Regional Economic Development in China in 2020

Jiuwen Sun

Regional Economic Development in China in 2020

palgrave
macmillan

Jiuwen Sun
Renmin University
Beijing, China

ISBN 978-981-95-3603-0 ISBN 978-981-95-3604-7 (eBook)
https://doi.org/10.1007/978-981-95-3604-7

Jointly published with China Renmin University Press Co., Ltd.

This Palgrave Macmillan imprint is published by the registered company Springer Nature Singapore Pte Ltd.
The registered company address is: 152 Beach Road, #21-01/04 Gateway East, Singapore 189721, Singapore

Foreword

There is a remarkable increase in the number and types of academic publications in China today. Encouraged by such academic prosperity, we, Renmin University of China (RUC), present our finely-designed collection of "Research Report Series," which is the concerted efforts of RUC scholars. In this collection, you will see application and countermeasure research that contains dynamic analyses, evaluations, and forecasts and explores solutions to major practical social problems for China's political and academic circles.

This collection of research papers feature academic-frontier-based approaches, innovative content, focus on current hotspots, and reference significance for the Chinese government and enterprises. These research papers are about the scientific research achievements made by RUC scholars: they record the breakthroughs made by the scholars and embody how they fulfilled their responsibilities of exploring challenging research topics and providing consulting services as think tanks; they facilitate the transformation and application of research achievements in humanities and social sciences by offering indicators and statistical data in relevant fields, evaluating current situations, and making forecasts. These research papers are about the academic reputation of RUC: over the years, RUC's high-quality research results have provided valuable information for the CPC, the Chinese government, as well as decision-makers of enterprises; these thematic, in-depth research papers released by RUC at regular intervals are widely acknowledged and recognized by readers.

RUC publishes this collection of research papers as a way to accumulate its scholastic achievements and spread its academic ideas. Famous for its excellence in humanities and social sciences, RUC insists on serving the government, the people, and society through its research activities. Since 2002, we have organized interdisciplinary research groups to publish a series of research results that faithfully reflect existing problems in the development of China's economy and society and those in China's humanities and social sciences: *Report on China Economic Development*, *Report on China Social Development*, and *Report on China Humanities and Social Sciences Development*. These three reports and *Report on China Law Development* published in recent years are known as the Four Reports of RUC. RUC has also published research papers from other disciplines, which have a certain influence on academic circles and society. Most of these research papers focus on major issues in political, economic, social, and cultural fields. They are often empirical and forward-thinking academic research that feature methods such as observing, analyzing, evaluating, and forecasting and provide reliable data and facts, in-depth analyses, and strategic forecasts. With an ambition to increase our influence and enhance our social service function, we have put these research papers into different groups and leveraged our publisher capacity to plan, edit, and release this collection of research papers. For your information, we plan to add new research papers to and remove outdated ones from this collection in the future according to changing situations to ensure the vitality of this series.

The research report series of RUC also facilitates discipline construction, strengthens education, and inspires innovation. First, boasting a continuous publication model, this series offers researchers a platform to share their research results and conduct academic exchanges. They also have chances to become stable members of our editorial team to enhance their own academic strengths as we would like to establish cooperation with competent researchers to improve our working ability. Second, this series requires a large amount of research-related materials, which gives students abundant opportunities to participate in academic research activities and increase their abilities. Third, to make this series focus on practical issues, we have to keep close contact with governments and enterprises and pay attention to actual conditions and needs of society. Such ties and interactions help to promote in-depth cooperation among universities, industries, enterprises, governments, academic circles, and foreign institutions, which effectively pool resources and remove barriers hindering innovation activities.

To adapt to the informatization, digitalization, and networking trends, this series of RUC is published in both paper and digital form. A corresponding literature database is developed to offer readers value-added services such as rich digital resources and convenient knowledge management tools and search tools for relevance search and cross-topic searches.

The research report series of RUC is the university's innovative endeavor to pool research resources and facilitate the transformation of research results. Given this, we will do our best to satisfy the needs of our readers and ensure the quality of each research report and that of the entire series by keeping up with the times and constantly following key issues, hotspots, and focus changes in our economic and social development. We are determined to build this series into a high-quality academic brand, creating academic core values and better assuming our social responsibilities as an academic institution.

Beijing, China Chen Yulu

Competing Interests The author has no competing interests to declare that are relevant to the content of this manuscript.

General Report: China's Regional Economy in 2020 and Its Medium- to Long-Term Development Ideas

For China, 2020 was a critical year: the CPC's first centenary goal of building a well-off society in an all-round way was scheduled to be achieved that year; in the meantime, a new endeavor was initiated, aiming to achieve socialist modernization by 2035 and develop China into a modern socialist country that is prosperous, strong, democratic, culturally advanced and harmonious by mid-21st century. In 2020, China saw historic opportunities created by a new round of scientific and technological revolution but had to face challenges brought by trade frictions and the outbreak of COVID-19 coronavirus infection. Based on a systematic study of domestic and foreign environments, China creatively put forward the strategy of pursuing a "dual-circulation" (the international circulation + the domestic circulation) development paradigm to blaze a new trail for China's economic growth and high-quality regional economic development.

I Background of China's Regional Economic Development

We analyze the background of China's regional economic development in 2020 based on the following six background factors.

(I) Continuous Impact of the COVID-19 Pandemic

The outbreak of COVID-19 coronavirus infection raged in China and other countries across the globe, creating serious problems. When the outbreak began, China adopted strict measures to curb the spread of the virus, such as population mobility reductions and postponed resumption of work and production. This had a negative impact on China's production activities and economy as well as its people's income and livelihoods. When the virus was under control, local governments used different methods to promote economic growth: some issued coupons to restore and expand consumption; some enhanced the resilience of their regional economies and improved their abilities to cope with risks by vigorously developing the digital economy.

Nevertheless, local governments had to shoulder the responsibility of preventing the virus from making a comeback in the following autumn and winter. While recovering their economies, they had to enhance their epidemic prevention measures and make up for the shortcomings in their public health systems, which had been exposed by the previous fight against the virus, such as those regarding emergency supplies and management. General Secretary Xi Jinping even published an article in *Qiushi* (Seeking Truth) journal to highlight the importance of enhancing our epidemic-prevention-and-control systems and mechanisms and improving our national public health emergency management system, which was entitled "Comprehensively Promote Law-Based Prevention and Control Measures and Improve Our National Public Health Emergency Management System."

(II) Digital Economy and New Impetus for Development

Given the rapid development of communication technology in China, Internet+ model emerged and matured. Because of mobility restrictions under epidemic situations, demand for cloud services, i.e. cloud office, cloud study, and cloud health services, surged. This led to the extensive popularization and application of digital-economy technologies such as the Internet of Things (loT), cloud computing, big data, artificial intelligence (AI), 5G, blockchain, etc.

At the same time, traditional industries were increasingly integrated with the digital economy and AI technology. This was a promising transformation given China's unremitting efforts to enhance its digital infrastructure and improve its computing and data-operation capacities. On July 8, 2020, *People's Daily* published an editorial announcing that the digital economy would start and fuel a new economic cycle.

(III) New Policies for Regional Economy

China's regional development strategy went through three stages after the state adopted the policy of reform and opening-up, reflecting the transition from unbalanced development to balanced development and then to coordinated development. Its regional development strategy was formulated based on the following five major regions: the Beijing-Tianjin-Hebei (BTH) region and the Yangtze River Delta (YRD) that pursues integrated development, the Yangtze River Economic Belt (YREB), the Guangdong-Hong Kong-Macau Greater Bay Area (Greater Bay Area or GBA), and the Yellow River Basin (YBR) as an ecological barrier and economic development zone.

In 2020, to overcome domestic and foreign challenges brought by the COVID-19 pandemic, China proposed the building of the Chengdu-Chongqing Economic Circle (CCEC) and promulgated plans to clarify the YBR's pursuit of ecological conservation and high-quality development and the strategic goals of China's new urbanization and its development campaign of its western region.

(IV) Trade Protectionism and Trade Frictions

There was a drastic shift in the US trade relationship with China when Donald John Trump, the 45th president of the United States, assumed power: the United States began pursuing trade protectionism, and anti-globalization trends emerged and grew strong. The United States even increasingly regarded China as a strategic competitor given its rapid economic development driven by its policy of reform and opening-up and its rising global influence. To choke off the development of China's economy, the United States used frequent trade frictions as methods to contain China.

The deeper reason for such trade frictions is the increasingly intense high-tech industry competition among different countries: given that the development of new industries needs sufficient purchasing power, countries become sensitive to trade deficits and want to ensure adequate market space for their own new industries by expanding their exports and reducing their imports. The developed countries represented by the United States, have pursued trade protectionism in formulating their recent trade policies, which has worsened the current international trade situation and caused frequent trade frictions.

(V) The Belt and Road Initiative (BRI) and the 15-Member RCEP

The Regional Comprehensive Economic Partnership (RCEP) was initiated by the Association of Southeast Asian Nations (ASEAN) in 2012. On November 15, 2020, the ten ASEAN countries and China, Japan, South Korea, Australia, and New Zealand signed the RCEP, which marked the establishment of the largest, most populous, and probably most promising free trade zone of today's world.

After years of development, the Belt and Road Initiative (BRI) gradually achieved its intended goals. It has already entered a phase of rapid advancement, which is aimed to achieve the following objectives: realizing orderly and free flow of economic factors, highly-efficient allocation of resources, and in-depth integration of markets; encouraging the BRI countries to achieve economic-policy coordination and carry out broader and more in-depth regional cooperation of higher standards; creating an open, inclusive, and balanced regional economic cooperation architecture that benefits all.

The BRI and RCEP will effectively expand China's westward and southward opening-up campaigns. This will accelerate the connections between China's central and western regions and eastern coastal areas, which is conducive to the success of its "dual-circulation" paradigm in promoting regional economic development.

(VI) Technology Restrictions and the Development of High-Tech Industries

Since 2018, trade frictions between China and the United States have spread to the high-tech sector. When ZTE and Huawei were subject to technical restrictions, a problem of strategic importance was exposed: some of China's critical technologies were still dependent on imported supplies. The United States' FY2019 National Defense Authorization Act (H.R. 5515) regarded China as a major strategic competitor of the United States and thus contained various measures to curb China's rise. Apart from intensifying its containment of China in the Indo-Pacific region, the United States has also increased its economic and technological restrictions on Chinese enterprises.

As the United States comprehensively tightens its control over exports, foreign investment, and government procurement, Chinese enterprises show a waning willingness to invest in the United States. The channels of acquiring advanced and new technologies from the United States by means of investment may be blocked, and the cost of technology learning of Chinese enterprises may increase significantly.

II Overall Structure of China's Regional Economy

The COVID-19 epidemic caused a direct, extensive, and profound impact on the overall regional economic development of China, resulting in a significant decline in the year-on-year provincial economic growth in the first quarter of 2020. The eastern coastal areas highly dependent on the export-oriented economy suffered severe losses, with its regional GDP growth falling by 4.7% year-on-year. Heilongjiang and Liaoning provinces in Northeast China that delayed the resumption of work and production for epidemic-control needs saw a year-on-year regional GDP growth decrease of 5.3%. The central and western regions were relatively less affected by the epidemic. Without counting Hubei province in, the year-on-year decrease rates of the central and western regions' GDP growth in the first quarter was 2.4 and 1.3% respectively. At the provincial level, the epidemic triggered the most serious decline in GDP growth in Liaoning, Shanghai, Heilongjiang, and Tianjin, causing a decline of 6 plus percentage points. The most seriously affected provincial administrative unit, namely Hubei province, saw its GDP in the first quarter decrease by 36.7% year-on-year. For more data, please see Fig. 1.

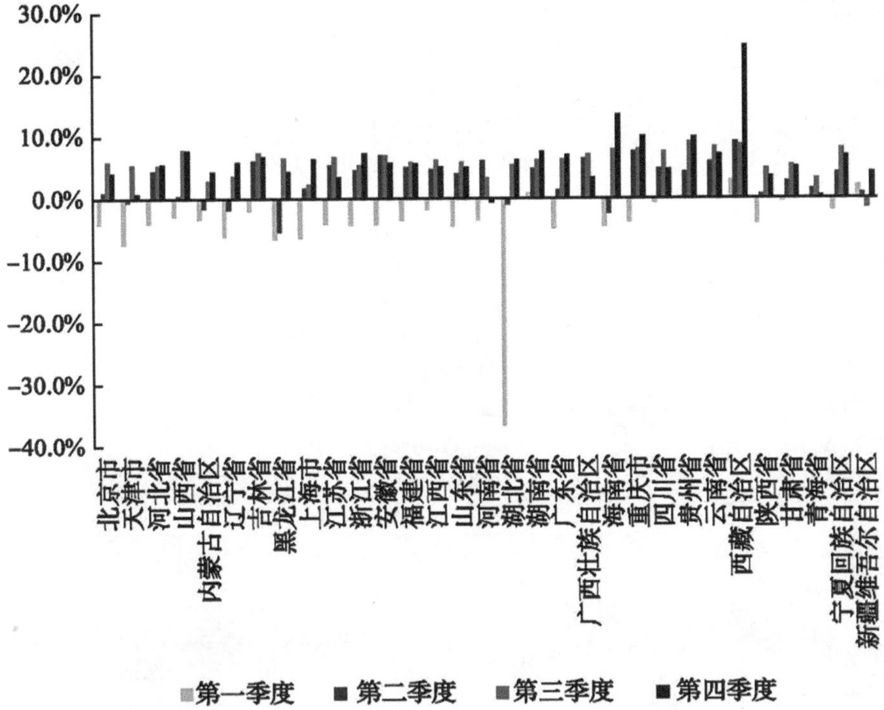

Fig. 1 Year-on-Year Provincial GDP growth rates in the Four Quarters of 2020

北京市	Beijing Municipality
天津市	Tianjin Municipality
河北省	Hebei Province
山西省	Shanxi Province
内蒙古自治区	Inner Mongolia Autonomous Region
辽宁省	Liaoning Province
吉林省	Jilin Province
黑龙江省	Heilongjiang Province
上海	Shanghai Municipality
江苏省	Jiangsu Province
浙江省	Zhejiang Province
安徽省	Anhui Province
福建省	Fujian Province
江西省	Jiangxi Province
山东省	Shandong Province
河南省	Henan Province

(continued)

(continued)

湖北省	Hubei Province
湖南省	Hunan Province
广东省	Guangdong Province
广西壮族自治区	Guangxi Zhuang Autonomous Region
海南省	Hainan Province
重庆市	Chongqing Municipality
四川省	Sichuan Province
贵州省	Guizhou Province
云南省	Yunnan province
西藏自治区	Tibet Autonomous Region
陕西省	Shaanxi Province
甘肃省	Gansu Province
青海省	Qinghai Province
宁夏回族自治区	Ningxia Hui Autonomous Region
新疆维吾尔自治区	Xinjiang Uygur Autonomous Region
第一季度	Q1
第二季度	Q2
第三季度	Q3
第四季度	Q4

Some provinces (provinces, otherwise written in this book as provincial administrative units, including autonomous regions and municipalities directly under the administration of the Central Government) maintained positive growth under the impact of the epidemic, but their growth slowed significantly. Tibet and Xinjiang delivered positive performances because of the long distances between them and Hubei. Situated right beside Hubei and placed under grave epidemic-control pressure, Hunan still managed to achieve the highest provincial GDP growth in the country in 2020, with a year-on-year growth of 0.87%. This achievement should be attributed to Hunan's advantageous industrial machinery manufacturing industry and stable foreign trade orders. Guizhou actualized positive growth for two reasons: its pillar industries, namely cigarette and beverage industries, were not susceptible to the epidemic; its newly established big data industry experienced explosive growth because of the surging demands under the epidemic conditions.

Most provinces saw their economies get back on track in the second quarter of 2020, basically getting back to the levels of the same period in the previous years. A total of 18 provinces recovered to their levels of the same period in 2019, but their cumulative year-on-year GDP growth rates, ranging from 0.35 to 3.38%, were still substantially lower than those of the same period in 2019. The year-on-year growth rate of GDP in Tianjin, Inner Mongolia, Liaoning, Hainan, and Hubei was still a negative number, but this did not negate the fact that Hubei's economy recovered rapidly in this time period: its year-on-year growth rate of GDP was −1.1%. In Beijing, Shaanxi, Shandong, Guangdong, and Hainan, the economy recovered relatively slower in the second quarter, but in the third quarter, they recovered to the same levels of the same period in 2019. The economic recovery in the second and third quarters was not just a rebound after an excessive recession. Instead, it suggested that the negative effects of the epidemic had gradually been offset and that the economic growth was steady.

Some localities were still in the stage of crisis recovery. In Tianjin, Shanghai, Liaoning, Inner Mongolia, Heilongjiang, and Hubei, GDPs failed to recover to the levels of the same period of 2019 even in the third quarter of 2020. In Tianjin and Hubei, this situation lasted throughout the fourth quarter of the year: their GDPs respectively recovered by 99.8 and 94.8%, in comparison with the numbers of the same period in the previous year.

Most provinces saw upturns in their economies, and for some of them, the upturns were big. Nevertheless, few changes occurred to the overall structure of China's regional economic development. Figure 2 shows absolute changes in provincial GDPs in each quarter of 2020: most provincial administrative units (China's provincial administrative units include provinces, autonomous regions, and municipalities directly under the Central Government, hereinafter collectively referred to as "provinces") had positive year-on-year growth rates and recovered their economies to levels of the same periods in 2019; Tianjin, Inner Mongolia, Liaoning, Shanghai, Heilongjiang, and Hubei, however, still faced a difficult economic convalescence; the Pearl River Delta (PRD), the YRD, and the BTH region, which quickly recovered from the impact of the epidemic and realized high economic growths, remained major contributors to China's economic growth with their output value accounting for a large proportion of the national total.

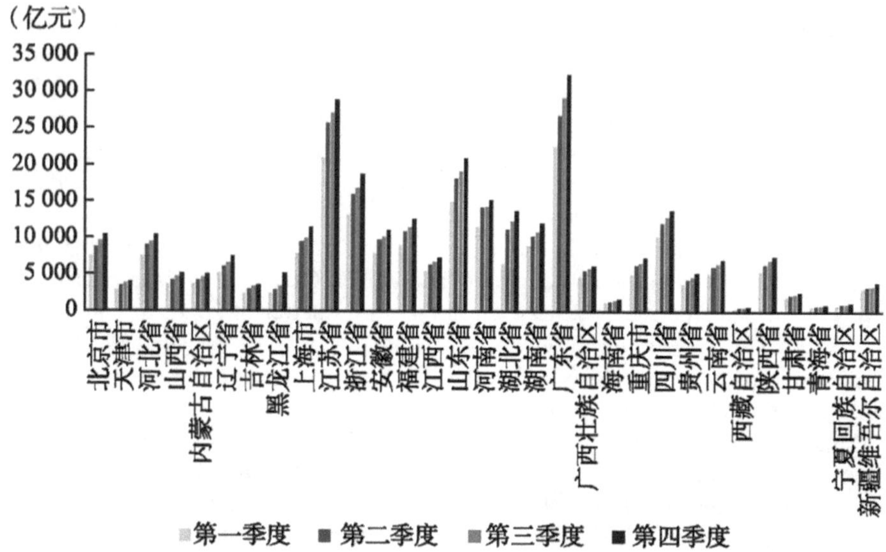

Fig. 2 Absolute Changes in Provincial Quarterly GDP of 2020 (*Data Source* The National Bureau of Statistics of China)

北京市	Beijing Municipality
天津市	Tianjin Municipality
河北省	Hebei Province
山西省	Shanxi Province
内蒙古自治区	Inner Mongolia Autonomous Region
辽宁省	Liaoning Province
吉林省	Jilin Province
黑龙江省	Heilongjiang Province
上海	Shanghai Municipality
江苏省	Jiangsu Province
浙江省	Zhejiang Province
安徽省	Anhui Province
福建省	Fujian Province
江西省	Jiangxi Province
山东省	Shandong Province
河南省	Henan Province
湖北省	Hubei Province
湖南省	Hunan Province
广东省	Guangdong Province
广西壮族自治区	Guangxi Zhuang Autonomous Region

(continued)

(continued)

海南省	Hainan Province
重庆市	Chongqing Municipality
四川省	Sichuan Province
贵州省	Guizhou Province
云南省	Yunnan province
西藏自治区	Tibet Autonomous Region
陕西省	Shaanxi Province
甘肃省	Gansu Province
青海省	Qinghai Province
宁夏回族自治区	Ningxia Hui Autonomous Region
新疆维吾尔自治区	Xinjiang Uygur Autonomous Region
第一季度	Q1
第二季度	Q2
第三季度	Q3
第四季度	Q4

Overall, the central region's contribution to the national GDP decreased slightly, the western region's contribution rose slightly, and the proportions of the eastern and northeastern regions in the national GDP did not change significantly. In the first three quarters of 2020, the GDP of the central region totaled 17.03 trillion yuan, accounting for 23.66% of the national total, a 0.32 percentage point decrease from the same period last year; Under external shocks, the northeastern region showed inadequate economic resilience and recovered relatively slower. In the first three quarters, its GDP reached 3.51 trillion yuan, which accounted for 4.88% of the national number, a 0.08 percentage points decrease compared with the same period of 2019; Located far away from Hubei, most of the western provinces were slightly affected by the epidemic. In the region, Yunnan and Guizhou recovered relatively faster. The regional GDP of the first three quarters amounted to 12.4 trillion yuan, which accounted for 17.23% of the national total, an increase of 0.28 percentage points year-on-year. The eastern region's proportion in the national GDP stayed almost the same as before: its GDP of the first three quarters was 39.04 trillion, accounting for 54.23% of the national total and increasing by 0.12 percentage points in comparison with the same period of the previous year.

In terms of industrial structure, the proportion of the manufacturing sector, or the secondary sector, rose rapidly during the epidemic, from 34.9% in the fourth quarter of 2019 to 60.5% in the second quarter of 2020. Nonetheless, its contribution to the national GDP began to decline in the third quarter. During this period, which was under the impact of the epidemic, the secondary sector became the most important driver of economic development for the following reasons: There was a rapid decline in the tertiary sector under epidemic conditions; The secondary sector experienced a period of rapid growth, which was caused by the surging demand for medical and health products such as masks, disinfectants, protective clothing, and respirators; In the second quarter of 2020, which featured the resumption work and production, efforts were concentrated on the manufacturing sector. For more detailed data, please refer to Fig. 3.

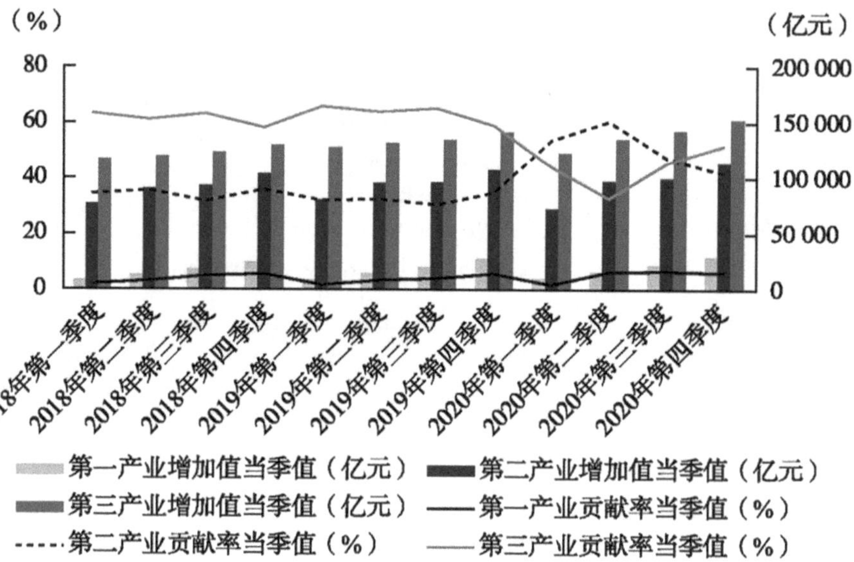

Fig. 3 Industry added values and contribution rates 2018–2020 (*Data Source* The National Bureau of Statistics of China)

第一产业增加值当季值 (亿元)	Quarterly added value of the primary industry (100 million yuan)
第二产业增加值当季值 (亿元)	Quarterly added value of the secondary industry (100 million yuan)
第三产业增加值当季值 (亿元)	Quarterly added value of the tertiary industry (100 million yuan)
第一产业贡献率当季值 (%)	Quarterly contribution rate of the primary industry (%)
第二产业贡献率当季值 (%)	Quarterly contribution rate of the secondary industry (%)
第三产业贡献率当季值 (%)	Quarterly contribution rate of the tertiary industry (%)
2018第一季度	Q1 2018
2018第二季度	Q2 2018
2018第三季度	Q3 2018
2018第四季度	Q4 2018
2019第一季度	Q1 2019
2019第二季度	Q2 2019
2019第三季度	Q3 2019

(continued)

(continued)

2019第四季度	Q4 2019
2020第一季度	Q1 2020
2020第二季度	Q2 2020
2020第三季度	Q3 2020
2020第四季度	Q4 2020

It should be noted that the fast growth in the secondary sector did not change the overall situation: the tertiary sector remained the biggest contributor to China's economy. The primary sector remained stable as it stayed almost immune to the effects of the epidemic. The secondary sector experienced a sharp decrease in the first quarter and then a rebound in the second one, basically recovering to its pre-epidemic status. The tertiary sector returned to its pre-epidemic levels in terms of economic aggregate and growth rate in the third quarter and grew further in the fourth quarter.

In autumn and winter, the possibility of epidemic recurrence increased, posing a potential threat to the development of the tertiary sector. Nevertheless, its impact on the tertiary sector was limited because the sound epidemic-prevention-and-control system effectively prevented a long-term nationwide epidemic outbreak.

III New Features of China's Regional Economy

When the world economy was facing recession, the outbreak of COVID-19 worsened the situation by giving a blow to economic globalization. As a response to such changes, China put forward its new development paradigm featuring dual circulation: taking domestic circulation as the main body, with domestic and international circulation reinforcing each other. Under this new paradigm, China planned to provide a guarantee for the stable and high-quality development of its regional economy by doing well in the following aspects: strengthening its cities' supporting role in regional economic development, accelerating its industrial transformation and upgrading, ensuring equal access to public services, and vigorously developing the digital economy and related industries. In general, there were three new spatial characteristics in its regional economy and four key points regarding the transformation of its growth drivers.

(I) Spatial Characteristics

1. **Multiple Fulcrums.** Multiple fulcrums describe the 21 city clusters (city groups consisting of large, nearly contiguous cities with many adjoining satellite cities

and towns) planned by China: further-enhanced mature clusters in the eastern regions such as the BTH region, the YRD, and the GBA; improved clusters in the central region and the middle reaches of the Yangtze River; recovered clusters in the northeastern region; and fast-developed clusters in the western region, including Chengdu-Chongqing, Central Guizhou, Central Yunnan, Central and Southern Tibet, Tianshan North-Slope, Lanzhou-Xining, Ningxia Riverside, and Guanzhong-Plain City Clusters. As the main driving force to reshape our country's economic geography, urbanization has constantly balanced our regional economic development, expanded our economic development space, and provided new impetus for our overall economic growth.

2. **The Artery-Based Economy.** The artery-based economy is embodied through the BRI, the integrated development of the BTH region, and the ecological conservation and high-quality development in the YRB. It refers to an effective method used by China to realize its industrial transfer through the influence of traffic arteries. These development plans and arrangements stand as proof that China develops its economy to benefit all of its provinces and stops sacrificing its resources and environment for economic development. China is determined to pursue sustainable development in the process of balancing its development.
3. **Reticulated System.** A reticulated system has been developed due to increasing spatial, industrial, and economic linkages between our cities, especially after the opening of our high-speed railways. With drastically shortened traveling times within China, the development of the central and western regions as well as the transformation and upgrading of the eastern economies have been accelerated.

(II) Transformation of Growth Drivers

1. **Urbanization and Urban System Development.** In a new development phase featuring new domestic and international environments, China has attached more significance to urbanization and reconsidered its role. The optimization of Industrial layouts in urban areas can improve factor allocation efficiency and provide new impetus for industrial development. The enhancement of urban facilities and public services and the reasonable increase of public consumption can help to expand domestic demand and build a strong domestic market. The development of urban areas can benefit the rural areas, helping counties and towns (townships) to tap their development potentials; the removal of barriers for factor flows can promote the synergistic development of urban and rural areas. The agglomeration and the diffusion effects of cities can be leveraged for better city clusters and urban systems and more balanced regional development.
2. **Coordination Between Urban and Rural Areas and Rural Revitalization.** Urban-rural relationships refer to the general connections and interactions between urban and rural areas under certain social and economic conditions. As a fundamental relation of human society, urban-rural relationships are rich

in connotations, involving economic, social, social class, ecological, environmental, cultural, and spatial factors. As all of its impoverished counties have been removed from the poverty list, China has made a historic adjustment to its major agricultural work. Despite that, efforts to consolidate the achievements of the poverty alleviation campaign will not slacken since they constitute the foundation for our rural revitalization strategy and the sustained development of our regional economy.

3. **Industrial Transformation and Upgrading.** China is the only country obtaining all the categories listed in the United Nations' industrial classification worldwide, but it still has to prevent problems such as excessive deindustrialization and industrial hollowing-out from happening during its process of industrial upgrading. Given the intensifying international competition and the COVID-19 pandemic, countries across the globe adopted policies to attract their enterprises to return, which created an increasingly apparent deglobalization trend. In response to this, China had to rely on industrial transformation and upgrade to ensure the smooth and safe operation of its domestic economy and safeguard its development interests: It had to take active measures to make up for shortcomings of its industrial chain and supply chain and increase investment to solve such weak links; It had to rely on its super-large-scale market and complete industrial system to build its unique advantages that facilitated the rapid large-scale application and iterative upgrading of new technologies, accelerated its transformation of scientific and technological achievements into real productivity, and improved and safeguarded its industrial chain.
4. **Domestic Circulation and "Dual-Circulation" Paradigm.** Facing drastic changes in the international environment, China needed to guarantee its economic security and expand its economic development space. Therefore, it decided to take its domestic market or domestic circulation as the mainstay and let the domestic and international markets reinforce each other, which was the "dual-circulation" paradigm. From an international perspective, the rise of trade protectionism and populism across the globe caused drastic changes in international investment trends and the global industrial chain, which created a challenging international environment for China. China had to ensure its independence in its key areas to effectively deal with an increasing number of external uncertainties. From a domestic perspective, China needed to shift from high-speed growth to high-quality development. The 19th National Congress of the CPC, held in 2017, marked the beginning of the new era of socialism with Chinese characteristics (hereinafter, the new era). The new era called for high-quality development under "dual-circulation" paradigm given the greatly-improved productivity and the increasingly prominent problem that insufficient and imbalanced development could hardly meet people's growing needs for a better life. It should be noted that the paradigm with domestic circulation as the mainstay does not suggest a passive shrinkage. Instead, it is a strategic choice to build, through in-depth reforms, vicious domestic and international circulations and a more resilient economic operating system with the domestic circulation serving as the mainstay and the international circulation playing an auxiliary role.

IV Breakthrough Points for China's Regional Economic Development

(I) The Joint Development of Economic Regions

Despite making miraculous economic achievements after adopting the policy of reform and opening-up, China has to face increasingly evident resource bottlenecks and environment constraints, such as water shortage, ecological environment fragility, insufficient resources and environment carrying capacity, and economic and social development's unavoidable conflict with resources and environment. Knowing that its high-quality development is underpinned by the coordinated development of its economic regions, China has promulgated a series of region-based strategies for its eastern, central, western, and northeastern regions: the construction of the YREB, the ecological conservation and high-quality development in the YRB, the construction of the CCEC, the integrated development of the YRD, the coordinated development of the BTH region, and the construction of the GBA. These new regional policies that play a regulatory role in allocating resources, enhancing transportation infrastructure, or optimizing regional coordination mechanisms make China's regional economy a new booster for its national economic growth. The new regional economic development paradigm features the coordination of different economic zones: to be more specific, it sees that the further development of the BTH region, the YREC, the GBA, and the integrated YRD facilitates the ecological conservation and high-quality development in the YRB and the growth of CCEC. The joint development of the economic zones calls for the establishment of effective systems and mechanisms for coordinated regional development, which are conducive to the integration of their transportation and information networks, the removal of institutional barriers, the collaboration between regions and industries, and the enhancement of their inter-regional economic and trade ties.

(II) The People-Oriented Urbanization: China's New-Type Urbanization

Optimized ways to realize the new-type urbanization with Chinese characteristics must be explored to improve the quality of urbanization in China. This is because of the following new changes occurring during the 14th Five-Year Plan period (2021–2025): (1) It is unlikely that China will rapidly increase its urbanization rate again because this rate has already exceeded 60%; (2) Because of the impact of the COVID-19 pandemic, the re-examination of our choice of urban scaling and designing has come to the foreground; (3) The urbanization of rural population transferred to urban areas calls for the establishment of a new-type urban-rural relationship regarding land-related issues. To cope with these changes, the previous development path featuring urban land expansion has to be replaced with a new one that highlights the improvement of urban culture and public services, i.e., the goal of urban development must be changed from building big cities to creating ideal cities.

Regarding the current stage of China's urbanization, it is essential to efficiently implement urban renewal plans and do well in the following aspects: promoting urban ecological restoration and functional improvement; coordinating urban planning, constructing, and managing activities; rationally deciding on the size, population density, and spatial structure of a city; and ensuring the coordination between towns and cities of different sizes in urban development. For distinct, attractive city images, efforts must be exerted to protect the historic and cultural relics in cities and improve urban management abilities. For better urban communities, old communities need to be renewed and relevant community services must be improved. To solve problems regarding flood control and drainage systems, cities will be developed or transformed into sponge cities (cities with abundant natural areas such as trees, lakes, and parks—or other good designs intended to absorb rain and prevent flooding) or resilient cities (cities boasting lifeline systems of roads, utilities, infrastructure, and other support facilities that are designed to continue functioning in the face of rising water, high winds, shaking ground, and other shocks and extreme conditions). For better urban governance, it is a must to optimize megacities' risk prevention and control systems and our urban systems that involve central cities (a central city constitutes the densely populated center of a metropolitan area), metropolitan areas (a metropolitan area is a major city together with its suburbs and nearby cities, towns, and environs over which the major city exercises a commanding economic and social influence), and city clusters (city groups consisting of large, nearly contiguous cities with many adjoining satellite cities and towns). The choice and development of central cities and city clusters is an issue of national significance. To be specific, China has three types of central cities that are the national-level, regional-level, and local-level central cities, and allows its metropolitan areas to exercise cross-administrative-region functions in economic regions that have no big city clusters. These central cities and metropolitan areas promote the rational allocation of resources within economic regions. When the central cities or city districts expand their influences for regional development, they also expand the space for China's new-type urbanization.

(III) Narrowed Inter-Regional Development Gaps

China's 9.6 million square kilometers of land can be roughly divided into three parts: the eastern coastal region, the central region, and the western region. The eastern region actualized rapid development in the process of undertaking international industrial transfer with its closer ties with the international market and its early-established and relatively complete market mechanism. The central and western regions lagged behind the eastern region in the early periods of China's reform and opening-up for several reasons: unfavorable geographical conditions, immature market mechanisms, unsmooth commodity circulation channels, and restrictions on the free circulation of land, technical capital, and other factors among different localities. Given the long-existing East-West gradient in economic development and the adverse influence brought by geographical locations and immature transportation conditions, inequality between the eastern region and the central and western regions becomes apparent. Besides, the inequality of development opportunities between regions also leads to a widening development gap between North and South China. Given this, narrowing the regional development gaps and promoting a balanced regional development have been recognized as important contents of China's national territory spatial planning (the spatial and temporal arrangements for the development and protection of a territorial space).

It is a must to stimulate the vitality of the factor market by improving transportation infrastructure. The actualization of the regional balance and free flow of production factors helps to narrow regional development disparities. The enhanced transportation infrastructure not only reduces inter-regional traveling times but also increases remote areas' abilities to pursue development.

It is necessary to optimize China's industrial spatial arrangements by adjusting its existing industrial layout to facilitate the country's domestic circulation. To do well in this, it is essential to grasp the four features of the current global industrial layout. (1) Production activities are distributed all over the world, and as service-related costs fall, the income gap between skilled and unskilled workers gradually becomes apparent; the cost of communication and trade also gets lower than before. (2) Production activities begin to cluster within their respective regions since such clusters better facilitate economic activities when the cost of trade decreases. (3) End product manufacturers and their intermediate product providers tend to congregate for joint development for two reasons: Compared with the transportation cost of end products, the transportation cost of intermediate products increases; Not all business services are tradable. (4) In the face of huge economic fluctuations, the pattern of the distribution of production activities worldwide is unstable.

It is essential to utilize both our land and sea resources. Regarding this point, the CPC Central Committee made the following statement in its report to the 19th National Congress of the CPC: "We will pursue coordinated land and marine development, and step up efforts to build China into a strong maritime country." During the 14th Five-Year Plan period (2021–2025, China will step up the utilization of its land and sea resources. To achieve this goal in the current international environment, it is necessary to properly handle the following issues: disputes over sea resources and the stabilization of our surrounding environment; the utilization of offshore resources and ocean development; the optimization of marine industrial structure and the adjustment of the relevant industrial layout; the balance between the aggregate and quality of the marine economy; the protection of marine resources and ecological environment; and the utilization intensity and time arrangements. On this basis, China will formulate its national marine strategies and marine economy policies.

(IV) Ecological Civilization and Ecological Security

Ecological security is a concept that has gained widespread attention in China in recent years, especially in the new era, when the country begins to pursue high-quality economic growth and when ecological security has an increasingly significant impact on its society and economy. Ecological security is a term used to analyze and evaluate natural and semi-natural ecosystems from the perspective of human utilization of natural resources and human living conditions. It reflects that economic and social development cannot be separated from the support of the ecosystem, which is especially true in the new era that highlights the pursuit of high-quality economic development. Given this, ecological security becomes a must-considered factor in our regional development.

It is necessary to promote ecological civilization through the formulation and implementation of national major development strategies. While pursuing development, various economic regions of China also ensure their ecological security since their development strategies include content about ecological protection and green development. A typical example is the ecological conversation and high-quality development strategy of the YRB. These economic regions seek development while taking into consideration resources and environment-carrying capacities. They give full play to their respective strategic advantages and aim to achieve the strategic goals of the country. They actively practice coordinated development and faithfully implement major policies. Therefore, a precise understanding of green development policies and principles, which are included in major regional development strategies, the strategy for coordinated regional development, and the functional zoning strategy, helps us to do well in the ecological conservation work when making our national territory spatial plans.

It is essential to grasp different economic regions' characteristics and functions and their roles in our national territory spatial plans. The 5th Plenary Session of the 19th CPC Central Committee proposed to divide our territorial space into three categories based on the environment-carrying capacities and different regional characteristics: urbanized areas, agricultural production areas, and ecological function areas. A thorough understanding of these three categories is of utmost importance in making national territory spatial plans. In this book, we defined them as follows. (1) Urbanized areas refer to the highly populated and economically active areas with clear boundaries, i.e. the 21 city clusters of China. (2) Agricultural production areas are responsible for the production of agricultural products and are built on the basis of protected basic farmland and ecological space, which are roughly equal to the eastern plains and the western oasis areas. (3) Ecological function areas are ecological security-related areas responsible for ecological security-related issues, i.e. China's entire territorial space excluding the areas included in the above two categories.

Contents

List of Figures

Chapter 1

Chapter 2

Chapter 3

Chapter 4

Chapter 6

Chapter 7

Chapter 8

Chapter 9

Chapter 10

List of Tables

Chapter 8

Chapter 10

Part I
New Environments for China's Regional Economic Development

Chapter 1
Domestic Environment

1 Regional Economic Development Under the Impact of the COVID-19 Epidemic

The spread of COVID-19 in China and other countries has caused unpleasant results. This section explores two issues: (1) the economic growths and economic recovery speeds of the Chinese economic regions and the structural changes in the regional economy under the epidemic situations; (2)environmental changes and new requirements for regional economic development with regularized epidemic prevent and control measures in place.

1.1 The Influence of the COVID-19 Epidemic on Regional Economic Growth

This subsection analyzes the influence of the COVID-19 epidemic on the economic growth of different Chinese localities, their differences in economic recovery, and the relationship between the epidemic and their economic growth, and on this basis, assesses the structural changes caused by the epidemic in the regional economy.

1. **Provincial Economic Growth and Recovery.** The impact of the epidemic was widespread and profound. It spread rapidly across China because the outbreak of this highly-contagious virus began around the Lunar New Year of 2020, when people moved on a large scale and at high frequency. To analyze the epidemic's impact on different regions, we compare the year-over-year provincial GDP growth rates of the first quarter of 2020 to those of previous periods in Fig. 1. As shown in the figure, the first quarter of 2020 saw most provinces suffer negative growth. For Qinghai, Guangxi, Yunnan, Gansu, Sichuan, and Jiangxi, the decline was within two percentage points because they were slightly affected by the

J. Sun, *Regional Economic Development in China in 2020*,
https://doi.org/10.1007/978-981-95-3604-7_1

epidemic outbreak that happened far away from them, and because their dependence on the international market was low. For provinces in eastern and central regions, including the five neighboring provinces of Hubei i.e. Anhui, Chongqing, Shaanxi, Jiangxi, and Henan, they experienced a moderate-degree impact and saw their provincial GDPs decrease by 2 to 5 percentage points year-on-year. The heavily-affected ones included Liaoning, Shanghai, Heilongjiang, and Tianjin, whose GDPs fell by more than 6 percentage points year-on-year. Among them, Liaoning and Heilongjiang should attribute the problem to their large proportions of traditional industries and low-degree industrial diversification; Shanghai ascribed the issue to its high dependence on the international market and the deglobalization trend caused by the COVID-19 pandemic; and Tianjin's bad performance was not what people generally expected. Among all provinces of China, Hebei suffered the most from the epidemic, with a year-on-year decrease of 36.75% in its GDP in the first quarter of 2020.

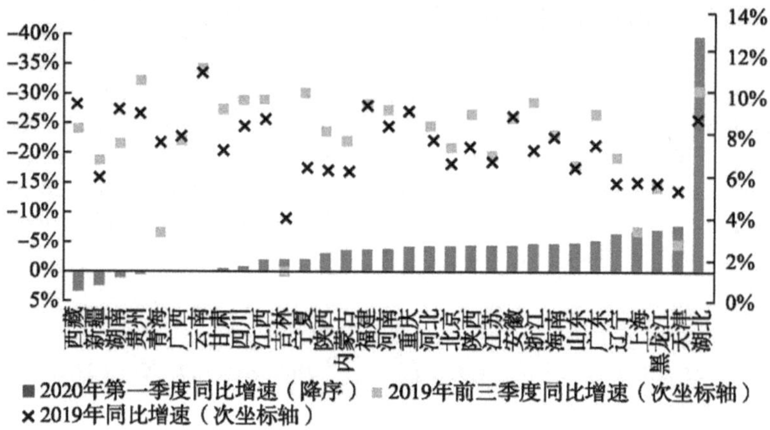

Fig. 1 Comparison between the year-on-year provincial GDP growth rates in Q1 2020 and those in previous periods (*Data Source* The National Bureau of statistics of China)

2020年第一季度同比增速(降序)	Year-on-year growth rates in Q1 2020 (descending order)
2019年前三季度同比增速(次坐标轴)	Year-on-year growth rates in the first three quarters of 2019 (secondary axis)
2019年同比增速(次坐标轴)	Year-on-year growth rate in 2019 (secondary axis)
西藏	Tibet
新疆	Xinjiang
湖南	Hunan
贵州	Guizhou
青海	Qinghai

(continued)

(continued)

广西	Guangxi
云南	Yunnan
甘肃	Gansu
四川	Sichuan
江西	Jiangxi
吉林	Jilin
宁夏	Ningxia
陕西	Shaanxi
内蒙古	Inner Mongolia
福建	Fujian
河南	Henan
重庆	Chongqing
河北	Hebei
北京	Beijing
陕西	Shaanxi
江苏	Jiangsu
安徽	Anhui
浙江	Zhejiang
海南	Hainan
山东	Shandong
广东	Guangdong
辽宁	Liaoning
上海	Shanghai
黑龙江	Heilongjiang
天津	Tianjin
湖北	Hubei

The situation was different in some provinces: they resisted the impact of the epidemic and achieved positive growth in the first quarter of 2020. Tibet and Xinjiang maintained positive growth due to their distance from Hubei. Although situated close to Hubei and faced with high epidemic-control pressure, Hunan still managed to keep its steady growth and realized an increase of 0.87% year-on-year, with its GDP growth ranked first in the country in this period: As a top-notch manufacturer of industrial machinery in China, Hunan gained benefits from its relatively stable foreign trade orders and the surging demand for machinery when the nationwide resumption of work and production began. Guizhou's GDP growth should be attributed to its dependence on cigarette and beverage industries: The two industries increased respectively by 22.3% and 5.8% year-on-year in the first quarter of 2020.

On the whole, China's provinces showed strong economic resilience: In the second quarter of 2020, most of them recovered to levels of the same period in the previous year. Table 1 shows the quarters in which the provinces recovered to levels of the same period in 2019 and relevant year-on-year GDP growth rates in the recovery periods. As shown in the table, a total of 18 provinces achieved economic recovery in the second quarter of 2019, but their year-on-year GDP growth rates only ranged from 0.35% to 3.38%, which were significantly lower than the numbers of the same period in 2019. Beijing, Shaanxi, Shandong, Guangdong, and Hainan recovered relatively slower: It was in the third quarter of 2020 that their GDPs finally returned to levels of the same period in 2019.

Inner Mongolia, Liaoning, Heilongjiang, and Shanghai did not recover until the fourth quarter, with their year-on-year quarterly GDP growth rates being 086%, 083%, 063%, and 1.43%. Tianjin failed to actualize its economic recovery in 2020: Its annual GDP of 2020 was 0.15% lower than that of 2019. Toward the end of 2020, Hubei was still in the recovery stage, but its recovery speed was fast. In the third and fourth quarters of 2020, its quarter-on-quarter GDP growth rates reached 5.44% and 6.26%, respectively. By the end of 2020, its annual GDP was equivalent to 94.8% of its annual GDP in 2019.

It is generally agreed that a region's industrial structure exerts a great influence on its economic resilience. Some studies further point out that a region's economic resilience is determined by its industrial structure, institutional environment, social capital, and cultural factors. Sun et al.'s (2017) findings indicate that industrial structure factors, including the diversity of industrial structure and the types of leading industries, determine a region's economic resilience: With increased diversity, the region gains a better ability to resist external shocks and locate new growth paths; A region without such diversity is vulnerable to external shocks and lacks the flexibility in finding new alternative industries. Generally speaking, differences in leading industries lead to differences in economic resilience, and regions highly dependent on heavy industry have relatively weak resilience. Liaoning, Inner Mongolia, and Heilongjiang, which had large proportions of traditional industries, did lag behind in recovering their economies. Tianjin's weak resilience could be reasonably explained by its high dependence on heavy industries. Shanghai's secondary sector was on a downward trend before the epidemic, but the epidemic drastically accelerated the decline. Under such circumstances, its economy recovered slowly when its tertiary sector was not much affected by the epidemic.

All in all, most provinces delivered good economic performance in 2020, and some provinces did exceptionally well. As shown in Fig. 2, a comparison of the year-on-year provincial GDP growth rates of 2019 and those of 2020, most provinces sustained positive growth. For some of them, their growth rates in 2020 were even higher than those in 2019. Affected by the COVID-19 epidemic, the year-on-year growth rate of national GDP was only 2.99%. This was an unavoidable result when most Chinese provinces had low growth rates. The bright side was that western provinces such as Tibet, Guizhou, Chongqing, and Yunnan were clearly in a period of rapid development: Their emerging service industry maintained a good momentum of development, and their GDP growth rates exceeded the national number by 2.5 percentage points, which revealed the high growth potential of the region.

Table 1 Economic recovery quarters for the provinces

Quarter	Province
Q2	Hebei (0.42%), Jilin (2.37%), Jiangsu (0.88%), Zhejiang (0.35%), Anhui (1.72%) Fujian (0.98%), Jiangxi (1.58%), Henan (1.52%), Hunan (2.96%), Guangxi (3.38%) Chongqing (2.15%), Sichuan (2.12%), Guizhou (2.43%), Yunnan (3.09%), Gansu (1.21%) Qinghai (0.86%), Ningxia (1.22%), Xinjiang (1.45%)
Q3	Beijing (1.23%), Shaanxi (2.05%), Shandong (2.07%), Guangdong (1.21%), Hainan (0.58%)
Q4	Inner Mongolia (0.86%), Liaoning (0.83%), Heilongjiang (0.63%), Shanghai (1.43%)

Note Percentages in parentheses represent the corresponding provinces' year-on-year GDP growth rates in the quarters when their economies recovered to levels of the same period in 2019

Data Source The National Bureau of statistics of China

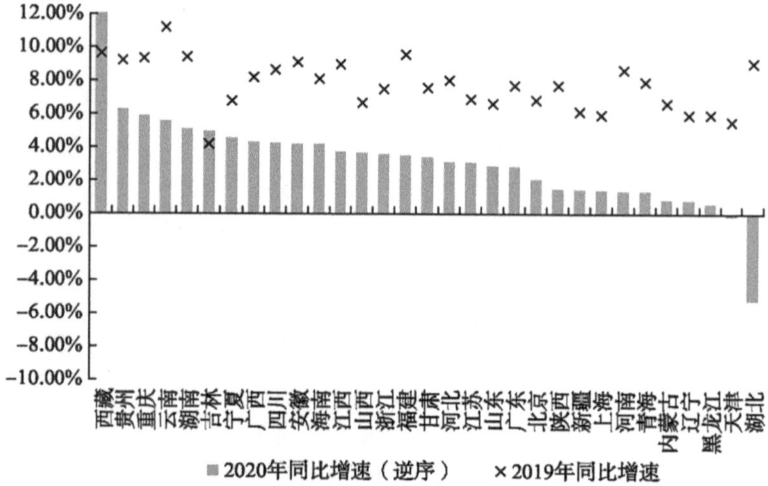

Fig. 2 Comparison between year-on-year provincial GDP growth rates in 2020 and those in 2019 (*Data Source* The National Bureau of statistics of China)

2020年同比增速(逆序)	Year-on-year growth rates in 2020 (descending order)
2019年同比增速	Year-on-year growth rates in 2019
西藏	Tibet
新疆	Xinjiang
湖南	Hunan
贵州	Guizhou
青海	Qinghai
广西	Guangxi
云南	Yunnan
甘肃	Gansu
四川	Sichuan
江西	Jiangxi
吉林	Jilin
宁夏	Ningxia
陕西	Shaanxi
内蒙古	Inner Mongolia
福建	Fujian
河南	Henan
重庆	Chongqing
河北	Hebei
北京	Beijing
山西	Shanxi

(continued)

(continued)

江苏	Jiangsu
安徽	Anhui
浙江	Zhejiang
海南	Hainan
山东	Shandong
广东	Guangdong
辽宁	Liaoning
上海	Shanghai
黑龙江	Heilongjiang
天津	Tianjin
湖北	Hubei

2. **The Influence of the COVID-19 Epidemic on Provincial Economic Growth.** To find out how much the COVID-19 epidemic has influenced the regional economic growth, we count the provinces' (excluding Hubei) cumulative numbers of confirmed COVID-19 cases in 2020, sort out the provinces' year-on-year GDP growth rates in 2020, and analyze these two sets of data. As shown in Fig. 3, the cumulative number of confirmed COVID-19 cases is negatively correlated with the GDP growth rate, i.e., provinces with more cumulative confirmed cases have lower GDP growth rates. Nevertheless, the correlation coefficient between the two variables is relatively small. To eliminate the influence of the provinces' different population sizes, this subsection divides the provincial cumulative numbers of confirmed cases in 2020 by the total provincial population counted at the end of 2019 to get the cumulative numbers of confirmed cases per 10,000 people, and compares these numbers with the year-on-year growth rate of provincial GDPs in 2020, as reported in Fig. 4. The results of a simple linear regression show that when the cumulative number of confirmed cases per 10,000 people increased by one, the GDP growth rate will drop by 7.13%. Given that the average cumulative number of confirmed cases per 10,000 people for all provinces (excluding Hubei) is less than 0.5, it is more reasonable to say that when the cumulative number of confirmed cases per 10,000 people increased by 0.5, the GDP growth rate drops by 3.57%. It is worth noting that as the provinces are still in their recovery period in 2020, their year-on-year GDP growth rates are also related to the drops in their GDPs caused by the epidemic and related hysteretic effects.

Therefore, further examination is needed to determine how much the cumulative number of confirmed cases per 10,000 people has influenced the provincial economic growth. Figure 5 shows the relationship between provincial GDP growth and the number of newly confirmed cases per 10,000 people in the fourth quarter of 2020. As can be seen from this figure, there is no significant correlation between the two variables. This shows that when epidemic-prevention-and-control measures are

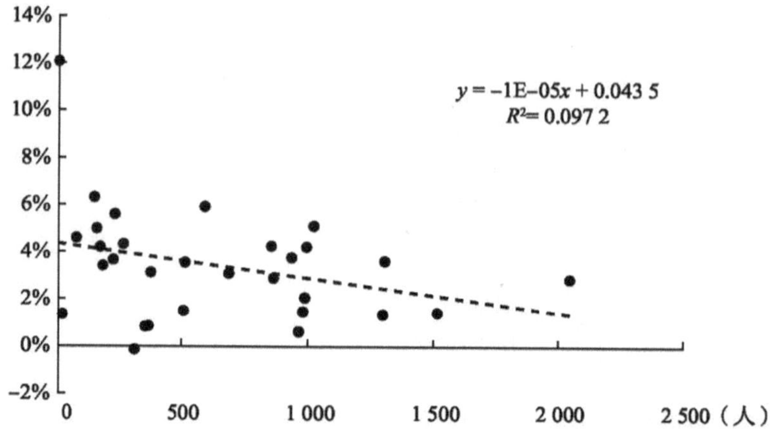

Fig. 3 Provincial cumulative numbers of confirmed COVID-19 cases and year-on-year provincial GDP growth rates in 2020 (*Note* This figure does not contain the data of Hubei) (*Data Source* Wind information and the National Bureau of statistics of China)

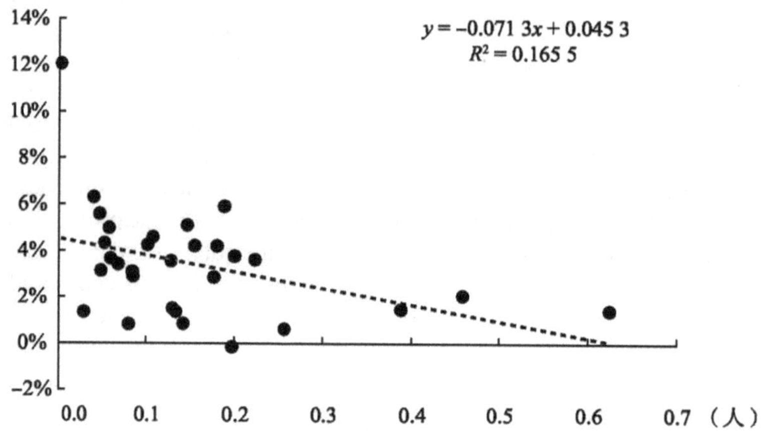

Fig. 4 Provinces' confirmed COVID-19 cases per 10,000 people and year-on-year GDP growth rates in 2020 (*Note* This figure does not contain the data of Hubei) (*Data Source* Wind information and the National Bureau of Statistics of China)

regularized, the virus does not have much influence on the GDP and that only short-term impacts of the epidemic cause sudden drops in GDP. Based on this discovery, we judge that in the short term, the epidemic can not directly cause changes to the existing regional economic structure.

In fact, the economic growth of a locality is affected by several things, such as its original GDP size, industrial structure, institutional environment, technological progress, and international environment. These things will not change significantly in a short time. Relevant research also shows that a structural change to the regional

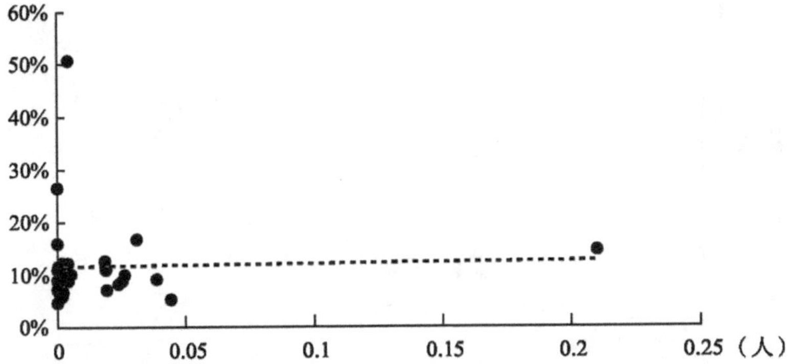

Fig. 5 Provinces' newly-confirmed COVID-19 cases per 10,000 people and Quarter-on-quarter GDP growth rates in Q4 2020 (*Note* This figure does not contain the data of Hubei) (*Data Source* Wind information and the National Bureau of statistics of China)

economy is a relatively long-term process. Figure 6 shows the relationship between provincial GDP growths in 2020 and those in 2019. As shown, there is a positive correlation between the two data groups. Even considering the impact of the epidemic, 31% percent of the GDP growth rates in 2020 are determined by the GDP growth rates in 2019; Provinces with faster GDP growths before the epidemic experience smaller downturns under the impact of the epidemic; Provinces with slower GDP growths before the epidemic see sharp economic downturns caused by the epidemic. This shows that the current structure of provincial economic growths is relatively stable. To further analyze the epidemic's impact on the structure of the regional economy, the following paragraph analyzes the GDP structure based on the division of the four major economic regions of China.

Overall, the year 2020 only saw a slight decline in the central region's GDP share and a small increase in the western region's GDP share, and the eastern and northeastern regions' proportions in the national GDP remained basically unchanged. Table 2 shows that the drop in the central region's GDP share comes from the sharp downturn in the economy of Hubei, which is a province in the central region. In 2020, the central region's GDP totaled 19.62 trillion yuan, accounting for 23.67% of the country's GDP, a 0.28 percentage point decrease from the same period of the previous year. The northeastern region recovered relatively slower than other regions due to its inadequate capacity to cope with external shocks. In 2020, its total GDP was 5.11 trillion yuan, accounting for 5.05% of the national GDP, a 0.05 percentage points year-on-year decrease. Most western provinces, especially Yunnan and Guizhou, were not much affected by the epidemic outbreak in the far-away central province and recovered quickly. The region's GDP reached 17.4 trillion yuan in 2020, accounting for 17.16% of the national total, a year-on-year increase of 0.24 percentage points. The eastern region's contribution to the national GDP did not change significantly in comparison with its performance in 2019. In 2020, its regional GDP totaled 54.79 trillion yuan, taking up 54.12% of the national total

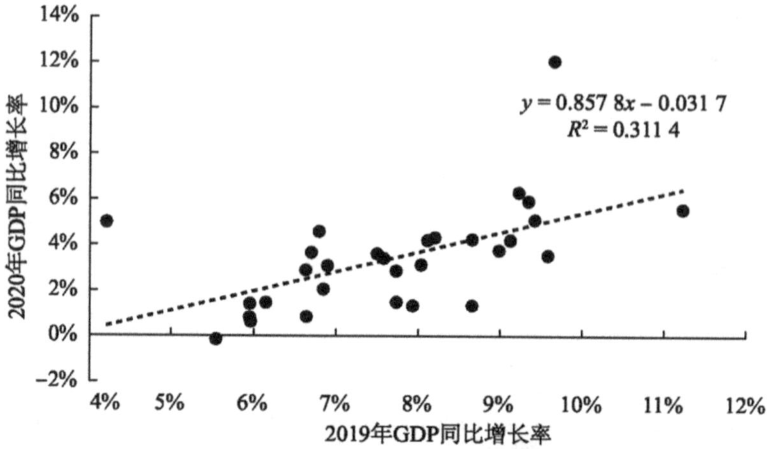

Fig. 6 Year-on-year provincial GDP growth rates in 2019 and 2020 (*Note* This figure does not contain the data of Hubei) (*Data Source* Wind information and the National Bureau of statistics of China)

and going up by 0.09 percentage points year-on-year. Sun (2020) pointed out that from a long-term perspective, the epidemic did not cause any fundamental changes to the structure of China's regional economy: From 2006 to 2019, the changes in the eastern, central, western, and northeastern regions' contributions to the national GDP ran between 2 and 3 percentage points, with the average annual change of about 0.5 percentage points. Therefore, we have reason to believe that the above-mentioned changes in the regions' proportions in the national GDP are normal fluctuations instead of significant structural changes caused by the epidemic.

Table 2 Regional contributions to the National GDP in 2020 and their Year-on-year changes

	Time period	The eastern region (%)	The central region (%)	The western region (%)	The northeastern region (%)
The national total	2020	54.12	23.67	17.16	5.05
	2019	54.03	23.95	16.92	5.10
	Year-on-year change in GDP share	0.09	−0.28	0.24	−0.05
The national total (excluding Hubei)	2020	56.55	20.24	17.93	5.28
	2019	56.67	20.24	17.75	5.35
	Year-on-year changes	−0.12	0.01	0.19	−0.07

Data Source The National Bureau of statistics of China

3. **The Influence of the COVID-19 Epidemic on the Structure of the Regional Economy.** The epidemic's impact on the structure of the regional economy becomes even more insignificant after excluding the data of Hubei. The final three lines in Table 2 show the four regions' GDP shares in 2020 after the removal of Hubei's data: compared with the statistical results involving Hubei's data, the changes in the regions' GDP shares are reduced to within 0.2 percentage points, with the eastern region's share decreased slightly and the central region's share increased slightly. The decrease (−0.12%) in the GDP share of the eastern region probably results from the slow economic recoveries in Shanghai and Liaoning. The almost negligible change (0.01%) in the central region's GDP share demonstrates the region's high economic resilience: it has the ability to offset the negative influence of the economic plunge of Hubei, which is deeply troubled by the epidemic. The rise (0.19%) in the western region's GDP share is not surprising: the region's rapid growth in recent years and the fact that it is situated far away from the areas severely impacted by the epidemic can explain this positive change. The decrease (−0.07%) in the northeast region's GDP share gets more evident when the statistical results do not involve the data of Hubei.

On the whole, whether the influence of Hubei's economic plunge is considered or not, there are no substantial changes in the structure of the regional economy. The changes in the regions' GDP shares are within 0.3 percentage points. The GDP share of the western region increased slightly, while that of the northeast region decreased slightly. As to the eastern and central regions, their proportions in the national GDP are different under different preconditions, but the changes are small.

1.2 New Requirements for Regional Economic Development Under the Conditions of Regularized Epidemic Containment Efforts

While implementing regularized epidemic containment measures within itself, China had to cope with increasingly complicated domestic and international situations. Internationally, the spread of the virus aggravated the global economic downturn and the deglobalization trend, which propelled China into pursuing a development paradigm featuring "dual circulation." This created new opportunities for its regional economy. Domestically, the epidemic exposed shortcomings in public service systems and urban development patterns, which needed reasonable solutions. Given the new opportunities and challenges emerging under the conditions of regularized epidemic containment efforts, new requirements for China's regional economy were put forward.

1. **The Pursuit of a Coordinated Regional Development for the New "Dual-Circulation" Paradigm.** The spread of the virus brought about a global supply chain crisis: the global supply chain was interrupted, and the localization rates

of production activities were increased. This severely impacted the process of economic globalization. Given the Chinese government's decision to regularize its epidemic containment efforts and the continuous global economic recession, China's economic sector could hardly follow its old development path featuring high dependence on the international market and felt the urge to improve its development quality and solve the problem of unbalanced development. It called for a new development strategy. Under such circumstances, the Standing Committee of the Political Bureau of the CPC Central Committee first proposed the following measures: advancing the supply-side structural reform; giving full play to China's advantages, namely, its super-large-scale market and the huge potential of its domestic demand; and pursuing a new development paradigm involving both domestic and international circulations. The 5th Plenary Session of the 19th CPC Central Committee re-emphasized these points when adopting the "Proposal of the CPC Central Committee to Formulate the Outline of the 14th Five-Year Plan for Economic and Social Development (2021–2025) and Long-Range Objectives Through the Year 2035." To pursue the new "dual-circulation" development paradigm, it is essential to take domestic demand as the starting point and foothold in our endeavor to remove obstacles in our production, distribution, exchange, and consumption links and ensure smooth circulations for our factors, industries, regions, and rural–urban systems.

For a long time, our regional development has been unbalanced and inadequate. With the beginning of China's reform and opening up, the eastern region has taken the lead in actualizing high economic growth by leveraging its advantages in location, policy support, and ties with the international market. Nevertheless, the central and western regions have developed relatively slowly during these years, and the northeast region has even experienced an economic recession in recent years. Now that China has entered a new normal in economic development, the gap between North and South China becomes increasingly apparent, further exacerbating the trend of regional economic differentiation. The unbalanced development of our regional economy has hindered the flow of our talents, capital, and industries. Given the central and western regions' huge population losses to the southeast coastal areas, some cities have become shrinking cities troubled by the problem of labor shortage. Because of the industrial upgrading and transformation in the eastern region, some traditional industries moved to the central and western regions, which to some extent, has boosted these regions' development. Nevertheless, the upgrading and transformation have not fundamentally changed the development pattern of the eastern region. As a result, the central and western regions still lag behind the eastern regions in industrial development.

For the success of the new "dual-circulation" paradigm, it is essential to coordinate the development of different regions. Some scholars have fully recognized the importance of balanced regional economic development in pursuing the new paradigm. Jia (2020) regarded the development campaign of the western region as one of the three major engines of domestic circulation in China. Guo (2020) pointed out that regional economic development disparities hinder the flow of people and goods among regions and that coordinated development of the regions should be promoted through an adjustment to the regional economic arrangements. The coordination among the regions helps different localities give full play to their comparative advantages and encourages them to actively participate in the domestic and international value chains and find development paths that are suitable for themselves. In this way, all regions in the country can mobilize their resources for high-quality economic development, improved living standards for the people, and an expanded domestic market. With a strong domestic market, our economy will have endogenous drivers, and our domestic circulation will be smooth. The central and western regions need to seize opportunities brought by this new paradigm. By enhancing their infrastructure such as transportation networks, ensuring equal accessibility to public services, and promoting the free flow of factors among the regions, they can develop new industries, discover new development drivers, and promote high-quality regional development.

2. **Efforts to Promote Equal Accessibility to Public Services for Enhanced Regional Risk Management Abilities.** As an essential part of the new-era regional-coordination mechanism, efforts exerted to ensure equal accessibility to public services are also important means to improve the regions' risk management abilities. The outbreak of the COVID-19 epidemic has fully exposed China's shortcomings in its public health service and risk management systems, such as the shortage of emergency supplies for epidemic prevention and control purposes. Regarding these issues, General Secretary Xi Jinping published an article in *Qiushi* (Seeking Truth) journal to highlight the importance of enhancing our epidemic-prevention-and-control systems and mechanisms and improving our national public health emergency management system, which was entitled "Comprehensively Promote Law-Based Prevention and Control Measures and Improve Our National Public Health Emergency Management System."

Public services are provided to meet the basic needs of our people, ensure the normal functioning of our society, and promote social equity and development. To meet our people's longing for a better life and realize the coordination of our economic regions, it is essential to adopt measures to ensure the effective release of our public service resources, equal accessibility to our public services, and improved living standards for all people. Such measures highlight social justice, enhance our people's sense of fulfillment, and embody the principle of building together and sharing together as a community. It must be noted that China has shortcomings not only in its public health service system but also in its education, medical care, old-age care, and public cultural service systems and that regional differences in such public services exist extensively in China. Efforts made to ensure equal accessibility to public services help people in different regions develop their abilities and improve their caliber, which is conducive to our regional economic development. Inequality in this regard shall be avoided as it creates obstacles to coordinated regional development. In addition, a region's public service system shall be complete since any shorting in any public service link or field can have a systematic impact that affects the economic operation of the region. A complete, efficient public service system helps to enhance a region's resistance to external shocks and improve its crisis response ability and economic recovery ability.

3. **Efforts to Improve the Carrying Capacity of Central Cities and City Clusters and Appropriately Limit the Size of Mega Cities.** The COVID-19 epidemic has also revealed problems in China's urban development patterns. Many scholars have studied these issues. Shi (2020) observed that Chinese cities' over-reliance on spatial agglomeration facilitated the massive spread of the virus and that since this traditional pattern used for a centralized utilization of economic and land resources faced great challenges in the "post-epidemic" era, it was essential to give full play to the knowledge spillover effect and pursue economic modernization and transformative development. Duan (2020) emphasized that it was necessary to change the traditional center-periphery model for urban development, avoid an excessive obsession with city-scale expansion, and pursue a networked and multi-center model. Li (2020), however, insisted that big cities with more active economic factors and more diversified economies have stronger innovation abilities and better economic resilience under epidemic conditions.

Overall, whether in a large-city- or small-city-based development model, work must be done to improve the carrying capacities of the central cities or city clusters. The 5th Meeting of the Central Committee for Financial and Economic Affairs clarified the necessity of enhancing the economic carrying capacity and population-carrying capacity of economically advantaged regions such as central cities and city clusters. The central cities and city clusters are the main engines of China's economic development. They have a strong appeal to economic resources and population but face grave problems such as land resource shortage, unreasonable spatial planning, resource bottlenecks, environment constraints, and the "big city disease." In particular, some cities suffer from problems caused by unreasonable planning. The population density is particularly high in some of their urban areas and very low in the rest, which leads to difficulties for commuters, low utilization efficiency of land resources, and great pressure on resources and the environment in some urban areas. These problems were particularly obvious during the epidemic: the high population density, long commuting times, and large-scale population movements facilitated the spread of the virus in cities. Therefore, it is a must to improve the carrying capacity of central cities and city clusters by reasonably planning their urban space, properly adjusting their division of administrative regions, and strengthening inter-city coordination. In the meantime, it is necessary to limit the scale of megacities and speed up the drive to relocate the non-core functions of megacities to their surrounding cities. Besides, the influence of central cities should be enhanced to promote the coordinated development of large, medium, and small cities.

2 Digital Economy for New Drivers of Regional Economic Development

With the continuous improvement of information and communication technologies, the "Internet + " new business models have gradually matured over the years. IoT, cloud computing, big data, artificial intelligence, 5G, blockchain, and other digital economy technologies have developed rapidly, continuously creating new business patterns and models. For instance, during the COVID-19 epidemic of 2020, China's digital economy contributed a lot to the country's epidemic containment by providing services such as online office, online education, online medical treatment, etc., which further expanded the scope of the digital economy, creatively solved problems, and helped to enhance the country's economic resilience. In the meantime, the digital economy became increasingly integrated with the country's industrial development. Given the continuously-consolidated digital infrastructure and computing and data-operation capabilities, industrial digitalization has broad prospects in China. Just as *People's Daily* remarked in an editorial published on July 8, 2020, the digital economy will start and fuel a new economic cycle. As the vigorous development of the digital economy is bound to have a profound impact on the future development of China's regional economy, which is the focus of this book, it is necessary to find answers to the following questions: What is the digital economy? What is the status quo of China's digital economy? How will the digital economy affect China's regional economic development?

2.1 The Concept and Status Quo of the Digital Economy

As early as the 1990s, the Organisation for Economic Co-operation and Development (OECD) first proposed the concept of the digital economy. Later, this concept became increasingly broad with the maturation of Internet technology and the increase in the number of its application scenarios. The "G20 Digital Economic Development and Cooperation Initiative," released at the 2016 G20 Hangzhou Summit, defines the digital economy as follows: "The digital economy refers to a broad range of economic activities that include using digitized information and knowledge as the key factor of production, modern information networks as an important activity space, and the effective use of information and communication technology (ICT) as an important driver of productivity growth and economic structural optimization." From this definition, it can be seen that the digital economy is not an independent industry but rather an interactive industrial system made up of multiple industries. It includes (1) information technologies such as 5G, big data, cloud computing, artificial intelligence, and IoT and communication facilities that underpin the development of the digital economy and (2) the digital transformation and upgrading of high-tech industries, traditional manufacturing industries, and other industries, such as optimizing their corporate organizations and production and trade patterns and reconstructing the traditional value chains. In summary, the digital economy includes not only the new digital industry but also the digitalization of the existing industries, involving all aspects of corporate production and economic activities.

Governments around the world have adopted corresponding policies to speed up the development of the digital economy. The UK government issued its Digital Charter in 2018 as a way of creating good conditions for the growth of its digital economy. In the same year, France introduced a series of policy documents on the digital economy, such as its 5G road map and program for the use of the digital economy for industrial transformation. The United States formulated policies such as its data science strategic plan and its national strategy for American leadership in advanced manufacturing, which made it clear that priority should be given to the development of smart and digital manufacturing. Germany released its "High Technology Strategy 2025" in 2018 to promote artificial intelligence applications. In recent years, the Chinese government has also introduced several policies to promote the development of the digital economy. In 2020, China's top-level administrative bodies including the National Development and Reform Commission (NDRC), the Ministry of Industry and Information Technology (MIIT), and the Office of the Central Cyberspace Affairs Commission (OCCAC) issued six policy documents (See Table 3), which explicitly took digital resources as a factor of production and proposed to accelerate the digital transformation of industries and develop new digitized industries for the creation of new drivers for the real economy. China's 14th Five-Year Plan also clearly proposes to develop the digital economy and promote the in-depth integration of the digital economy and the real economy. The Plan also contains a series of related arrangements such as those about data resource protection, relevant international rule-making, and a national unified platform for data sharing.

China's fasting-developing digital economy is now second only to the United States in size. According to estimates by the China Academy of Information and Communications Technology (CAICT), the global digital economy reached 31.8 trillion yuan in 2019 and grew 5.3% year-on-year, constituting an important driver of world economic growth. Table 4 shows some statistical results about the development of the top 10 digital economies in the world in 2019: the United States ranked first in the world, accounting for 41.09% of the world's total; China ranked second, with its digital economy totaling US $5.2 trillion and lagging far behind the United States; the other countries' digital economy did not cross the threshold of US $2.5 trillion. As to the digital economy's contribution to GDP, the United States and Germany saw over 60% of their GDPs come from the digital economy; Japan, France, and South Korea relied on the digital economy to realize more than 40% of their GDPs; for China, the number stood at only 36.17%. In comparison with other countries, China still had to increase the penetration rates of its digital economy in different sectors. Regarding the growth rate of the digital economy, China topped the list with an increase as high as 16.5%, showing strong growth momentum. With the support of our existing policies and the in-depth development of industrial digitalization, China can further tap its potential for digital economy development in the future.

Table 3 Policy documents on the digital economy issued by China in 2020

Serial number	Document title	Issuing authorities	Main contents
1	The Outline of the 14th Five-Year Plan for Economic and Social Development (2021–2025) and Long-Range Objectives Through the Year 2035 of the People's Republic of China	The State Council	The development of the digital economy must be boosted, including digital industrialization and industrial digitalization. The digital economy and the real economy shall be integrated, and a digital industrial cluster with international competitiveness shall be built. Efforts must be exerted to strengthen the development of digital society and digital government and improve the use of digital and intelligent technologies in public service and social governance sectors
2	Notice on Promoting the Accelerated Development of the Industrial Internet	The MIIT	A total of 20 measures are proposed regarding the new-type infrastructure construction, the integration of innovations and applications, security systems, drivers for innovative development, an ecological industrial layout, and industrial policy supports
3	Guideline to Fuel the Use of Digital Technologies Among Small and Medium-Sized Enterprises	The MIIT	Small and mid-size enterprises (SMEs) shall be encouraged to resume their work and production and increase their potential for development by adopting digital, network, and intelligent technologies
4	Opinions on Building a More Complete System and Mechanism for Market-Oriented Allocation of Factors	The State Council	Put efforts into the tasks of developing our data factor market, promoting the sharing of our government data, increasing the value of our social data resources, and strengthening the integration and security protection of our data resources

(continued)

Table 3 (continued)

Serial number	Document title	Issuing authorities	Main contents
5	Plan to Boost the Development of the New Economy by Promoting the Use of Cloud, Data, and Intelligence Technologies	The NDRC and The OCCAC	New business forms of the digital economy shall be explored, and the digital transformation of enterprises shall be pushed forward. A data supply chain and a digital ecosystem covering the upstream and downstream of our industrial chain need to be developed. The classic paradigm, which consists of equipment digitalization, production line digitalization, workshop digitalization, factory digitalization, enterprise digitalization, industrial chain digitalization, and digital ecosystem, shall be pursued
6	Opinions on Supporting the Healthy Development of New Business Forms and Models and Activating Consumer Markets for Expanding Employment	13 authoritative bodies including the NDRC	The digital transformation of our industries must be accelerated, and new drivers for our real economy shall be strengthened. The platformization of relevant industries must be encouraged, and the digital transformation of traditional enterprises needs to be pressed ahead

Source Information available on the Internet

2.2 *Digital Economy Development in Various Localities of China*

1. **Related Policies Documents.** Provinces and cities across China have actively provided strong policy support for digital economy development. Table 5 summarizes the digital economy-related policy documents issued in recent years in various provinces and municipalities. As shown in the table, provinces of China have introduced plans for digital economy development since 2016, expounding upon the strategic focus of digital economy development, the digital transformation of manufacturing and service industries, digital government, and guarantee mechanisms for digital economy development. Leveraging their advantages in digital infrastructure, Zhejiang and Fujian first promulgated provincial digital economic development plans in 2016. Guizhou was the first western province to make plans for its digital economic development and achieve remarkable results: the growth rate of its digital economy remained the highest in China for five consecutive years. Later, Guangxi, Qinghai, Inner Mongolia, and Tibet also issued their special plans for digital economy development, and other provinces quickly stepped up their efforts to study and formulate their plans for digital

Table 4 The development of digital economy in the Top 10 countries with the strongest digital economy

Country	Scale of the digital economy (USD)	Contribution to national GDP (%)	Contribution to the global digital economy (%)
The United States	13,065,200,000,000	60.96	41.09
China	5,195,400,000,000	36.17	16.34
Germany	2,438,000,000,000	63.39	7.67
Japan	2,394,900,000,000	47.12	7.53
The United Kingdoms	1,760,600,000,000	62.27	5.54
France	1,169,800,000,000	43.08	3.68
South Korea	799,500,000,000	48.68	2.51
India	585,600,000,000	20.53	1.84
Canada	45,400,000,000	26.15	1.43
Mexico	290,300,000,000	23.07	0.91
Brazil	390,100,000,000	21.20	1.23
Italy	389,400,000,000	19.46	1.22

Source The "New Picture of Global Digital Economy (2020)" released by the China academy of information and communications technology (CAICT)

economy development during the period of the 14th Five-Year Plan to promote the orderly and healthy development of the digital economy. To promote the innovation and development of the digital economy in different localities, China officially launched the plan to build six national digital economy innovation and development experimental zones in Heibei (Xiongan New Area), Zhejiang, Fujian, Guangdong, Chongqing, and Sichuan at the 6th World Internet Conference in 2019. As of 2020, all relevant provinces except Hebei have introduced their plans for their national digital economy innovation and development experimental zones. It should be noted that Hainan, Ningxia, and Xinjiang have not released any supportive policies.[1]

2. **The Status Quo.** To further analyze the situation of digital economic development in different provinces, we measure the situation of digital economic development for each of them. As there are few studies on the measurement of provincial digital economic development, we choose to adopt the methods proposed by Zhao et al. (2020) and Liu et al. (2020): We decompose the evaluation of digital economy development into three dimensions, namely informatization, Internet development, and digital transaction development and select 13 indicators for the measurement. To truthfully reflect the regional differences in digital economy development and conduct comparative studies across time periods, all data are

[1] The information comes from this Chinese website https://www.chinanews.com/gn/2019/08-09/8921787.shtml.

Table 5 Policy documents on the digital economy issued by Chinese localities from 2016 to 2020

Serial number	Province	Document title	Time of issue
1	Beijing	Beijing's Action Plan for Promoting Innovation and Development of the Digital Economy (2020–2022)	September 7, 2020
2	Tianjin	Tianjin's Action Plan for Promoting Innovation and Development of the Digital Economy (2019–2023)	June 3, 2019
3	Hebei	Policies on Promoting the Accelerated Development of the Digital Economy	October 1, 2020
4	Shanxi	Shanxi Province's Policies on Promoting the Accelerated Development of the Digital Economy	August 26, 2019
5		Shanxi Province's Opinions on Promoting the Accelerated Development of the Digital Economy	August 26, 2019
6	Inner Mongolia	Digital Inner Mongolia Development Plan From 2018 to 2025 (Draft)	January 11, 2019
7		Opinions of Inner Mongolia Autonomous Region People's Government on Promoting the Development of the Digital Economy	December 31, 2019
8	Liaoning	Liaoning's Guideline on Developing the Digital Economy	October 18, 2020
9	Jilin	Opinions on Accelerating the Replacement of Old Growth Drivers with New Ones and Promoting High-Quality Development Through the Development of Digital Jilin	July 17, 2018
10	Heilongjiang	Digital Heilongjiang Development Plan (2019–2025)	June 4, 2019
11	Shanghai	Shanghai's Opinions on Promoting the High-Quality Development of the Real Economy by Accelerating the Development of the Digital Economy	October 18, 2019
12	Jiangsu	Opinions of the General Office of the Provincial Government on the In-Depth Development of the Digital Economy	October 8, 2020
13	Zhejiang	Work Plan for the Development of Zhejiang National Digital Economy Innovation and Development Experimental Zone	December 23, 2020
14		Five-Year Plan for the Multiplying Growth of the Digital Economy in Zhejiang Province	September 14, 2018
15		The 13th Five-Year Plan for the Informatization Development in Zhejiang Province (The Development Plan of Digital Zhejiang 2.0)	November 15, 2016
16	Anhui	Notice of Anhui Provincial People's Government on Printing and Distributing Several Policies to Support the Development of the Digital Economy	October 23, 2018

(continued)

Table 5 (continued)

Serial number	Province	Document title	Time of issue
17		Guiding Opinions on Accelerating the Construction of Digital Jianghuai	September 12, 2018
18	Fujian	Development Plan for National Digital Economy Innovation and Development Experimental Zone (Fujian)	February 14, 2020
19		Opinions on Accelerating the Innovation Development of the Industrial Digital Economy of the Province	February 6, 2018
20		Fujian Province's 13th Five-Year Plan for the Development of Digital Fujian	May 17, 2016
21	Jiangxi	Jiangxi Province's Opinions on Pursuing the Digital Economy Development Strategy	February 18, 2019
22		The Three-Year Action Plan for the Development of Digital Economy (2020–2022)	April 16, 2020
23	Shandong	Shandong Province's Opinions on Supporting the Development of the Digital Economy	July 12, 2019
24		Plan for the Development of Digital Shandong (2018–2022)	February 27, 2019
25	Henan	The 2020 Work Plan for the Development of the Digital Economy in Henan Province	May 31, 2020
26		Henan Province's Plan to Accelerate the Development of the Digital Economy	June 27, 2019
27		Henan Province's Three-Year Action Plan for Intelligent Manufacturing and Industrial Internet	April 18, 2018
28	Hubei	Measures to Accelerate the Development of the Digital Economy and Create New Economic Growth Engines	June 17, 2020
29	Hunan	Hunan Province's Digital Economy Development Plan (2022–2025)	January 18, 2020
30	Guangdong	Work Plan for the Development of Guangdong National Digital Economy Innovation and Development Experimental Zone	November 28, 2020
31		Guangdong Province's Regulations on Promoting the Digital Economy (Draft)	November 4, 2020
32		Guangdong Province's Action Plan for Developing Digital Industrial Clusters (2019–2025)	May 6, 2020
33		Measures to Speed Up the Building of Top Digital and Innovative Cities	March 30, 2020
34		Guangdong Province's Digital Economy Development Plan (2018–2025)	April 10, 2018

(continued)

Table 5 (continued)

Serial number	Province	Document title	Time of issue
35	Guangxi	Measures to Accelerate the Development of the Digital Economy in Guangxi	October 15, 2020
36		Guangxi Province's Digital Economy Development Plan (2018–2025)	September 17, 2018
37		Three-Year Action Plan for the Development of the Digital Economy in Guangxi (2018–2020)	August 30, 2018
38		Measures to Speed Up the Development of the Digital Economy in Guangxi	August 29, 2018
39	Chongqing	Work Plan for the Development of Chongqing National Digital Economy Innovation and Development Experimental Zone	June 22, 2020
40	Sichuan	Work Plan for the Development of National Digital Economy Innovation and Development Experimental Zone (Sichuan)	October 22, 2020
41		Opinions of Sichuan Provincial People's Government on Accelerating the Highly Integrated Development of the Digital Economy and the Real Economy	November 16, 2018
42		Opinions of Sichuan Provincial People's Government on Accelerating the Development of the Digital Economy	August 6, 2019
43	Guizhou	Opinions of the Provincial People's Government on Promoting the Innovative Development Featuring Big Data, Cloud Computing, and Artificial Intelligence Technologies and Accelerating the Development of Digital Guizhou	June 21, 2018
44		Notice of Guizhou Provincial People's Government on Printing and Distributing the Plan for Implementing the Cooperation of Ten Thousand Enterprises and Boosting the Development of the Digital Economy	February 11, 2018
45		Opinions of the CPC Guizhou Provincial Committee and Guizhou Provincial People's Government on Accelerating the Development of the Digital Economy	March 21, 2017
46		Guizhou Province's Digital Economy Development Plan (2017–2020)	February 7, 2017
47	Yunnan	Digital Yunnan Three-Year Action Plan From 2019 to 2020 (Draft)	July 10, 2019
48		Yunnan Province's Three-Year Action Plan for Developing the Digital Economy and Promoting the Industrial Transformation and Upgrading	November 22, 2018

(continued)

Table 5 (continued)

Serial number	Province	Document title	Time of issue
49	Tibet	Digital Economy Development Plan of Tibet Autonomous Region (2022–2025)	May 21, 2020
50	Shaanxi	Key Points in the Work of Developing the Digital Economy in Shaanxi Province in 2020	July 2, 2020
51	Gansu	Special Action Plan for Promoting the Development of Digital and Information Industries in Gansu Province	June 11, 2018
52	Qinghai	Opinions on Accelerating the Development of the Digital Economy in Qinghai Province	2019
53		Qinghai Province's Digital Economy Development Plan (2019–2025)	2019

Note There is no information available on the Internet specifying the dates for the release of the two policy documents by Qinghai province
Source of Information This table is based on the information available on the Internet

processed by a linear standardization, and the maximum and minimum values are based on the data of the reference period which is the year 2017. As to the method of weighting indicators, the existing research mainly adopts the principal component analysis method, which is suitable for the determination of the weighting of parallel indicators. Considering the hierarchy in our indicator system, we adopt the method of hierarchical average weighting, that is, the weight of each-level indicator = 1 / the number of the indicators at the level. The specific indicators selected and their weights are shown in Table 6. Among them, the optical cable density, the base station density, and the network access density are calculated by dividing the absolute number of each province by the area of each province.

Despite that each province sees remarkable growth in its digital economy, the provinces can still be roughly divided into three groups based on their different digital economy development levels. According to their provincial digital economy indexes of 2019, the 31 provinces are arranged in descending order in Fig. 7. Overall, a three-tier system can be used to describe the provincial difference in digital economy development. Guangdong is the first-tier province that outshines its peers in terms of digital economy development. Its digital economy index is as high as 1.09, 0.27 higher than the second-ranked province. The second-tier provinces include Shanghai, Beijing, Zhejiang, and Jiangsu. Their differences in digital economy indexes are very small, running between 0.74 and 0.81. The third-tier group has similar characteristics: provinces in this group do not differ much from one another in terms of the digital economy index. Shandong ranks first in this group, with its index 0.19 lower than that of Jiangsu. The third-tier provinces can be subdivided into three sub-groups. The first-tier sub-group consists of Shandong, Sichuan, Henan, Hebei, Fujian, and Anhui, with their digital economy indexes running between 0.4 and 0.6. The third-tier sub-group contains only Qinghai and Tibet, whose digital economy indexes are less

Table 6 Evaluation indicator system for the development of the digital economy

First-level indicator	Weight	Second-level indicator	Weight	Measurement indicator	Weight
Informatization Development index	0.333 3	Informatization infrastructure	0.166 7	Optical cable density	0.055 6
				Base station density	0.055 6
				Employment in the information sector	0.055 6
		Informatization influence	0.166 7	Telecommunication services revenue	0.055 6
Network Development index	0.333 3	Fixed terminal Network infrastructure	0.083 4	Network access density	0.083 4
		Mobile terminal Network infrastructure	0.083 4	Mobile phone penetration	0.083 4
		Fixed terminal Network influence	0.083 4	Broadband penetration	0.083 4
		Mobile terminal Network influence	0.083 4	Mobile network penetration	0.083 4
Digital transaction Development indicator	0.333 3	Digital transaction Infrastructure	0.166 7	Proportion of enterprise websites	0.055 6
				Proportion of computers used by enterprises	0.055 6
				Proportion of e-commerce	0.055 6
		Digital transaction influence	0.166 7	E-commerce sales	0.083 4
				Online retail sales	0.083 4

Note This evaluation indicator system is based on the findings of Liu (2020)

than 0.2. The remaining provinces belong to the second-tier sub-group whose digital economy indexes come between 0.2 and 0.4. The huge development gaps shown in this figure reflect a major problem we face in developing the digital economy, which is referred to as the "digital divide." As for the changes in the digital economy index, Guangdong's digital economic growth is the most significant, nearly achieving double growth: its index growth is 0.5 between 2018 and 2019. Among the second-tier provinces, Jiangsu and Zhejiang are the best performers, with their indexes increased by 0.33 and 0.29, respectively. Three third-tier provinces, namely Henan, Shandong, and Sichuan, also deliver good performance, with their index growths running between 0.25 and 0.26. The increase of Heilongjiang, Jilin, Ningxia, Qinghai, Hainan, and Tibet is relatively slower: their indexes grow by less than 0.1. Although all provinces have actualized significant growth, there is a big gap in their growth amount. Provinces with higher digital economy development levels grow faster than those that lag behind, further widening the "digital divide." Therefore, attention must be paid to the polarization issue against the background of universal growth. Efforts must be made to accelerate the development of the provinces that lag behind and help those with moderate development levels make breakthroughs. Only in this way can we gradually change this polarization phenomenon and bridge the "digital divide."

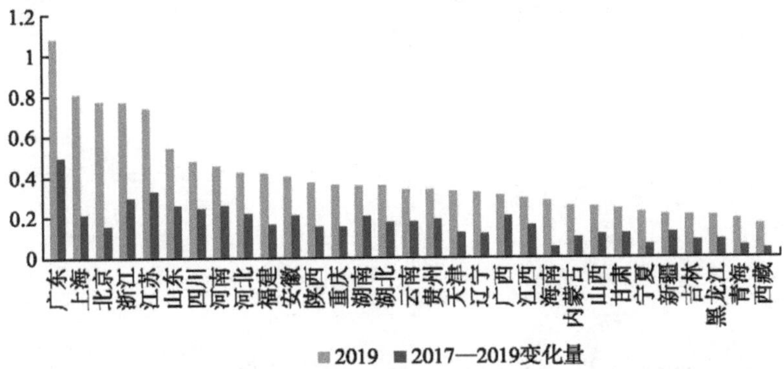

Fig. 7 Provincial digital economy indexes in 2019 and changes From 2017 to 2019 (*Source of Information* EPS data platform)

2019	Indexes for 2019
2019–2017变化量	Differences between 2017 and 2019
西藏	Tibet
新疆	Xinjiang
湖南	Hunan
贵州	Guizhou
青海	Qinghai
广西	Guangxi
云南	Yunnan
甘肃	Gansu
四川	Sichuan
江西	Jiangxi
吉林	Jilin
宁夏	Ningxia
陕西	Shaanxi
内蒙古	Inner Mongolia
福建	Fujian
河南	Henan
重庆	Chongqing
河北	Hebei
北京	Beijing
山西	Shanxi

(continued)

(continued)

江苏	Jiangsu
安徽	Anhui
浙江	Zhejiang
海南	Hainan
山东	Shandong
广东	Guangdong
辽宁	Liaoning
上海	Shanghai
黑龙江	Heilongjiang
天津	Tianjin
湖北	Hubei

The development of the digital economy in China's four major economic regions is also unbalanced. There is an obvious development gap within the eastern region, and the region also contains the top five provincial digital economies shown above. The central and western regions lag behind. To further examine the development gaps among and within the four major economic regions, we calculate the variability coefficient within each region, take the average of the provincial digital economy indexes within a region to represent its digital economy development level, and then calculate the coefficient of variation of the four regions, as shown in Table 7. These four regions are arranged as follows according to their digital economy development levels: the eastern region, the central region, the western region, and the northeastern region. The gap between the top-ranked eastern region and the other regions is big. The data for the year 2017 show small differences among the central, western, and northeastern regions. Nevertheless, the central region realizes fast growth and gradually leaves the western and northeastern regions behind after 2017. It is worth noting that although provinces with better digital economies realize bigger growth, the eastern region's growth rate is not so high as that of the central region. Therefore, the inter-region variation coefficient gradually shrinks year by year. As to the intra-region development gaps, the biggest gap is found in the eastern region, the second biggest one, in the western region, and the smallest one, in the central region. The eastern region's intra-region coefficient of variation is even close to its inter-group coefficient of variation. The intra-region variation coefficients for the central and western regions increase slightly year by year. This shows that although the inter-regional development gap is gradually narrowing, the gaps within the regions are still widening. It is a must to diminish the development gaps among different localities and promote the balanced development of the digital economy.

Table 7 Four major economic regions' performances in digital economy and their coefficients of variation

Index	Region	2017	2018	2019
Mean value	The eastern region	0.387 7	0.501 0	0.619 4
	The central region	0.166 0	0.262 6	0.353 9
	The western region	0.154 9	0.227 1	0.290 7
	The northeastern region	0.149 3	0.200 7	0.245 7
Intra-region coefficient of variation	The eastern region	0.443 2	0.416 4	0.416 9
	The central region	0.170 2	0.198 3	0.207 8
	The western region	0.295 4	0.299 8	0.318 0
	The northeastern region	0.269 4	0.222 4	0.227 1
Inter-region coefficient of variation		0.499 6	0.431 0	0.413 5

2.3 *The Digital Economy Causing Structural Changes to China's Regional Economy*

The essence of the digital economy is the economics of a two-sided market. Digital technologies break the space constraints between market entities to ensure efficient communication for them. These technologies remold our production sector to create new value chains. They help to timely provide customized products or services, which makes our social division of labor more component-based. They can change our traditional ways to organize activities in our production, market, and public service sectors, ensuring the efficient allocation of our social resources and adding new impetus to our economic development. The development of the digital economy, in return, creates broad prospects for the development of the digital industry for our economic regions. This is because the digital economy needs the support of a series of technologies such as 5G, big data, cloud computing, IoT, and artificial intelligence. In a word, in the era of the digital economy, data has become a new factor of production; intelligent algorithms have become a new mode of production; and the impact of the digital economy on our economic and social activities is widespread, all-inclusive, and profound. As a new driver of our economy, the digital economy can cause substantial changes to the structure of our regional economy.

As mentioned in the previous sub-section, the widening of the existing inter-regional development gaps in the development of the digital economy has caused a grave problem called the "digital divide." If this trend continues, the new round of industrial revolution fueled by the digital economy will further expand the inter-regional development gaps. Localities that have achieved a leading position in the development of the digital economy can achieve nonlinear, enormous growth, further expanding their development advantages. Those that have lagged behind cannot realize such rapid development, thus failing to keep up the pace and further widening the development gaps. Nonetheless, it must be noted that the problem of the "digital divide" is not unsolvable. If underdeveloped areas leverage their unique advantages, they can still seize development opportunities and realize the rapid development of the digital economy. In this way, they have a chance to overcome the difficulties caused by their geographical locations, resource bottlenecks, and environment constraints. They can rely on the development of the digital economy to achieve leapfrog development, thus changing the existing regional economic structure and promoting balanced development.

First, underdeveloped localities usually have low-cost land and labor resources and strong policy supports, offering a fault-tolerate, favorable environment for the development of new products, new services, and new modes. Take the western region as an example. The land costs there are low, which is attractive to data center runners who need large space to meet the cooling requirements of large servers. These large servers are important facilities that underpin the rapid development of the digital economy. Apart from its cheap land, the region also has rich power resources supplied stably and at reasonable prices (the northwestern region is rich in wind power resources and the southwestern region is rich in hydro-power resources). The electricity bills there exclude the costs of long-distance power transmission, which provides a reliable guarantee for the stable operation of big data service centers in the region. These advantages support the region's participation in digital economy development.

Second, the digital economy can help underdeveloped regions to solve problems related to public services equalization. By offering high-quality online public services such as online education and medical care services to the population in underdeveloped areas, the digital economy solves the problem of insufficient public service supplies for the regions, which is caused by their low population density, insufficient scale merit, and lack of funds. This will meet the people's growing needs for a better life. Such a change in the mode of public service supply may change the current mode of population mobility: it may reduce the population loss of the central and western regions, enrich their local labor resources, and help to provide high-caliber talents for the development of their local high-tech industries, which will boost their local economic development.

Third, by developing the digital economy, underdeveloped regions can upgrade their traditional industries and traditional production modes, promoting the digital transformation of their traditional industries. The digital economy can help these regions directly contact the consumer markets, which is conducive to the consumption-driven improvement of products and services. Under such circumstances, products that meet market demands will be manufactured, avoiding problems caused by product life cycles, and traditional industries will be reshaped and upgraded. This point is especially applicable in the northeastern region. Just as Dong (2019) has pointed out, by developing the digital economy, the northeastern region can upgrade its traditional industries, create new economic drivers to replace the old ones, and promote the networked, intelligent, and flexibility-improved transformation of its traditional manufacturing sector. Such efforts will inject new vitality into its regional economic development. In addition to the economic benefits, the digital economy can also promote the development of e-government and smart cities, enhance urban governance capabilities, and improve the regional business environment. This can reduce the institutional costs of enterprise development, reduce the burdens for enterprises, and improve the efficiency of economic operations.

Fourth, digital economy development facilitates the integrated development of a region. Many scholars have proved this point theoretically. Fan (2020) observed that the digital economy created favorable conditions for the establishment of an all-factor regional interest balance mechanism, and Wang et al. (2020) proved the digital economy's contribution to the integrated development of the YRD by analyzing the panel data of 41 cities in the YRD region from 2011 to 2018. Existing research also shows the digital economy's positive influence on the integrated development of urban and rural areas. He et al. (2020) have explained the digital economy's positive role in promoting the development of China's inclusive finance (China inclusive finance is aimed to remove barriers that exclude people from participating in the financial sector and use financial services to improve their lives), strengthening the inclusiveness of economic growth, and narrowing the gap between urban and rural areas. In practice, the "Outline of the Integrated Regional Development of the Yangtze River Delta" issued by Chinese authorities explicitly puts forward the requirements of creating a digital YRD and building new-generation information infrastructure, highlighting the development of the digital economy as a method to boost the integrated development of the delta. As the development of the CCEC is emphasized as a national strategy, Chongqing's Liangjiang New Area and Chengdu's Tianfu New Area have jointly held an industrial service summit for the CCEC, with the two cities reaching a consensus on promoting regional integration through the digital economy.

3 New Policies for China's Regional Economic Development

In recent years, the CPC Central Committee has attached great importance to the balance of regional economic development and successively put forward major strategic measures such as the coordinated development of the BTH region and the development of the YREB and the GBA. Given the impact of the COVID-19 epidemic and the great changes in the international and domestic situations in 2020, the Chinese government has proposed the strategy of developing the CCEC and issued strategic plans to promote the ecological conservation and high-quality development of the YRB, the development campaign of the western region, and new urbanization.

3.1 Ecological Conservation and High-Quality Development of the YRB

As an ecological corridor connecting the Qinghai-Tibet Plateau, the Loess Plateau, and North China Plain, the YRB is an important ecological barrier as well as a place of strategic importance in China. In the meantime, it is also China's important energy provider, chemical raw material manufacturer, and major industrial base that is rich in resources such as coal, oil, natural gas, and non-ferrous metals: its coal reserves account for more than half of the country's total. On September 18, 2019, during a visit to Henan, General Secretary Xi Jinping announced a national strategy: the ecological conversation and high-quality development in the YRB. The main objectives of this strategic plan are to strengthen the environmental protection work in the YRB, ensure the long-term stability of the Yellow River and the economical and intensive use of water resources in the YRB, and promote the high-quality development of the YRB and the conservation of the Yellow River traditional culture. On December 16, 2019, General Secretary Xi Jinping highlighted the need to study the issues regarding the ecological conservation and high-quality development of the YRB in his article published in *Qiushi* (Seeking Truth) journal: "A Plan to Promote Mutual-Enforcing and High-Quality Development of Economic Regions." On August 31, 2020, the Politburo of the CPC Central Committee held a meeting to review the "Outline of the Yellow River Basin's Ecological Protection and High-quality Development Plan", which marked that the ecological conservation work and high-quality development of the YRB entered a new phase.

3.2 The Joint Development of the CCEC

The Chengdu-Chongqing area is situated in the Sichuan Basin in the upper reaches of the Yangtze River. This place is adjacent to Shaanxi, Gansu, Yunnan, Guizhou, Qinghai, Tibet, Hunan, and Hubei provinces, boasting the advantages of locating on the main traffic arteries from south to north and those from east to west. In addition to its superior geographical location, this area has rich natural resources, a large carrying capacity, and a relatively strong economy. It also has a vast hinterland space and a high potential for economic growth. As early as 2016, the NDRC approved the "Development Plan for Chengdu-Chongqing City Cluster", proposing a spatial arrangement featuring "one axis, two belts, two cores, and three zones". The arrangement is aimed to develop the Chengdu-Chongqing axis, the riverside belt cities, and the Chengdu-Deyany-Mianyang-Leshan belt cities and enhance the central roles of Chengdu and Chongqing. The 6th meeting of the Central Committee for Financial and Economic Affairs, held on January 3, 2020, explicitly proposed the goal to develop the CCEC. At the meeting, General Secretary Xi Jinping made an important speech, announcing that the CCEC would be built into an economic growth pole for the high-quality development of the western region and emphasizing the strategic importance of the CCEC. On October 16, 2020, the Politburo of the CPC Central Committee held a meeting to review the "Outline of the Development Plan for the Chengdu-Chongqing Economic Circle." The meeting recognized the CCEC's importance for the mutual-enforcing and high-quality development of China's economic regions and its significance to the building of the new "dual-circulation" development paradigm, which takes domestic circulation as the mainstay, with domestic and international circulations reinforcing each other.

3.3 New Plan for the Development Campaign of the Western Region

The year 2019 marks the 20th anniversary of the development campaign of the western region. In the two decades, the western region made significant progress, but the problems of unbalanced and insufficient development were obvious. In 2020, when the development campaign of the western region entered its third decade and when domestic and international environments got complex, further measures were adopted to strengthen the development campaign. On May 17, 2020, the "Guiding Opinions of the CPC Central Committee and the PRC State Council on Promoting the Development Campaign of the Western Region in the New Era and Starting a New Chapter" was released. The document highlights a new development mode including environmental protection efforts, opening-up, and high-quality development, and puts forward overall requirements such as quality improvement, efficiency improvement, and driver replacement. Based on this, it clarifies the specific objectives of the development campaign of the western region in the new era: By 2020, together with the whole country, the region has to achieve the goal of building a well-off society in an all-around way; By 2035, the western region will basically realize socialist modernism, with its public service abilities, infrastructure, and the standards of living generally comparable to those of the eastern region. By promoting mutual-enforcing development, the western and eastern regions will advance and thrive together. This will enhance the stability of China's ethnic and frontier areas and ensure the harmonious coexistence between people and nature.

3.4 People-Oriented New Urbanization

Urbanization is the only way to realize modernization. It helps to settle issues regarding agriculture, rural areas, and rural people and address the main social contradiction (the contradiction between the people's ever-growing demands for a good life on the one hand and unbalanced and insufficient development on the other) in China's new era. It constitutes a powerful engine for our high-quality economic development. General Secretary Xi Jinping published an article entitled "Several Major Issues

Concerning the National Medium- and Long-Term Strategies for Economic and Social Development" in *Qiushi* journal in 2020. In the article, he pointed out that it was necessary to promote people-oriented urbanization and make our cities healthier, safer, and more liveable, ensuring that our cities can provide high-quality living space for our people. The 5th Plenary Session of the 19th CPC Central Committee emphasized the necessity of building a new paradigm for our land space development and protection work and that of promoting coordinated regional development and people-oriented new urbanization, further clarifying China's urbanization goals during the 14th Five-Year Plan period.

References

Dong, J. (2019). Accelerating digital northeast construction to help northeast China revitalize faster. *China Price, 2019*(12), 31–34.

Duan, J. (2020). *Houyiqing shidai chengshi fazhan yu zhili ying shixian sige tupo* (Four breakthroughs should be achieved in urban development and governance in the post COVID-19 pandemic era). *Governance, 2020*(Z4), 40–43.

Fan, H. (2020). Explore and establish the total factor balance mechanism of inter-regional interests. *Regional Economic Review, 2020*(01), 1–3.

Guo, Q. (2020). *"Shuangxunhuan" xin fazhan geju de xianshi luoji yu shixian lujing* (Realistic logic and realization path of the new development paradigm featuring "dual circulation"). *Seeker, 2020*(06), 100–107.

He, Z., Zhang, X., & Wang, G. (2020). Digital finance, digital divide, and multidimensional poverty. *Statistical Research, 37*(10), 79–89.

Jia, G. (2020). China's central and western development strategy and the dual circulation pattern with the domestic large-scale circulation as its main body. *Journal of Guangxi Normal University (Philosophy and Social Sciences Edition), 56*(06), 103–113.

Li, X. (2020). The epidemic prevention and control on a regular basis and the new regional economic structure. *The Chinese Banker, 2020*(10), 72–75.

Liu, J., Yang, Z., & Zhang, S. (2020). Research on the measurement and driving factors of China's digital economy. *Shanghai Journal of Economics, 2020*(06), 81–96.

Shi, F. (2020). Research on resilient transformative development of China's urban economy in the post-COVID-19 pandemic era. *Study & Exploration, 2020*(09), 105–111.

Sun, J., & Sun, X. (2017). Research progress of regional economic resilience and exploration of its application in China. *Economic Geography, 37*(10), 1–9.

Sun, J. (2020). *Xinguan feiyan yiqing dui zhongguo quyu jingji fazhan de yingxiang chutan* (Preliminary study on the influence of the COVID-19 epidemic on regional economic development in China). *Regional Economic Review, 2020*(02), 8–11.

Wang, Q., Wu, J., & Hu, A. (2020). Digital economy and integration of the Yangtze River Delta: Analysis based on spatial panel data model. *Journal of Zhejiang University of Technology (Social Sciences Edition), 19*(01), 39–46.

Zhao, T., Zhang, Z., & Liang, S. (2020). *Shuzi jingji chuangye huoyuedu yu gaozhiliang fazhan: Laizi zhongguo chengshi de jingyan zhengju* (Digital economy, entrepreneurial activity, and high-quality development: Empirical evidence from Chinese cities). *Management World, 36*(10), 65–76.

Chapter 2
International Environment

1 The Impact of US-China Relations and Trade War on China's Regional Economy

1.1 Background for Trade Frictions

Just as Xue et al. (2018) have pointed out, the United States has been adept at using trade wars to intervene against its competitors. It adopted such measures against the USSR, its major competitor after World War II: It imposed a strategic embargo, trade bans targeted at key sectors, and export quotas; suspended investments in key areas; and put pressure on the USSR from political, diplomatic, and military perspectives. What it did impacted the fragile economy of the USSR. The United States even fought trade wars with its allies. When it was troubled by the stagnation of the 1970s and when the European economies recovered from World War II and became its competitors in trade, the United States declared a trade war against its European allies, affecting the trade of agricultural products and steel and even public procurement activities. The United States also launched a trade war against Japan. With Section 301 of the Trade Act of 1974, it imposed sanctions on textiles, color TV sets, steel, cars, and semiconductors made in Japan. After that, it gradually expanded the trade war to influence Japan's foreign exchange market and financial market. Dong (2018) pointed out that the US-Japan trade war impacted Japan's currency exchange rates, trade, and economy and that it propelled Japan to adopt an ultra-loose monetary policy, which caused its real estate bubble and indirectly triggered its economic recession.

It is obvious that in the US-led global trading system and international monetary system featuring the hegemony of the US dollar, the United States has accumulated extensive experience in suppressing competitors through trade wars. In the current round of trade frictions, the United States targets China. Before gaining the presidential power, Donald J. Trump made many unrealistic accusations against China and

J. Sun, *Regional Economic Development in China in 2020*,
https://doi.org/10.1007/978-981-95-3604-7_2

China-US trade. After taking office, he adopted a series of policies to exert pressure on China. Since 2018, the United States has successively conducted unconventional trade investigations against China. In January of that year, Trump announced additional tariffs on solar panels and washing machines, targeting China. In February, the U.S. Secretary of Commerce Wilbur Ross issued a preliminary decision to impose anti-dumping duties on cast iron pipe parts from China. In March, Trump imposed additional tariffs on \$60 billion worth of goods from China under Section 301. In April, the United States issued the list of industries influenced by these additional tariffs and a prohibition on suppliers selling chips to ZTE on April 17. In May, China and the United States began a series of negotiations during which, despite the news about the easing tension, there was ultimately no agreement reached by the two sides. In June, the United States released a list of \$50 billion worth of goods that would trigger additional tariffs and announced that the first part of the list, which was worth \$34 billion, would be subject to additional tariffs starting from July 6. Liu (2018) observed that on July 6, 2018, the United States officially imposed tariffs on these Chinese goods and threatened to impose a fresh 10% tariff on \$200 billion worth of goods from China. In August of the year, the White House issued a statement that it wanted to increase the tariffs imposed on the \$200 billion worth of goods from China from 10% to 25% and released the second list of imported Chinese goods, which were worth \$16 billion, influenced by the additional tariffs that would take effect on August 23. In September, Trump announced a 10% additional tariff on \$200 billion worth of Chinese imports. In the same month, the United States also announced that tariffs would be increased to 25% from January 1, 2019, and threatened to impose more drastic tariff collection measures immediately if China reacted against U.S. industries such as agricultural products. On the evening of December 1, 2018, the leaders of China and the United States met in Buenos Aires, Argentina: the two sides decided to stop escalating trade restrictions, increasing tariffs, or introducing new tariffs on other goods. On February 24, 2019, Trump tweeted that he would delay the March 1 tariff increase on Chinese goods. On May 6, 2019, when everyone thought that the trade frictions were coming to an end, the United States abruptly announced that starting from May 10, 2019, the tariff rate on the \$200 billion worth of listed Chinese goods would be increased from 10% to 25%. This marked the beginning of the second round of trade frictions.

1.2 Literature Review

Dawley et al. (2010) have defined that the term "economic resilience" was generally used to describe an economy's ability to resist and recover from shocks. In the existing research, Martin (2012) has explained the connotations of economic resilience by dividing it into the following aspects: (1) the ability to withstand the adverse impact of external shocks; (2) the speed and extent of economic recovery in the aftermath of shocks; (3) the ability to pool internal resources to adapt to the environment in the recovery process; (4) the ability to make innovation in accordance with changing external conditions. In conclusion, the core of economic resilience is the ability to resist and recover from external shocks, which is closely related to the characteristics of an economic system and is conducive to the sustainable development of the system.

After the 2008 Financial Crisis, the research on economic resilience increased rapidly, making it a hot research topic. Some studies found big differences in the economic resilience of different regional economies across the globe. Buckman et al. (2015) discovered that within the EU, regions with higher population density and more high-tech industries boasted higher economic resilience. Caro (2014) revealed that in Italy, regions highly dependent on manufacturing industries were more economically resilient. Zhang et al. (2012) analyzed China's provincial economies and concluded that the economic resilience of Beijing and Shanghai was better than that of Sichuan. Other studies focused on the regional differences in shock resistance and post-crisis recovery. Fingleton et al. (2012) found that the economic regions of the UK did not differ much from one another in terms of post-crisis recovery, but their differences in the ability to resist shock were huge.

Noticing that economic resilience varied greatly from region to region, many researchers explored the reasons for the differences. They generally agree that factors such as industrial structure, institutional and cultural environment, and social networks affect regional economic resilience. First, existing literature generally considers the industrial structure as the most important factor influencing the resilience of a regional economy while recognizing the influence of industrial diversity and prime industries on economic resilience. Brown et al. (2017) concluded that regions with diversified industries showed better resilience in the face of shocks and those relying on a single industry had insufficient economic resilience. Davies (2011) observed that regions with heavy industries as their prime industries lacked economic resilience, such as the traditional industrial region in Detroit and that in northeast China. Second, the institutional environment is also regarded as an important reason. Whitley (2000) believed that the economy would be more resilient in a region with less government intervention and more favorable policies because such an environment enabled market players to maintain a keen grasp of information and allowed them to make flexible responses to market fluctuations and take targeted measures. Third, the influence of the cultural environment is widely recognized. Glaeser (2012) held that regional culture and customs had a noticeable influence on the resilience of a regional economy and that a more open, diversified, and entrepreneurial environment helped enhance regional economic resilience.

1.3 Economic Resilience of Regional Economy

The Sino-US trade frictions impinged on both the U.S. and Chinese markets. Figure 1 shows the impact of the trade frictions on the U.S. and Chinese stock markets. As shown in Fig. 1, stocks in the two markets fell sharply in February 2018 after substantial frictions between China and the United States, and the U.S. and Chinese stock markets showed similar fluctuations until June. In June, when the United States decided on the tariffs, there were very different changes in the US and China stock markets: the U.S. stock market, after a short-term setback, began to rise; the Chinese stock market continued falling. This situation, however, did not last long. At the end of September, when the United States announced a substantial increase in tariffs on products imported from China, stock markets of the two countries plunged. They kept falling until December, when the heads of the two states met and decided to suspend tariff escalation and other trade restrictions. Unfortunately, in May 2019, when the United States declared a trade war against China again, the two countries' stock markets crashed again. These events showed how much trade frictions impacted both Chinese and American economies. Therefore, we should be more prudent in our analyses and take active measures to resist the impact of trade frictions.

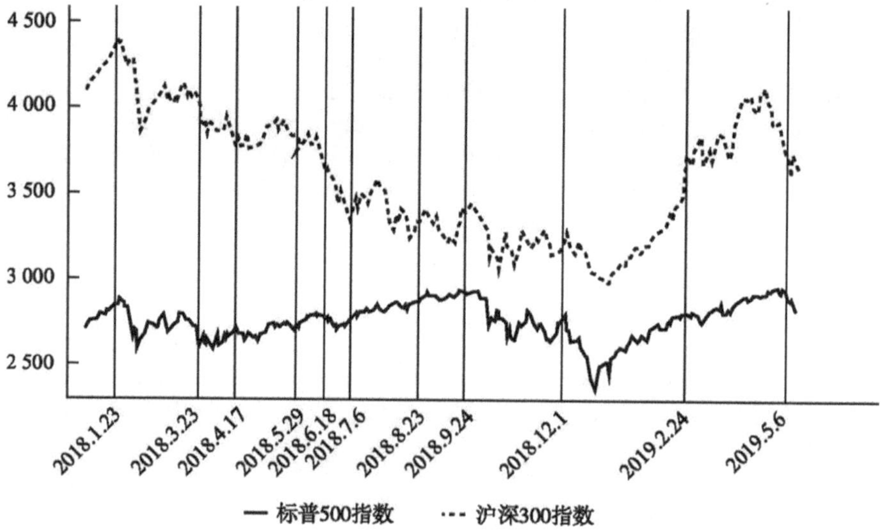

Fig. 1 How trade frictions impacted the U.S. and Chinese stock markets

2018.1.23	January 23, 2018
2018.3.23	March 23, 2018
2018.4.17	April 17, 2018
2018.5.29	May 29, 2018
2018.6.18	June 18, 2018
2018.7.6	July 6, 2018
2018.8.23	August 23, 2018
2018.9.24	September 24, 2018
2018.12.1	December 1, 2018
2019.2.24	February 24, 2009
2019.5.6	May 6, 2019

(continued)

(continued)

标普500指数	S&P 500 Index
沪深300指数	CSI 300 Index

External shocks not only directly impact on regional economies but also indirectly influence them through inter-regional circulations of trade and factors. Given continuously-decreased trade costs and constantly-enhanced market integration, a role-specialized division of labor is formed among regions: regions are interlinked with one another through production and trade ties. Under such circumstances, the influence of external shocks can be transmitted among regions through their input-output connections. In a word, to accurately estimate the economic resilience of a region, we have to consider the direct and indirect effects of external shocks. This chapter uses the dynamic spatial panel data model to control the direct and indirect effects of macro shocks for the estimates of regional economic resilience.

We adopt the method proposed by Vega et al. (2016) to estimate the impact of macro shocks on a regional economy. A macro shock is introduced as a "common factor" in the dynamic spatial panel data model, and estimates are made through the following equation.

$$y_{rt} = \beta_1 * y_{rt-1} + \beta_2 \sum_{j=1}^{N} w_{rj} * y_{rt} + \beta_3 \sum_{j=1}^{N} w_{rj} * y_{rt-1} + \beta_4 * y_{Nt} + \beta_5 * y_{Nt-1}$$
$$+ \gamma_2 * X + \varepsilon_{rt}$$

In the equation, the explanatory variable y_{rt} is regional economic growth rate; y_{Nt} is the macroeconomic situation at the national level, reflecting the intensity of a macro shock;y_{rt-1} is the regional economic growth rate in the previous year, which describes the lag of economic growth in time; X is the control variable; w_{rj} is the element corresponding to the spatial weights matrix; ε_{rt} is a random perturbation term. Among them, β1 = α1 depicts the correlation of regional economic growths in time; β2 = α2 is the spatial autoregressive coefficient; and β3 = α3 is the lagged spatial autoregressive coefficient.

To estimate the resilience of a regional economy, we need to further calculate the impact of a macro shock on a regional economy. Its calculation formula is $\gamma_{1r} = \beta_4 / (1\text{-}\alpha_2) = \beta_5 / (\text{-}\alpha_1\text{-}\alpha_3)$, and its variance can be obtained by calculating the quotient of random variables. By using the method of measurement, we avoid the strong subjectivity of using an index system and can get a more objective sensitivity γ_{1r} of regional economic growth to external macroeconomic shocks. The larger it is, the greater the impact of the macroeconomic shock on the regional economy is, and the weaker the economic resilience is. On the contrary, the smaller the γ1r is, the less sensitive the regional economy is to macro shocks, and the stronger the economic resilience is. The results are shown in Table 1.

The results show that after the 2008 Financial Crisis, China faced an overall downward trend in its economy., but its four major economic regions delivered quite different performances. Among them, provinces in the northeast region, especially Heilongjiang, had the lowest economic resilience, followed by the central region. The

Table 1 Basic estimation results and robustness test of the spatial weights matrix

Ranking	Province	Estimated coefficient	Standard deviation	Ranking	Province	Estimated coefficient	Standard deviation
1	Guizhou	– 5.30	1.77	17	Liaoning	0.81	1.56
2	Tibet	– 3.22	1.86	18	Sichuan	0.95	1.47
3	Chongqing	– 2.70	3.60	19	Hebei	1.15	3.81
4	Yunnan	– 2.18	1.27	20	Jiangxi	1.42	1.20
5	Shanghai	– 2.05	1.96	21	Hunan	1.46	1.20
6	Shandong	– 1.66	1.79	22	Gansu	1.55	4.56
7	Guangdong	– 1.04	1.97	23	Anhui	1.61	1.54
8	Beijing	– 0.88	23.51	24	Qinghai	2.12	1.48
9	Fujian	– 0.43	4.03	25	Zhejiang	2.19	2.05
10	Henan	– 0.03	0.16	26	Ningxia	2.36	1.16
11	Tianjin	0.01	0.01	27	Hainan	2.43	1.64
12	Inner Mongolia	0.01	0.00	28	Guangxi	2.73	2.11
13	Jilin	0.17	0.69	29	Heilongjiang	3.41	3.61
14	Jiangsu	0.25	0.66	30	Xinjiang	4.26	2.60
15	Shaanxi	0.54	0.67	31	Shanxi	5.29	2.07
16	Hubei	0.80	1.64				

eastern region was the most resilient one, and the western region had the most obvious intra-regional disparities in terms of economic resilience. Most provinces in southwest China such as Guizhou, Yunnana, and Chongqing were resilient enough to resist the macro impact, but provinces in northwest China such as Xinjiang, Qinghai, and Shaanxi and the southern province of Guangxi had insufficient economic resilience. We replace the spatial weights matrix with the inverse distance matrix for a robustness test, and the results support the conclusion of the basic regression (see Table 2).

Given the on-going trade frictions, the impact mostly comes from foreign trade. Under such circumstances, when estimating the intensity of the impact on a region, it is essential to take into account the region's dependence on foreign trade while considering its economic resilience. As a region's dependence on foreign trade rises, the intensity of the impact on it increases, and vice versa. Therefore, we make a distribution map of regional economic resilience and trade (foreign trade) dependence and mark the mean value of trade dependence, as shown in Fig. 2.

Table 2 Estimation results and robustness test of inverse distance matrix

Ranking	Province	Estimated coefficient	Standard deviation	Ranking	Province	Estimated coefficient	Standard deviation
1	Guizhou	– 5.10	1.55	17	Hubei	0.39	1.00
2	Chongqing	– 2.47	2.84	18	Inner Mongolia	0.39	0.30
3	Yunnan	– 2.32	1.15	19	Shaanxi	0.41	0.52
4	Tibet	– 1.95	1.78	20	Gansu	0.78	28.19
5	Shandong	– 1.67	1.38	21	Hunan	0.86	0.87
6	Shanghai	– 1.18	1.80	22	Anhui	1.05	1.38
7	Guangdong	– 0.74	1.85	23	Jiangxi	1.14	1.07
8	Fujian	– 0.47	2.55	24	Guangxi	1.31	2.33
9	Henan	– 0.41	0.72	25	Zhejiang	1.52	1.75
10	Beijing	– 0.35	1.24	26	Qinghai	1.57	1.41
11	Tianjin	– 0.09	0.29	27	Hainan	1.85	1.34
12	Sichuan	– 0.05	1.41	28	Ningxia	2.25	0.99
13	Jiangsu	0.11	0.40	29	Heilongjiang	2.41	2.58
14	Hebei	0.15	0.58	30	Xinjiang	3.62	-2.32
15	Jilin	0.32	0.65	31	Shanxi	3.82	1.76
16	Liaoning	0.35	0.96				

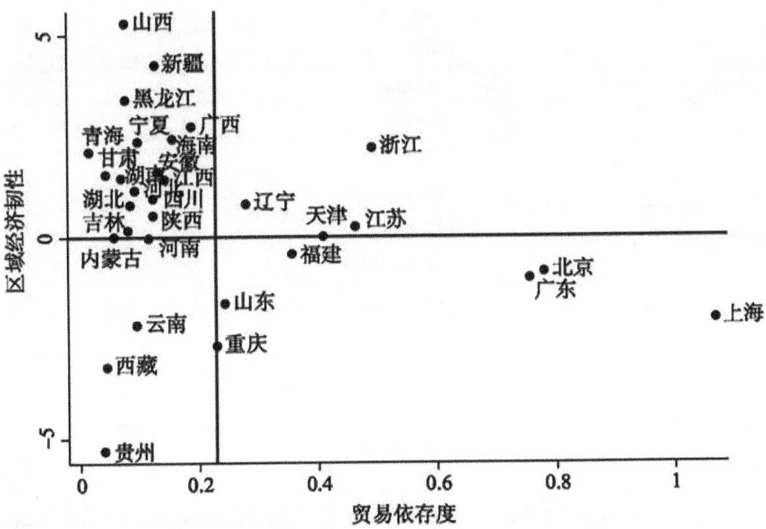

Fig. 2 Regional economic resilience and trade dependence distribution map

区域经济韧性	Regional economic resilience
贸易依存度	Foreign trade dependence
西藏	Tibet
新疆	Xinjiang
湖南	Hunan
贵州	Guizhou
青海	Qinghai
广西	Guangxi
云南	Yunnan
甘肃	Gansu
四川	Sichuan
江西	Jiangxi
吉林	Jilin
宁夏	Ningxia
陕西	Shaanxi
内蒙古	Inner Mongolia
福建	Fujian
河南	Henan
重庆	Chongqing
河北	Hebei
北京	Beijing
山西	Shanxi
江苏	Jiangsu
安徽	Anhui
浙江	Zhejiang
海南	Hainan
山东	Shandong
广东	Guangdong
辽宁	Liaoning
上海	Shanghai
黑龙江	Heilongjiang
天津	Tianjin
湖北	Hubei

In Fig. 2, the regions in the first quadrant had high trade dependence and low economic resilience and thus suffered a more intense impact. These regions are Jiangsu, Zhejiang, Tianjin, and Liaoning, all located in the eastern region. Given that Jiangsu, Zhejiang, and Tianjin are China's major economic development zones and that they are among the provinces that are hit the hardest by trade frictions, it is a must to take proactive measures to stabilize their economic growth during trade frictions. In addition, as Liaoning, the largest economy of northeast China, is also among the most vulnerable provinces, trade frictions will worsen the economic situation in northeast China. Most Chinese provinces are in the second quadrant: they have low trade dependence and low economic resilience. Provinces in this quadrant include some central and western provinces and two northeastern provinces, namely Heilongjiang and Jilin. These provinces are hardly affected by trade frictions, but they are not resilient enough to handle the impact by themselves. Once these provinces are trapped by severe difficulties caused by the impact, national and systemic risks can easily be triggered. During trade frictions, it is essential to ensure that the adverse effects of the impact do not spread to these vulnerable provinces. Provinces in the third quadrant have low trade dependence and strong economic resilience, including Chongqing, Guizhou, Yunnan, and Tibet. They are the least vulnerable provinces to trade frictions. The remaining provinces are in the fourth quadrant, most of them being eastern provinces. These provinces have high trade dependence and low economic resilience. They will be hit hard by trade frictions, but their industrial, institutional, and cultural advantages can help them resist these shocks. Their capacities for offsetting those adverse effects determine whether China can successfully overcome the impact.

2 The Influence of the BRI on the Four Economic Regions

2.1 Background of the BRI

The BRI has entered the phase of rapid advancement. The following objectives are set for this phase: realizing orderly and free flow of economic factors, highly-efficient allocation of resources, and in-depth integration of markets; encouraging the BRI countries to achieve economic-policy coordination and carry out broader and more in-depth regional cooperation of higher standards; creating an open, inclusive, and balanced regional economic cooperation architecture that benefits all. These goals are step by step turned into reality. The "Vision and Actions on Jointly Building Silk Road Economic Belt and 21st-Century Maritime Silk Road" makes it clear that the Silk Road Economic Belt focuses on bringing together China, Central Asia, Russia, and Europe (the Baltic); linking China with the Persian Gulf and the Mediterranean Sea through Central Asia and West Asia; and connecting China with Southeast Asia, South Asia, and the Indian Ocean. From a domestic perspective, this plan mainly affects the country's northwestern and southwestern provinces. The 21st-Century Maritime Silk Road is designed to go from China's coast to Europe through the South China Sea and the Indian Ocean in one route, and from China's coast through the South China Sea to the South Pacific in the other. This plan profoundly influences the development of China's coastal provinces. Since launching its regional economic plan in 1999, China has achieved remarkable progress in territorial space development and regional economy, creating universal prosperity. In particular, our accelerated marketization process has enabled our economic regions to build increasingly strong ties with the global economy, which underpin the progress of our BRI. So far, provinces such as Fujian, Xinjiang, Yunnan, and Inner Mongolia have already become frontiers of opening-up under the BRI.

First, for China, the BRI is the continuation and improvement of its regional economic development strategy. The imbalance of regional development has remained a unsolved problem in the country for a long time. From adopting the reform and opening-up policy to the beginning of the 21st century, the regional development strategy has undergone a transformation: the tolerance for unbalanced development has been replaced by the pursuit of coordinated development. The unbalanced development strategy implemented in the early stages of the reform and opening-up has greatly accelerated China's economic development, narrowed the gap between China and the developed countries, and created a "Chinese miracle." Nevertheless, the unbalanced development strategy has widened the gap between the eastern and western regions. When financial crises caused marked decreases in the international trade, China faced declining external demand and timely adopted strategies such as the development campaign of the western region, the revitalization of the northeast industrial base, and the rise of central China. These strategies constituted the initial architecture of China's regional development. In the 13th Five-Year-Plan period, the strategy of pursuing coordinated development in the BTH region and that of developing the YREB were put forward and implemented, significantly boosting the development in these two regions. Proposed and implemented on the basis of the above-mentioned development strategies, the BRI further improves China's regional development arrangements and thus has a guiding significance.

Second, the BRI helps to overcome new challenges that emerge when the Chinese economy enters its new normal, which is different from the high-speed growth pattern exhibited in the past. In the third decade of the 21st century, the western region stands at a new historical point and enters a new development phase. Because of the development campaign of the western region and the rise of central China, the relative gap between the eastern region and the central and western regions has been narrowed in recent years, but the absolute gaps between them are still widening. Their disparities in education, medical care, infrastructure, etc. are still obvious, indicating that their differences in development quality are still apparent. Chinese authorities regard the narrowing of inter-regional development gaps as a historic task. Nowadays, the country gradually adapts to the economic new normal where high-speed economic growth is replaced by medium-high-speed growth, and its industrial structure is constantly optimized and upgraded. In the meantime, it has to face increasingly apparent risks accumulated over the years and economic slowdowns caused by the worsening international economic situation and slower domestic economic growth. Risks such as overcapacity in some industries, local debt issues, fiscal sustainability, and deflation also need special attention. Under such circumstances, the BRI offers us opportunities to solve these problems.

2.2 The Influence of the BRI on the Eastern Region

The eastern region plays a significant role in China's economic development, especially in its manufacturing industries and foreign trade. Leveraging the industrial effect, cost effect, and employment effect brought by the BRI, the eastern region has the chance to further improve its economic strength.

1. **Industrial Effect.** The industrial transfer and upgrade carried out under the BRI will exert two types of industrial effect on the eastern region. (1) The implementation of the BRI means lower transport costs and expanded foreign markets, which will provide favorable conditions for the transfer of labor-intensive industries. The 65 BRI countries will become not only receivers of low-end manufacturing industries transferred out of China but only providers of huge markets. (2) The implementation of the BRI promotes a smooth transfer of the lower-end manufacturing industries out of the eastern region, which helps to pool resources and create new market space for the development of emerging industries. To be more specific, when labor-intensive industries in the eastern regions move out smoothly, the huge surplus of labor, land, and capital can ensure a good start for the development of capital-intensive industries. This enables the emerging industries to be started up smoothly and the industrial transformation and upgrading of the eastern region to be concluded successfully.
2. **Cost Effect.** Thanks to the infrastructure enhancement efforts under the BRI, the costs of production, trade and communication between China and the world are greatly reduced. That is, the implementation of the BRI makes China's exchanges with foreign countries increasingly frequent. (1) The implementation of the BRI greatly reduces the operating costs of Chinese transnational corporations (TNCs) and increases their international competitiveness. (2) As a major exporter and importer, China can reduce the import and export costs through the implementation of the BRI and thus improve the overall living standards of its people. (3) The implementation of the BRI facilitates the communication between China and BRI countries and reduces the cost of communication.
3. **Employment Effect.** (1) Despite that the implementation of the BRI will lead to the transfer of industries, the parent companies stay inside China will offer more unproductive employment opportunities such as management jobs; (2) If more subsidiaries are set up abroad, more jobs in sectors such as legal services, engineering consulting, and public relations will be created in China. (3) The implementation of the BRI will lead to an increase in China's international communication, which will bring China a large amount of foreign-market opportunities, boost the development of related Chinese industries, and create more jobs; (4) The transfer of low-added-value, low-end manufacturing industries out of the eastern region because of the implementation of the BRI will create favorable conditions for the upgrading of the region's industrial structure, which will bring about an increase in the employment of high-tech industries and greatly improve the employment rate of high-caliber working population in China.

2.3 *The Influence of the BRI on the Western Region*

Among the four major economic regions of China, the western region is situated the closest to Central Asia and Europe and destined to play a significant role in the BRI cooperation. Because of the BRI, this region has become a new frontier of our opening-up. It will realize progress by embracing the idea of opening-up, exploring creative institutions, and finding its new strategic position.

1. **New Frontier of Opening-Up.** In close proximity to Central Asia and Europe, the western region enjoys the natural advantage of opening up to the Eurasian continent. Western provinces such as Tibet, Gansu, Qinghai, Ningxia, and Xinjiang border Nepal, India, Russia, Mongolia, and Central Asian countries and have traditional trade ties with these foreign countries. Border trade in these areas is based on the differences and complementarities between natural resources and economies. Provinces in southwest China such as Guangxi, Chongqing, Sichuan, Guizhou, and Yunnan have the geographical convenience to cooperate with ASEAN countries. The development of the Silk Road Economic Belt have turned many remote inland provinces into important nodes in international logistics channels, creating new competitive advantages for these provinces. In September 2016, China approved the establishment of new free trade pilot zones (Pilot FTZs) in three western provinces, namely, Sichuan, Chongqing and Shaanxi (hereinafter referred to as FTAs). They are the first batch of Pilot FTZs in the western region. Through reform and trials, Shaanxi actively explores new paths for inland opening-up, trying to build a top-notch inland reform and opening-up base in the western region. Sichuan strives to become the best inland open economy for the coordinated development of coastal, riverside, and border opening-up bases. Chongqing FTZ shoulders the heavy responsibility of advancing the development campaign of the western region. Other western provinces such as Gansu, Ningxia, and Inner Mongolia offer an indispensable support for the opening-up of the western region with their enhanced comprehensive strengths and abundant natural resources. The western region plays an increasingly significant role in the BRI cooperation and China's opening-up. These inland areas have already become new frontiers of China's westward opening-up.
2. **New Changes to the Opening-Up.** ***(1) The western region will rely on itself instead of coastal provinces to pursue opening-up.*** Since the 12th Five-Year-Plan period, the western region has realized enormous economic progress: the average annual growth rate of its regional GDP is 157 %, and that of its fixed asset investment is 208%; its basic economic development model, industrial structure, and growth drivers have changed profoundly. Under the BRI, the western region adopts the policy of opening up to achieve continuous optimization, upgrading, and innovation of its economic structure, and it is continuously stepping up its opening-up efforts.

   (2) The western region sees growth in its imports and exports. From the beginning of the 12th Five-Year-Plan period to now, the average annual growth rate of its imports and exports is 294%. The quality of trade has improved significantly over the years, achieving the balance between imports and exports. The BRI's contribution to infrastructure enhancement, institutional betterment, and people-to-people exchanges have changed the old mode of foreign exchange in this region, which features the inflow of capitals, technologies, managerial experience and professionals. Now, this region places emphasizes people-to-people exchanges and economic and trade cooperation with the BRI countries, ramping up its opening-up efforts in medical, educational and financial sectors.

(3) The western region now pursues mutually-beneficial and win-win development instead of immediate interests. The BRI connects the Asia-Pacific and European economic circles and highlights the joint development, close ties, and shared opportunities, interests, and future of its member countries. This constitutes a realistic basis for the western region's win-win cooperation and joint progress with the BRI countries. With its comparative advantages in energy source, technology, capital, and market, the western region can carry out in-depth win-win cooperation in various fields with the BRI countries. They can jointly build their trade flows, industrial belts, networks, and cultural circles.

3. **Institutional Innovation.** The BRI is helping the western region pool resources and enhance their regional advantages. Local governments in the region have high expectations for the development dividends of the BRI, which is a major strategy for China's new round of opening-up. Nevertheless, in comparison with the policies issued in the early stages of the opening-up, the BRI boasts different connotations, characteristics, support systems, and policy tools. The opening-up under the BRI call for the building of a fair environment for domestic and foreign investors, and the traditional preferential treatments provided through land resources, taxation, factor prices, and financial subsidies are increasingly restricted. Together with the domestic policies aimed to deepen reform, the BRI will propel the western provinces to stop depending on the traditional paths and form reasonable expectations for the BRI-related policies. The local governments are encouraged to focus on transforming their government functions, innovating their institutions, optimizing their environments, providing better services, and cultivating needs in their opening-up endeavors. They are required to accelerate the construction of an open economic system and remove barriers in policy coordination, facilities connectivity, trade flows, financial integration, and people-to-people exchanges. They are expected to develop internationally-accepted institutions and supervision modes, facilitate the free flow of factors, and let the market play a decisive role in resource allocation. In a word, these governments will have to find a better and more reasonable way to play their roles.
4. **New Strategic Orientation of Opening-Up.** Most western provinces of China are located within the core area of the BRI. The BRI provides them with development opportunities as well as new status. Shaanxi is now positioned as the new starting point and bridgehead of the Silk Road Economic Belt; Xinjiang, as a principal participant and a core zone of the Economic Belt. Gansu shoulders the task of building the prime section of the Economic Belt by leveraging its advantages in location and featured industries. Ningxia and Qinghai, inland residential areas for

ethnic minority groups, become supporters and strategic bases for the BRI. Inner Mongolia takes advantage of its proximity to Russia to build itself into the frontier of China's northward opening-up. As the intersection of the Silk Road Economic Belt and the YREB, southwestern provinces that boast outstanding advantages in connecting domestic and foreign economic circles, become influential economic and cultural centers. Sichuan utilizes national strategies in its pursuit of all-around opening-up. Chongqing strive to become a new top-notch base for opening-up through the Chongqing-Xinjiang-Europe International Railway and its bonded port areas and free trade zones (FTZs). Yunnan makes great efforts to build a platform for economic and cultural exchanges and cooperation with South Asia and Southeast Asia under the BRI. Such repositioning for the opening-up drive under the BRI will enable the western provinces to give full play to their comparative advantages, develop a high-standard open economy, and comprehensively upgrade their opening-up.

2.4 The Influence of the BRI on the Northeastern Region

The BRI-related development and exchanges offer new opportunities for the northeastern region, the oldest industrial zone of our country.

1. **Expanded Cooperation Coverage.** With the advancement of the BRI, the northeastern region conducts cooperation and exchanges with an increasing number of BRI countries and regions, and the scope of cooperation continues to expand. Data show that by 2018, Heilongjiang Province has established 93 friendly partnerships with local governments and municipalities in 35 countries. The province has also built a mechanism for regular exchange visits with five Russian provinces and friendly relations between its chambers of commerce and their pears in Russia, Japan, South Korea and other countries. For Heilongjiang, the coverage of cooperation continues to expand, so does its circle of friends. Jilin has established friendly, cooperative relations with local governments in 83 countries and 382 multinational companies. Liaoning has 17 friendly provinces and 69 sister cities in 27 BRI countries. They bring the province a number of cooperation projects, realizing increasingly in-depth inter-regional cooperation and exchanges.
2. **Deepened Trade Cooperation.** Data shows that in 2017, the imports and exports between the northeastern region and the BRI countries amounted to $61.69 billion, realizing a year-on-year increase of 22% and accounting for 4.3% of the total China-BRI countries trade volume. The region's exports to the BRI countries amounted to $24.79 billion, increasing by 7.7 percentage year-on-year and taking up 3.2% of China's total exports to the BRI countries. Its imports from the BRI countries totaled $36.9 billion, achieving a year-on-year increase of 33.9 percentage points and accounting for 5.5% of China's total imports from the BRI countries. The recovery of import and export trade has reversed the trend of decline lasting from 2015 to 2016 in this region. Among the three northeastern provinces, Liaoning outshone its peers in the growth rate and trade scale

when promoting foreign trade with the BRI countries. Its total import and export volume reached $42.9 billion, realizing an increase of 29.6 percentage points compared with 2016 and accounting for 69.5% of the regional total. These data directly or indirectly show that with the advancement of the BRI, the trade cooperation between the northeastern region and the BRI countries is continuously deepened, and the foreign trade structure is constantly optimized.

3. **Major Channel for Opening-Up.** The BRI highlights the following five aspects: policy coordination, facilities connectivity, unimpeded trade, financial integration, and people-to-people bond. Regarding facilities connectivity, Heilongjiang made the following breakthroughs in the BRI-related construction in 2018. The China-Russia Tongjiang-Nizhneleninskoye Bridge was basically completed, and the Heihe-Blagoveshchensk Heilongjiang highway bridge was about to open. The opening of Heixiazi Island Highway Gateway was approved by the PRC State Council and pushed forward by the joint efforts of China and Russia. The China-Russia natural gas pipeline project was pressed ahead, and the preparation work for the Dongning Boundary Highway Bridge was accelerated. These projects enhanced the connections between Heilongjiang and the BRI countries and regions. In 2018, the total cargo volume at ports across the province reached 41.423 million metric tons, a year-on-year increase of 37.1 percentage points. The Harbin-Europe and Harbin-Russia train service recorded 144 trips, and 13 routes to Russian destinations were opened. The above-mentioned data show how the BRI-related construction projects have increased Heilongjiang's connections to BRI countries and regions and opened up economic channels for the province's opening-up drive. The BRI has also boosted infrastructure construction across Liaoning province. In June 2018, the Liaoning-Europe maritime international transport corridor opened up its second Arctic route; in December, Jilin and Liaoning proposed to jointly build the Changbaitong (Dan) Corridor and the open economic belts along the Yalu River and the Tumen River. Liaoning-Manzhouli-Europe international transport corridor, Liaoning-Mongolia-Europe international transport corridor, and Liaoning-Europe maritime international transport corridor have been built step by step. Port of Dalian and Port of Yingkou have been constantly improved for integrated development, and the function of the five regional airports have been continuously enhanced. As of 2018, Liaoning's China-Europe freight-train service has recorded 394 trips, transporting 32,000 twenty-foot equivalent unit containers of goods, and a new trans-Eurasia logistics node has been planned at Shenyang Port to further promote Liaoning's opening-up drive. Jilin's infrastructure construction has also gained impetus from the BRI. In 2015, the opening of the Changchun-Manzhouli-Europe Train accelerated the integration of Jilin into the BRI. Since then, the Changchun-Manzhouli-Europe freight-train service has transported 43,000 twenty-foot equivalent unit containers of goods, which is worth nearly 13.8 billion yuan. This train service connects nearly 100 sites in Russia and 30 sites in 10 European countries, consolidating Jilin's ties with the BRI countries and regions. In a word, the BRI has played a vital role in building physical channels for the opening-up drive of the northeastern region.

4. **Policy Support.** The northeastern region also enjoys the BRI-related policy support. To facilitate the advancement of the BRI, Liaoning introduced the "Master Plan for Liaoning Comprehensive Pilot Zone for the BRI," providing policy support for the BRI cooperation-related economic and trade investment, connectivity, open platforms, and financial services. In the meantime, the province adopted more than 40 open policies to facilitate the participation of general manufacturing industries, telecommunications sector, electric vehicle industry in the BRI-related cooperation. To actively respond to the regional opening-up plans of Russia and other countries, Heilongjiang introduced a series of policies and plans: the "Development Plan for 'China-Mongolia-Russia Economic Corridor' Heilongjiang Land and Maritime Silk Road Economic Belt," issued in 2015; "Several Opinions on Building 'Longjiang Silk Road Belt' to Accelerate the Pace of Opening-Up" and "Heilongjiang Province's Scheme to Implement the Special Plan for Promoting Scientific and Technological Innovation Cooperation Under the BRI", which were released in 2016; "Several Policies to Support Suifenhe-Dongning Key Development and Opening-Up Pilot Zone," promulgated in 2017; the "Overall Plan for the Development of Suifenhe-Dongning Key Development and Opening-Up Pilot Zone" and the "Plan for the Construction of Heilongjiang Eurasian Logistics Hub", issued in 2018. These policies and plans boosted Heilongjiang's cooperation with the BRI countries and regions. Jilin actively participates in the BRI. It has formulated and implemented a series of policies for this purpose, such as the "Development Plan (2018–2025) for the China-Mongolia-Russia Open Economic Belt." These policies has promoted the province's cooperation and communication with the BRI countries and optimized the environment for its provincial development.

2.5 *The Influence of the BRI on the Central Region*

As the inland central area of China, the central region has never been well integrated into international trade. It relies on mainly the eastern region to sell its product to the international market. With the BRI, this embarrassing situation will change.

1. **Opportunities to Develop Top-Notch Inland Open Economy.** To boost the development of their inland open economy, the five provincial capitals in the central region, namely Zhengzhou, Wuhan, Changsha, Nanchang, and Hefei, actively embrace the BRI. On June 14, 2017, President Xi Jinping met with Luxembourg Prime Minister Xavier Bettel: in Beijing. At the meeting, Xi Jinping emphasized the need to deepen financial and capacity cooperation between the two sides within the framework of the BRI and expressed China's support for the Zhengzhou-Luxembourg "Aerial Silk Road." Under such guidance, Henan province, situated in the hinterland of China, pursues opening-up and develops cross-border free trade through air transportation, online business platforms, and cross-border e-commerce. Wuhan and Changsha seize opportunities brought by the BRI, and Nanchang builds ties with the BRI regions in the west, central south

and southwest of China. The International Container Terminal of Hefei Port in Hefei City has opened two regional container shipping routes: the Hefei-Wuhu-Shanghai route and the Hefei-Nanjing-Shanghai route.

2. **Opportunities Brought by the Brand of China-Europe Freight Trains.** The Vision and Actions on Jointly Building Silk Road Economic Belt and 21st-Century Maritime Silk Road contains the following statement: "We should set up coordination mechanisms in terms of railway transport and port customs clearance for the China-Europe corridor, cultivate the brand of 'China-Europe freight trains,' and construct a cross-border transport corridor connecting the eastern, central and western regions." The China-Europe freight train services are available in several provincial capitals in the central region and some other places in China, but on the whole, the business situation is not good. The main reason is that these freight trains that transport goods to Europe often have to return to China without carrying much cargo. Among the localities offering this train service, Wuhan does the best in solving this problem because it provides tailored services for enterprises of different sizes. The Wuhan-Xinjiang-Europe international rail route (hereinafter referred to as the Wuhan-Europe International) offers less than container load (LCL) shipping services to small and micro enterprises, shared train services to small and medium-sized enterprises, and customized train services for large enterprises. The Wuhan-Europe International also transports European raw materials, especially Russian timber, back to Wuhan. These materials will then be shipped to cities in the northeastern and southeastern regions through the Yangtze River and sea transportation, such as Dalian, Yantai, Xiamen, and Shanghai. In this way, the Wuhan-Europe International connects resource networks in Russia and other European countries and promotes the joint development of the central region, especially the middle reaches of the Yangtze River. Through cooperation and integration of resources, this region pools its strength to enhance the brand of China-Europe freight trains.
3. **Opportunities Emerging in Expanded International Capacity Cooperation.** In 2015, the PRC State Council issued the "Guiding Opinions on Promoting International Production Capacity and Equipment Manufacturing Cooperation." According to the policy document, China plans to mainly cooperate with developing countries that have high compatibility with China's equipment and production capacity, strong desires for cooperation, and good cooperation conditions and foundations; and explore the markets in developed countries step by step. The document also lists steel, non-ferrous metal, building materials, railway, electricity, chemicals, textile, automobile, communications, engineering and machinery, aerospace, ship and marine engineering as key industries for cooperation. These industries are the central region's advantageous industries. In recent years, they have stepped up their efforts to go global. The six provinces in the central region have established hundreds of overseas enterprises of strategic emerging industries, including 156 enterprises set up by Hunan. Hubei is promoting the Beidou navigation system at home and abroad, and Thailand is the first overseas user of the system.

4. **Opportunities Generated by the Interaction Between the YREB and the BRI.** Among the six provinces in the central region, Anhui, Jiangxi, Hubei, and Hunan are situated within the YREB. The superposition effect of the YREB and the BRI development strategies can be realized through the following three aspects.

(1) ***Infrastructure Development.*** The central region shall actively construct a comprehensive transportation system consisting of air, rail, and water transportation services and underground pipelines. It shall promote the development of the YREB by increasing the facility connectivity and connects it to the Silk Road Economic Belt in the west and to the 21st-Century Maritime Silk Road in the east by enhancing transportation channels in the upper, middle, and lower reaches of the Yangtze River. While strengthening its own infrastructure, the central region should actively participate in the infrastructure development projects of the BRI countries. The central provinces, especially Hubei and Hunan, have rich experience in infrastructure development and competitive infrastructure-related industries. Relevant enterprises such as those engaging in construction or machinery business can go goal by participating in foreign infrastructure development projects. By seizing the opportunities of international capacity cooperation brought by the BRI and the Asian Infrastructure Investment Bank (AIIB), the central region can leverage its industrial advantages to realize international cooperation featuring complementary partnerships and win-win results and boost its local industrialization and urbanization.

(2) ***Industrial Transformation and Upgrading in the YREB.*** In its "Guiding Opinions on the Development of the Yangtze River Economic Belt Based on the Golden Waterway," the PRC State Council proposes the goal of nurturing world-class industry clusters. The four central provinces located within the YREB must faithfully pursue this goal as they are positioned as top-notch provincial economies in this Belt. According to the State Council, the YREB shall focus on industries such as electronic information, high-end equipment manufacturing, and automobiles, which are the advantageous industries of Hubei, Hunan, Anhui, and Jiangxi. There are 40 100-billion-yuan industry clusters in Hubei, which provides foundations for world-class industry clusters. Given the existing industrial resources, development environment, and opportunities in these central riverside provinces, it is highly possible to see several world-class industry clusters emerge here within five to ten years.

(3) ***Open Cooperation Corridor.*** The YREB is going to be an open cooperative corridor connecting the eastern, west, northern, and southern provinces of China. Leveraging the golden waterway provided by the Yangtze River, cooperation between key industrial parks and key ports can be strengthened, and the BRI-related trade and investment, especially the cooperation with EU countries in the fields of high-end technology, high-tech equipment, new energy, and new materials, can be utilized to drive the industrial transformation and upgrading in the YREB. Hubei needs to enhance the functions of its "Wuhan-ASEAN" and "Wuhan-Japan-South Korea" shipping channels and China-Europe fright

train services. The Wuhan-Xi'an high-speed railway line needs to be completed for the enhancement of the connection between the YREB and the Silk Road Economic Belt. The YREB also needs to expand its influence eastward to forming a fast-track link with the core area of the 21st-Century Maritime Silk Road.

3 The Influence of Technology Restrictions on Chinese High-Tech Industries

3.1 Background of the Technology Blockade

Since 2018, the U.S.-China trade war has spread to the high-tech sector. ZTE encountered a technology embargo. Huawei 5G technology promotion was frustrated. Chinese high-tech industry faced acute choke points. Over the year, the Bureau of Industry and Security (BIS) of the U.S. Department of Commerce has remained committed to building export control tools for various commodities and technologies to prevent exports of critical technologies to China. The United States has also created a common interest community with Europe through the Wassenaar Agreement to limit the transfer of European high technology to China. In addition, the United States has also introduced FY2019 National Defense Authorization Act, the Foreign Investment Risk Review Modernization Act of 2018 (FIRRMA), and the Export Control Reform Act of 2018.

The FY2019 National Defense Authorization Act (hereinafter referred to as the "Act") explicitly declares China as a strategic competitor of the United States and proposes to formulate a comprehensive government strategy to contain China. Besides stepping up the military containment of China in the Indo-Pacific region, it calls for more economic restrictions on Chinese enterprises. First, it request to restrict the U.S. government purchases of Chinese equipment and services for telecommunications and video surveillance purposes. Article 889 of the Act prohibits U.S. government agencies from entering into, extending, or renewing a contract with Huawei or ZTE or using their products, equipment, services, and critical technologies. Under the pretext of protecting public and national security, the Act bans the use of video surveillance products or telecommunications equipment produced by Hytera Communications Corporation, Hangzhou Hikvision Digital Technology Company, or Dahua Technology. It is worth noting that the United States does not just target certain Chinese enterprise. Instead, it targets the entire Chinese communications manufacturing industry and its fast-growing high-tech industry. Second, it prohibits the U. S. military from importing sensitive materials from China, North Korea, Russia, and Iran. The materials involved are samarium-cobalt magnets, neodymium-iron-boron magnets, tungsten metal powder, and tungsten heavy alloy, which are used in manufacturing industries related to national defense science and technology, such as the manufacturing of aviation equipment, weapons systems, supercomputers, satellites, etc. According to statistics released by the U.S. Geological Survey (USGS), the United Stated relied heavily on China for rare earth, with 78% of it rare earth imports coming from China in 2017. The adoption of these restrictions on the U.S. Department of Defense, which was aimed to make the United States' strategic material supply chain more independent from China, constituted a turning point in the development of the U.S. rare earth industry: In 2017, China exported 3,350.85 metric tons of rare earth permanent magnet materials to the United States, accounting for only 11.2 % of the total exported amount. The Act has little effect on China's rare earth industry. Nevertheless, as the U.S. military actively reduces its dependence on the Chinese rare earth magnets, we lose an essential bargaining chip in the Sino-US economic and trade negotiations.

The Foreign Investment Risk Review Modernization Act of 2018 (FIRRMA) is issued to prevent Chinese enterprise from buying key U.S. technologies. On August 23, 2018, Trump hosted a roundtable discussion on FIRRMA. At the roundtable, many members of Congress noted that this act was aimed to address threats from China. Main contents of FIRMA include the following.

First, FIRMA expands the scope of covered transactions subject to the Committee on Foreign Investments in the United States (CFIUS). The covered transactions originally referred to all foreign investments that could result in foreign control of a U.S. business. Now FIRMA authorizes CFIUS to review more transactions: foreign acquisitions of real estate that is located within or in close proximity to an U.S. air or maritime port, military installation, or another national-security-sensitive property of the United States Government; direct and indirect foreign investments in U.S. businesses that own critical infrastructure, critical technologies, or sensitive personal data of U.S. citizens, which may be exploited to get important non-public information or enable foreign investors to intervene in critical decisions of U.S. enterprises or transfers of intellectual property rights or critical technologies to specific joint venture projects, etc. That is, the CFIUS that is originally responsible for reviewing mergers and acquisitions (M&A) investments is authorized to review certain greenfield investment and joint venture projects. In addition, FIRMA gives a comprehensive definition of the term "critical technologies." It means any of the following: defense articles or defense services included on the United States Munitions List set forth in the International Traffic in Arms Regulations (ITAR); items included on the Commerce Control List set forth in Supplement No. 1 to part 774 of the Export Administration Regulations (EAR); chemicals, biological weapons, nuclear equipment and components, and certain drugs and toxins; emerging and foundational technologies covered by the Export Control Reform Act (ECRA). This expands the definition of "critical technologies": it refers to not only technologies in the military category but also emerging, cutting-edge technologies in nearly all fields.

Second, FIRRMA expands the scope of transactions reviewable by the CFIUS to address national security concerns, but does not give the term a clear definition. Besides its 12 criteria for consideration, the CFIUS is required to consider "whether a covered transaction involves a country of special concern that has a demonstrated or declared strategic goal of acquiring a type of critical technology or critical infrastructure that would affect United States leadership in areas related to national security."

The "country of special concern" here clearly implies China. Apart from this requirement, there are other contents targeting China, such as those about the accumulation of critical technologies, the U.S. citizens' sensitive information, the U.S. national network security, etc. This gives the CFIUS greater discretion and more chances to launch reviews on the pretext of national security.

Third, FIRRMA requires reports on foreign investments from China in the United States. The Secretary of Commerce is requested to submit to the U.S. Congress and the CFIUS a report on foreign investment transactions made by entities of the People's Republic of China (PRC) every two years. The report shall include a breakdown of the Chinese investments by government and non-government organizations and content about whether the investments align with the objectives outlined by the PRC's Made in China 2025 plan. Although the document does not stipulate discriminatory treatments for Chinese investments, but in fact, it creates prejudice against Chinese investments.

Fourth, FIRRMA extends the review time and improve the efficiency of CFIUS reviews. Beside stipulating simpler review application process and longer review times, it introduces a mandatory declaration system and authorizes the President to take any action with respect to a covered transaction related to national security issues. It further enhance the CFIUS's review capabilities and efficiency.

The ECRA is aimed to restrict technology exports to China. On August 1, 2018, the BIS updated the EAR Entity List to add 44 more entities in China. Among them, 17 companies were accused of illegally purchasing goods and technologies for unauthorized military use in China, and 27 companies were listed for potential risk of using or transferring U.S. goods for military purposes in China. The United States believed that these Chinese enterprises carry out or participate in activities that harm its national security or diplomacy interests and thus conducted strict reviews on their U.S. trade partners' exported products and technologies. This list update mainly affected military research institutions and involved products and critical areas in information technology such as electronic components, communication equipment, LED, and photovoltaics, targeting leading Chinese enterprises in various fields. On October 30, 2018, the United States Department of Commerce added Fujian Jinhua Company to the EAR Entity List, increasing the number of listed Chinese enterprises to 219, second only to listed Russian companies that amount to 335. The ECRA systematizes and codifies regulations and practices of export controls currently implemented by federal agencies of the United States, including the aforementioned EAR. It provides a permanent legislative basis for the current export controls over items used both for civilian and military applications. The ECRA has two main characteristics. First, it expands the scope of application for US export control laws and tighten export controls on "emerging and foundational technologies." It proposes to establish an interdepartmental committee led by the President to qualify "emerging and fundamental technologies." Nonetheless, in the absence of clearly-stipulated qualification standards, nearly all technologies deemed crucial to the national security of the United States can be qualified, including artificial intelligence, biopharmaceuticals, intelligent manufacturing, and next-generation information technologies. ECRA requires the Secretary of Commerce to impose commercial export controls on the export of qualified technologies. This means that the exports, re-exports, or transfers of the so-called "emerging and fundamental technologies" to countries under arms embargoes need special permissions from the US Department of Commerce. This makes it difficult for China to purchase advanced technologies through commercial channels. Second, the ECRA tightens licensing reviews of countries under arms embargo by the United States, namely the PRC, Belarus, Myanmar, Iran, Cuba, North Korea, Syria, and Venezuela. Among them, only China is a major trade partner of the United States. Therefore, this rule is considered a "tailored" export control requirement for China.

3.2 The Actual Impact

The increasing difficulties for Chinese enterprises to invest in the United States has affected their global businesses. These changes are reflected in data released by the United States: China's new investments in the country peaked at $27.43 billion in 2016, and $26.49 billion of it was invested in U.S. enterprises through mergers and acquisitions, accounting for 96.6% of the total amount; Due to harsh restrictions on Chinese investments, China invested only about $13.97 billion to the United States in 2017, only 50.9% of the 2016 amount; In the first half of 2018, Chinese enterprises' direct investments to the United States totaled only $1.8 billion, a year-over-year decline of more than 90%. As the United States is determined further tightens its export controls and restrictions over foreign investments and government procurements, Chinese enterprises' willingness to invest in the United States will decline further. The investment-related channels to obtain advanced and emerging technologies from U.S. enterprises may be completely blocked, and the cost of technology learning for Chinese enterprises will increase significantly.

As shown in Fig. 3, China's high-tech industries' expenditure on the introduction of foreign technologies increased steadily after 2014, peaking at 47.54 billion yuan in 2016, but it fell directly back to the level of 2012 in 2017; the expenditure picked up later but it only recovered to the level of 2016 in 2019. These changes suggested that protectionism and technical blockades in Western countries appeared earlier than 2018 when the U.S. official documents about technology restrictions were issued.

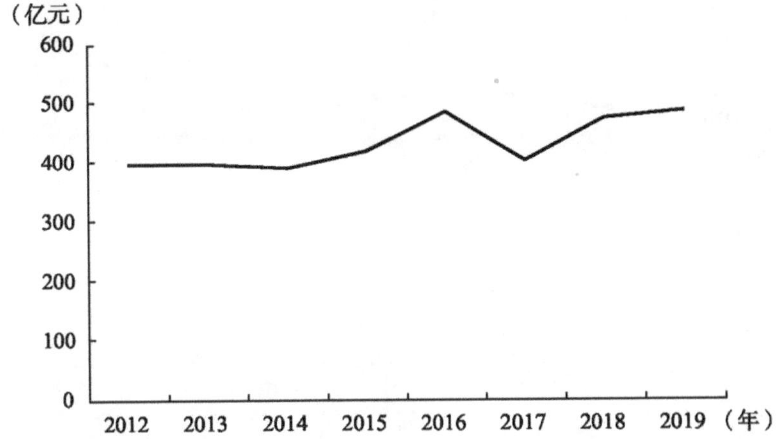

Fig. 3 Chinese high-tech industries' expenditure on the introduction of foreign technologies 2012–2019

Similar trend appears in international trade, as shown in Fig. 4. China's total high-tech products exports remain higher than the total imports. Both imports and exports were stable in the period from 2012 to 2016, but they entered a period of overall rapid growth after 2016. These changes reflected that China's high-tech industries quickly adjusted their trade strategies amid the technology blockade. They looked for alternatives among other trade partners and proactively broke through the blockade, turning the crisis into a development opportunity. Unfortunately, the new international cooperation has not fundamentally changed the overall situation. China still has to rely on technology imports, and its high-tech industries have to carry out independent innovations to make breakthroughs and realize leapfrogging development. According to Fig. 5, Chinese high-tech industries' spendings on domestic technology purchases remained stable from 2012 to 2017 but increased drastically after 2017. Up to now, the expenditure has already doubled. This indicates that fast-improved domestic technologies have become strong competitors and substitutes for foreign technologies and that China's independent innovation is entering a phase of vigorous development.

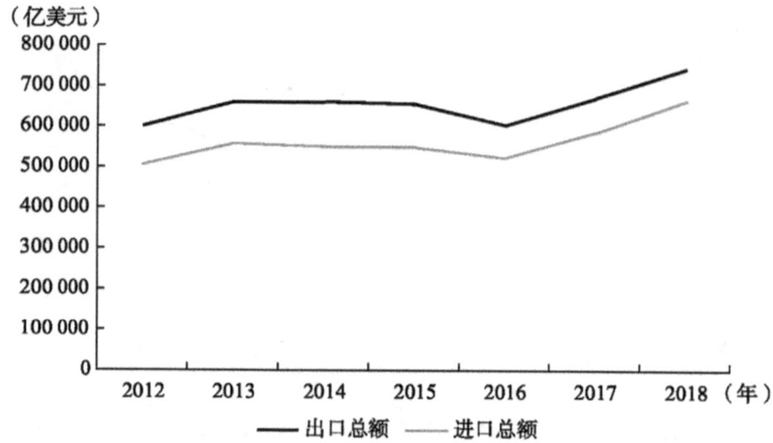

Fig. 4 Imports and exports of high-tech products 2012–2018

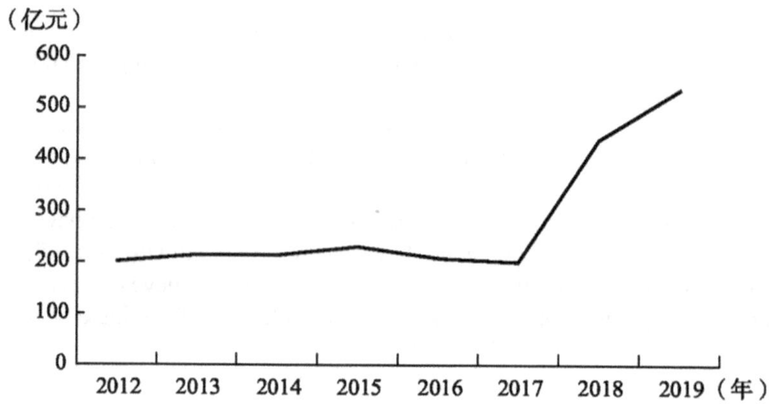

Fig. 5 Chinese high-tech industries' expenditure on domestic technology purchases 2012–2019

(亿元)	(100 million yuan)
(年)	(year)
出口总额	Total exports
进口总额	Total imports
亿美元	(100 million USD)

With substitutes for foreign technologies, imports and exports have not been much affected. Nevertheless, this does not mean that the technology blockade does not bring any difficulty to our country's high-tech industry. The main adverse effects are embodied in the changes in the operating revenues of Chinese high-tech industries. The operating revenue is a very important business indicator related to cash flows and the realization of profits. As shown in Fig. 6, the operating revenues of Chinese high-tech industries grew at a relatively steady pace before 2016, but the increase slowed down significantly in 2016 with the regularization of technology blockade measures. In 2018, the increase almost completely stopped. This situation is very unfavourable for the long-term development of our high-tech industries. In view of their high demand for capital, most high-tech industries may fail to survive in this harsh environment when their operating revenues stop increasing.

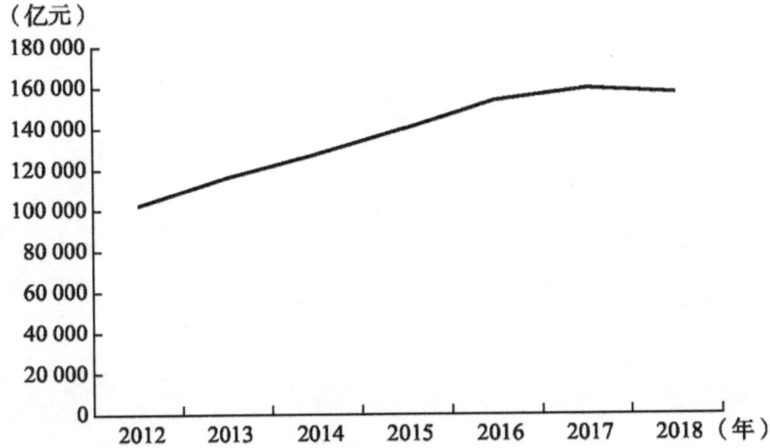

Fig. 6 Operating revenues of Chinese high-tech industries 2012–2018

(亿元)	(100 million yuan)
(年)	(year)

Part II
New Impetus for China's Regional Economic Development

Chapter 3
Strategy for the Ecological Conservation and High-Quality Development of the Yellow River Basin Economic Zone

1 Background

The Yellow River, the second largest river of China, originates in the Yueguzonglie Basin on the northern flank of the Bayan Har Mountains, which is situated at an elevation of about 4,500 meters. It flows 5,464 kilometers across north China to the Gulf of Bohai in Dongying City, Shandong province, with an average annual runoff of 58 billion cubic meters. The YRB is located in the north-central part of China, between 95°53′ and 19°05′ E and 32°10′ and 41°50′ N. It is 1,900 kilometers from east to west and 1,100 kilometers from north to south, covering an area of 795,000 square kilometers. From west to east, the Yellow River flows across nine provinces: Qinghai, Sichuan, Gansu, Ningxia, Inner Mongolia, Shaanxi, Shanxi, Henan, and Shandong (hereinafter collectively referred to as the YRB provinces), connecting 58 prefectural-level administrative regions and 362 county-level administrative regions. With the exception of Henan, Shandong and Sichuan, their provincial capitals are located with the YRB. Among the three, the provincial capitals of Henan and Shandong are situated near the YRB.

Among the nine provinces that the Yellow River flows through, Qinghai, Gansu, Inner Mongolia, Shaanxi provinces have relatively larger catchment areas of the river, as shown in Table 1. Among the nine provinces, Qinghai has the largest catchment area, and Shandong has the smallest one. The catchment area of the Yellow River in Ningxia accounts for 77.4% of the total land area of the province: this percentage far exceeds those of the other eight provinces. Therefore, it is fair to say that the development and environmental protection efforts dedicated to the YRB are of great significance to Ningxia.

The Huang-Huai-Hai Plain, Fenwei Plain, and Hetao Irrigation Area in the YRB have long been main agricultural production areas of China, providing important guarantees for its national supplies of grains and meat. Nevertheless, it was not until the founding of the PRC that the region started to improve its weak industrial foundations. Over the years, the development of energy industries, major industrial

J. Sun, *Regional Economic Development in China in 2020*,
https://doi.org/10.1007/978-981-95-3604-7_3

Table 1 Catchments of the yellow river in different provinces

Province	Qinghai	Sichuan	Gansu	Ningxia	Inner Mongolia	Shaanxi	Shanxi	Henan	Shandong
Catchment area (km^2)	152,250	16,960	143,240	51,393	150,962	133,300	97,138	36,164	13,633
Share in provincial area (%)	21.08	3.49	33.64	77.40	12.76	64.85	61.99	21.66	8.63

Data source http://www.huanghenet.com/

bases, and emerging cities in the YRB has underpinned the development of the region. The energy industries refer to the hydro-power industry in the upper reaches of the Yellow River, the coal energy industry in the middle reaches, and the oil and natural gas energy industries in the lower reaches. The YRB is hailed as China's "energy basin": Its raw coal production accounts for more than half of the national total; its oil production takes up about 1/4 of the national total. This region is rich in other mineral resources such as rare earths, gypsum, siliceous rock, bauxite, aluminum, refractory clay, etc. Besides, the concentrated distribution of mineral resources in this region is highly conducive to relevant mining industries.

The YRB can be divided into three parts. The western part stands at high altitudes and is snowy all the year round. The central part features Loess landforms and has a prominent problem of soil erosion. The eastern part mostly consists of alluvial plains at low altitudes, which are vulnerable to flooding. Atmospheric circulation and monsoon circulation lead to uneven precipitation in the YRB. Most places in the basin lack precipitation, and the water resource per capita data of the basin is less than 30% of the national average.

In the lower reaches of the Yellow River, the riverbed is several meters above the cities and farmlands around it. In the area, the Yellow River serves as a natural boundary between the Huaihe River Basin and the Haihe River Basin and a water resource for major cities and industrial and mining enterprises such as Zhengzhou, Kaifeng, Xinxiang Zhongyuan oil field, Shengsheng oil field, etc. Some places outside the YRB also rely on or plan to rely on the downstream and midstream of the Yellow River for water supplies: Tianjin City; parts of Heibei Province; Qingdao City; the region along the canal that transports water to Qingdao City; Shanxi Energy Base; and four lakes in Shandong Province, namely Nanyang Lake, Dushan Lake, Zhaoyang Lake, and Weishan Lake.

The Yellow River is the most important water source in the northwestern and northern regions of China. With only 2% of China's total water resources, it irrigates 15% of the nation's arable land and offers water to 12% of its population. It guarantees water supplies for not only the YRB but also nearby water-deficient areas including industrial, agricultural, and urban areas. As the average sediment discharge of the Yellow River has remained high for years, the problem of water pollution is severe. Even in its water functional zone, the water quality compliance rate is only 60%. With this water quality problem, plus with the high utilization rate of the water resources in the region, the YRB faces severe water shortages.

Table 2 shows the ecological status of each province in the YRB in 2019. Among the nine provinces and regions of the YRB, Shaanxi boasted the highest forest coverage rate, which was 43.06%, followed by Sichuan (39.60%), and Qinghai had the lowest rate: 7.26%. As to water resource per capita, Qinghai topped the list with the number reaching 15,182.49 cubic meters per person, and Henan Province ranked last, with only 175.21 cubic meters per person. Among the nine provinces, Shandong had the largest urban green space, which amounted to 252,338.37 hectares.

The Yellow River connects the Qinghai-Tibet Plateau, the Loess Plateau, and the North China Plain, flowing through ecologically fragile areas such as the soil-erosion areas of the Loess Plateau and five sandy deserts and desertified lands. In

Table 2 The ecological status of provinces in the YRB

Project province	Number of prefectural-level cities (whole number)	Forest area (% of land area)	Urban green space (hectare)	Per-capita water resource (m^3 per person)
Shanxi	11	22.8	51,445.59	261.31
Inner Mongolia	9	–	69,069.06	1,765.47
Shandong	16	17.95	252,338.37	194.06
Henan	17	24.94	115,269.26	175.21
Sichuan	18	39.60	124,157.92	3,288.94
Shaanxi	10	43.06	59,615.67	1,279.84
Gansu	12	–	29,186.1	1,233.54
Qinghai	2	7.26	7,309.06	15,182.49
Ningxia	5	–	26,215.91	182.21

fact, most of the ecological functional areas that it flows through are ecologically fragile. The Loess Plateau is widely known for its severe soil-erosion problem and weak ecosystem and economic foundations, and the lower reaches of the Yellow River is troubled by floods as well as insufficient water supplies. These are urgent ecological problems in the YRB.

The YRB must pursue high-quality development. The term refers to the pursuit of a high-quality, efficient, and sustainable development that benefits all: It features an upgraded development mode, an optimized economic structure, new growth drivers, and other measures to improve total factor productivity and build a modern economic system. Given the overall ecological fragility of the YRB and high flood risks in the lower reaches of the Yellow River, the contradiction between human and land in the region is acute. Therefore, when boosting the development of the YRB, it is essential to strike a balance between ecological protection and high-quality development. On one hand, the ecological protection underpins the high-quality development of the YRB. On the other hand, the high-quality development facilitates the ecological protection efforts and the sustainable development of the region.

On September 18, 2019, Xi Jinping, General Secretary of the CPC Central Committee, President of the PRC, and Chairman of the Central Military Commission, presided over a symposium on the ecological protection and high-quality development of the YRB and delivered a keynote speech. He emphasized that the efforts to promote the ecological protection and high-quality development of the YRB would improve people's living standards and make the Yellow River benefit humanity.

The Politburo of the CPC Central Committee held a meeting on August 31, 2020 to review the "Outline of the Yellow River Basin Ecological Protection and High-Quality Development Plan" and the "Comprehensive Report on the Fifth Round Inspection by the 19th CPC Central Committee." The meeting stressed the need to improve the ecological environment of the YRB, optimize its water resource allocation, and take water resources as the maximum rigid restriction. It called for effective measures to promote overall high-quality development of the region, accelerate the transformation of its development drivers, build its modern industrial system with distinctive advantages, and optimize its urban development pattern.

2 The Spatial Correlation and Interaction in the YRB: Metropolitan Areas in the Yellow River Ji-Shaped Bend

2.1 *Metropolitan Areas in the Yellow River Ji-Shaped Bend*

A metropolitan area is a spatial organization of urban forms. This term was first introduced by a French geographer named Jean Gottmann to describe large urban configurations. A metropolitan area refers to a major city together with its suburbs and nearby cities, towns, and environs over which the major city exercises a commanding economic and social influence: the central city and its peripheral areas participate in a regional division of labor and cooperation with a 1-hour commuting circle. Besides, a metropolitan area is often the core area of an urban cluster or city cluster. Generally speaking, metropolitan areas have the following characteristics. First of all, a metropolitan area must have at least one economically advanced central city. Secondly, there are close economic and social ties between the central cities and peripheral urban communities. Thirdly, a metropolitan area usually features a circular, multi-layer spatial structure. Fourthly, there is usually a trend of economic integration in a metropolitan area. It is worth noting that metropolitan areas are different from city clusters. The former is lesser to the latter in terms of size and development level, but the former is often the central part of the latter. The "Guiding Opinions of the National Development and Reform Commission on Cultivating and Developing Metropolitan Areas" defines the two terms as follows: A metropolitan area refers to the urban forms under the influence of a central city (a super-large or mega city); A city cluster consists of several metropolitan areas and central cities.

As for the selection and qualification of central cities, the PRC State Council issued the "Notice on Adjusting the Standards for Urban Scale Division" in November 2014, stipulating that a metropolitan area must have a big central city whose urban population is over 3 million or at least over 2 million. Li (2019) proposed two basic criteria for central cities: having an urban population of more than 3 million; being a municipality directly under the Central Government of the PRC, a provincial capital, or a sub-provincial-level city.

The are many cities in the Yellow River Ji-Shaped Bend ("Ji" refers to the Chinese character 几, whose upper part resembles the shape of the middle part of the Yellow River), which is a zone bordering five provinces, namely Gansu, Ningxia, Inner Mongolia, Shaanxi Shanxi, and Shanxi. Among these cities, there are three provincial capitals: Taiyuan (the capital of Shanxi Province), Hohhot (the capital of Inner Mongolia), and Yinchuan (the capital of Ningxia). Other major cities include Ningxia's Wuzhong and Zhongwei, Inner Mongolia's Wuhai, Bayannur, Baotou,

Ordos, Shaanxi's Yulin, and Shanxi's Shuozhou, Xinzhou, and Lvliang. Many cities in this region have good economic performance, but their exchanges and cooperation are insufficient. The inter-metropolis bounds in the Yellow River Ji-Shaped Bend can be strengthened through the development of metropolitan areas there, which is conducive to the coordinated development of the region. In the following paragraphs, we discuss and measure the strength of the urban spatial interaction among metropolitan areas in the Yellow River Ji-Shaped Bend.

2.2 Data Source

The data about urban populations and GDPs of cities at and above the prefectural level come from China City Statistical Yearbooks and China Statistical Yearbooks (CSYD) released between January 1, 2009 and December 31, 2018. Urban highway mileage data comes from www.chinahighway.com.

Spatial interaction refers to the mutual transmission process of commodity, population, labor, capital, technology, and information. It has a great influence on the establishment and change of inter-regional economic relations. On the one hand, it strengthens inter-regional links, facilitates mutual exchange of needed products, and provides more development space and opportunities. On the other hand, it can lead to inter-regional competition for resources, factors, and development opportunities, which can cause damage to some regions involved.

Spatial interaction theory holds that the spatial interaction among geographical entities depends on their mutual complementarity, accessibility, and disturbance, and that the strength of the interaction declines with increasing distance. In addition to the influence of distance, the size of urban population and economic scale also have remarkable influences on the strength of spatial interaction. The Gravity Model is most widely used in measuring the strength of the urban spacial interaction. We modify the model by taking into consideration the influence of geographical variables (adjacency, geometric distance, and accessibility) and economic variables (population size and economic scale) on the strength, and get the following calculation formula.

$$Fxy = \frac{\sqrt{P_x V_x P_y V_y}}{d_{xy}^{a_{xy}}}$$

In the formula, Fxy represents the strength of spatial interaction between two cities; Px and Py respectively represent the two cities' (cities at and above the prefectural-level) population sizes or total urban populations; Vx and Vy represent the economic scales or GDPs of the two cities; d is the distance between the two cities, calculated by highway mileage; a_{xy} is the friction coefficient of d, mainly referring to the differences in accessibility and generally ranging between 0.5 and 3. Based on existing literature and research, we make the following provisions on the value of a to facilitate relevant operation: a is 2.25 for two cities that are adjacent and connected by high-grade highways (national highways or expressways) or railways; a is 2.5 for two cities that are adjacent but not connected by high-grade highways (national highways or expressways) or railways; a is 2.75 for two cities that are connected by high-grade highways (national highways or expressways) or railways but not adjacent to each other; a is 3 for two cities that are not adjacent or connected by high-grade highways (national highways or expressways) or railways. For the three pairs of cities in Shanxi Province, namely Taiyuan-Xinzhou, Taiyuan-Lvliang, Taiyuan-Shuozhou, a is 2.25, 2.25, and 2.75, respectively. For the four pairs in Inner Mongolia, namely Hohhot-Ordos, Hohhot-Baotou, Hohhot-Bayannur, Hohhot-Wuhai, a is 2.25, 2.25, 2.75 and 2.75. As to the two pairs in Ningxia, Yinchuan-Wuzhong and Yinchuan-Zhongwei, a is 2.25 and 2.75.[57]

When studying the strength of the inter-provincial interaction, we use the distances (highway mileages) between provincial capitals to express the distances between provinces. As Taiyuan (provincial capital of Shanxi), Hohhot (provincial capital of Inner Mongolia), and Yinchuan (provincial capital of Ningxia) are not in proximity of Yulin (a city in Shaanxi), we determine that in these inter-regional pairs, a is 2.75. When calculating the total economic aggregate of an involved province, we calculate the economic aggregate of its involved cities, i.e. the economic aggregate of its involved metropolitan areas. As for the cities included in the calculation, there are four cities in Shanxi, namely, Taiyuan, Xinzhou, Shuozhou and Lyuliang; three cities in Ningxia, i.e. Yinchuan, Wuzhong and Zhongwei; five cities in Inner Mongolia, including Hohhot, Wuhai, Bayannur, Baotou, and Ordos; and Yulin City of Shaanxi.

2.3 *Strength of Inter-Provincial Interaction Between Metropolitan Areas in the Yellow River Ji-Shaped Bend*

From 2009 to 2018, the inter-provincial spatial interaction among metropolitan areas in the Yellow River Ji-Shaped Bend was significantly enhanced. As shown in Table 3 and Fig. 1, the strength of interaction between any two provinces increased to varying degrees: the largest and second largest increases are respectively found in the interaction between Shanxi and Inner Mongolia and that between Inner Mongolia and Shaanxi; the interaction strength between Shanxi and Inner Mongolia grew the fastest, followed by that between Inner Mongolia and Shaanxi.

Table 3 Strength of inter-provincial interaction between metropolitan areas in the yellow river Ji-shaped bend

Year region	2009	2010	2011	2012	2013	2014	2015	2016	2017	2018
Shanxi and Inner Mongolia	0.0355	0.0407	0.0494	0.0560	0.0598	0.0627	0.0647	0.0687	0.0645	0.0725
Shanxi and Shaanxi	0.0068	0.0086	0.0108	0.0129	0.0141	0.0152	0.0161	0.0168	0.0262	0.0296
Shanxi and Ningxia	0.0043	0.0049	0.0059	0.0069	0.0075	0.0080	0.0084	0.0092	0.0105	0.0121
Inner Mongolia and Shaanxi	0.0231	0.0272	0.0334	0.0389	0.0435	0.0467	0.0498	0.0539	0.0513	0.0569
Inner Mongolia and Ningxia	0.0054	0.0063	0.0078	0.0090	0.0101	0.0109	0.0116	0.0125	0.0119	0.0132
Shaanxi and Ningxia	0.0021	0.0027	0.0034	0.0042	0.0049	0.0054	0.0058	0.0063	0.0076	0.0111

Data source EPS data

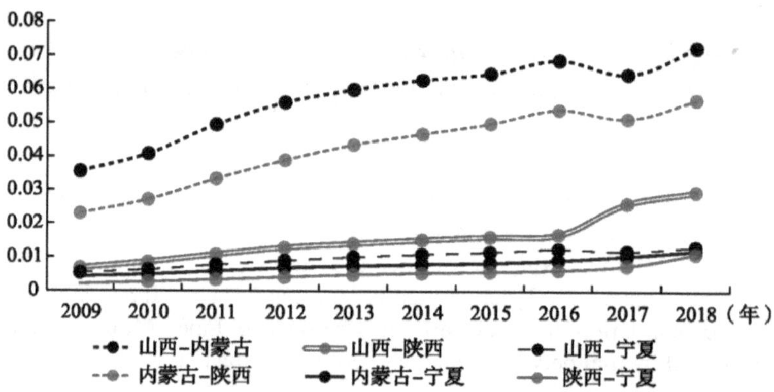

Fig. 1 Interaction strength. *Data source* EPS data (China Regional Economic Database)

山西-内蒙古	Shanxi-Inner Mongolia
山西-陕西	Shanxi-Shaanxi
山西-宁夏	Shanxi-Ningxia

(continued)

(continued)

内蒙古-陕西	Inner Mongolia-Shaanxi
内蒙古-宁夏	Inner Mongolia-Ningxia
陕西-宁夏	Shaanxi-Ningxia
(年)	(year)

In 2017, the interaction between Shanxi and Inner Mongolia and that between Inner Mongolia and Shaanxi decreased in strength. Ningxia had low spatial interactions with other provinces because of its low urban economic aggregates and long distances from other provinces. The interaction strength between Shanxi and Inner Mongolia was the highest among the pairs of provinces. This was because the involved cities in Shanxi had relatively strong economic performance and Inner Mongolia had the most cities located in the Yellow River Ji-Shaped Bend among the provinces involved in this calculation. The relatively short distance between the provincial capitals of the two provinces also contributed to the high interaction strength. The relatively long distance between Yulin and the provincial capital of Shanxi led to relatively low interaction strength between the two province. The reason distances significantly influence the spatial interactions was that cities in this region mainly relied on highways to connect with one another. The Yellow River cannot be used as their shipping channel because of the limitation of natural conditions.

2.4 *Strength of Intra-Provincial Interaction Between Metropolitan Areas in the Yellow River Ji-Shaped Bend*

Based on their high intra-provincial inter-metropolis interaction strengths, the provincial capitals of Shanxi, Inner Mongolia, and Ningxia were undoubtedly central cities. As shown in Table 4, from 2009 to 2018, the provincial capitals' interaction with other cities within their respective provinces increased significantly: The interaction strength between Taiyuan and Xinzhou remained the highest, followed by that between Yinchuan and Zhongwei and that between Hohhot and Ordos.

Although both located in Inner Mongolia, Hohhot and Bayannur had the lowest interaction strength. This was a result of Bayannur's small economic aggregate and inconvenient transportation. It was clear that distances had a strong influence on the strength of intra-provincial spatial interaction among cities. This suggests that the improvement of traffic conditions helps to strengthen inter-cities ties.

2.5 *Conclusions and Suggestions*

From 2009 to 2018, the inter-provincial spatial interaction among metropolitan areas in the Yellow River Ji-Shaped Bend increased remarkably. The strength of

Table 4 Strength of intra-provincial interaction between metropolitan areas in the yellow river Ji-shaped bend

Year region	2009	2010	2011	2012	2013	2014	2015	2016	2017	2018
Taiyuan and Shuozhou	0.000 38	0.000 41	0.000 51	0.000 59	0.000 60	0.000 61	0.000 55	0.000 58	0.000 63	0.000 72
Taiyuan and Xinzhou	0.016 98	0.019 91	0.023 42	0.026 44	0.028 40	0.029 05	0.030 15	0.032 51	0.037 23	0.044 30
Taiyuan and Lvliang	0.001 84	0.002 29	0.002 90	0.003 09	0.003 07	0.002 85	0.002 93	0.003 11	0.003 57	0.004 37
Yinchuan and Zhongwei	0.007 72	0.009 33	0.011 48	0.013 54	0.015 53	0.016 41	0.017 62	0.019 44	0.022 02	0.025 28
Hohhot and Ordos	0.003 29	0.003 79	0.004 76	0.005 33	0.005 92	0.006 49	0.006 97	0.007 47	0.006 29	0.006 70
Hohhot and Bayannur	0.000 01	0.000 01	0.000 01	0.000 01	0.000 02	0.000 02	0.000 02	0.000 02	0.000 02	0.000 02
Hohhot and Baotou	0.000 12	0.000 13	0.000 17	0.000 19	0.000 21	0.000 23	0.000 26	0.000 27	0.000 22	0.000 23
Hohhot and Wuhai	0.001 49	0.001 78	0.002 29	0.002 63	0.002 88	0.003 12	0.002 81	0.002 89	0.002 38	0.002 70

interaction between any two provinces increased to varying degrees: the interaction strength between Shanxi and Inner Mongolia had the biggest growth, followed by that between Inner Mongolia and Shaanxi; the interaction strength between Shanxi and Inner Mongolia grew the fastest, followed by that between Inner Mongolia and Shaanxi. With high intra-provincial inter-metropolis interaction strengths, the provincial capitals of Shanxi, Inner Mongolia, and Ningxia became central cities of their metropolitan areas. From 2009 to 2018, their interaction with other cities within their respective provinces increased: Taiyuan and Xinzhou outshone the other pairs of cities in terms of interaction strength, Yinchuan and Zhongwei ranked the second, and Hohhot and Ordos, the third. It is worth noting that this calculation include only Yulin City for Shaanxi Province. Due to the city's limited economic aggregates, Shaanxi's interaction with other provinces appear weak in the calculation results.

For its future development, the YRB needs an overall layout. It needs efforts to enhance its central cities' ties with their peripheral cities and towns, which will boost the development of the peripheral areas, optimize the division of labor within metropolitan areas, and promote cooperation and the construction of transportation networks and modern information networks. Such ties are conducive to the coordinated development of the region. As to specific methods to build such ties, we propose the following suggestions. (1) Eliminate the barriers to the inter-provincial and inter-city flows of factors. (2) Strengthen major infrastructure facilities, enhance transportation links, and increase accessibility. (3) Remove administrative barriers to inter-province and inter-city interactions. (4) Give full play to the role of market in allocating resources.

In 2020, faced with complicated domestic and international situations and the impact of the COVID-19 epidemic, most cities in the Yellow River Ji-Shaped Bend achieved economic growth by coordinating economic development plans with epidemic-prevention-and-control efforts and implementing effective policies. In the future, these cities shall faithfully pursue the "dual-circulation" development paradigm, strengthen inter-city economic ties, and take multiple measures to promote economic and social progress.

3 High-Quality Development of the YRB

3.1 Connotations of High-Quality Development

High-quality development has economic, social, and ecological connotations. It is a systematic, multi-dimensional, long-term development process driven by innovation, and it is green, efficient development that is realized without destroying the ecological balance. It highlights steady economic growth, healthy economic structure, and effective inter-regional coordination. It emphasizes environmental sustainability and people-oriented social development that benefits all.

The YRB has a fragile ecosystem due to the restriction of natural resources. While developing economy, it has to step up its ecological protection efforts to ensure its survival. Only when its ecology and environment is well protected can it development be sustainable. For this region, pursuing economic growth at the cost of environment will only leave hidden dangers for its future development. The YRB is a target area of China's poverty-alleviation campaign. Besides the task of ecological protection, provinces in the region also shoulder the responsibility of improving people's living standards through high-quality economic development.

In a word, the high-quality development of the YRB shall be based on systematic and coordinated arrangements. They shall put environmental protection and ecological balance in the first place, give full play to the decisive role of the market in allocating resources, and encourage innovations. They shall strengthen inter-city links while increasing the influence of the central cities as a way of realizing regional coordinated development. To investigate and evaluate the high-quality development of the YRB, we adopt the following five first-level indicators: ecological conservation, environmental protection, economic development, innovation-driven progress, and social betterment.

3.2 Research Method and Indicators

1. **Research Method.** To evaluate the high-quality development of the YRB as objectively as possible, we use the entropy weight method (EWM) to measure nine YRB provinces' high-quality development status from 2011 to 2019. The data comes from China City Statistical Yearbooks and CSYD, and provincial indicators come from provincial statistical bureaus. The original data are sorted out in the following way: there are a total of m evaluation objects and n evaluation indicators; in this chapter, m = 9. The specific evaluation steps are as follows.

(1) Due to the indicators' differences in dimension and order of magnitude, it is necessary to standardize each indicator;

(2) Calculate y_{ij}, i.e. the proportion of x^{ij} (the ith evaluation object's jth) in the jth indicator:

$$y_{ij} = \frac{x_{ij}}{\sum_{i=1}^{m} x_{ij}}$$

(3) Calculate the information entropy e_j of the jth index:

$$e_j = -\mathrm{K}\sum_{i=1}^{m} y_{ij}\ln y_{ij}$$

In the formula, $K = 1/\ln m$ is a non-negative constant, $0 \le ej \le 1$;

(4) Calculate the difference coefficient dj of the jth:

$$dj = 1 - ej$$

(5) Determine wj i.e. the weighting of the jth indicator:

$$w_j = \frac{d_j}{\sum_{j=1}^{n} d_j} = \frac{1-e_j}{n-\sum_{j=1}^{n} e_j}$$

(6) Calculate the high-quality development index Ui of the ith evaluation object:

$$U_i = \sum_{j=1}^{n} y_{ij}w_j \ (i = 1, 2, 3, \ldots, m;\ j = 1, 2, 3, \ldots, n)$$

2. **The Evaluation Indicator System of High-Quality Development in YRB.** Given the ecological priority and the essence of high-quality development in the YRB, this indicator system aims to evaluate the high-quality development status of the region under the constraints of its ecology and environment and

thus gives more weight to environmental and ecological indicators. It involves five dimensions of high-quality development (ecological conservation, environmental protection, economic development, innovation-driven progress, and social betterment), 5 second-level indicators, and 17 third-level indicators.

(1) ***Economic Development.*** Per capita GDP of the region represents the regional economic development status. The proportion of the secondary sector's added value in the regional GDP and that of the tertiary sector's added value in the regional GDP, both expressed as percentages, represent the status of the regional industrial structure. The proportion of foreign investment in the regional GDP, expressed as a percentage, reflects the degree of economic opening-up.
(2) ***Innovation-Driven Progress.*** The number (whole number) of professional and technical personnel in public (state-owned) enterprises and institutions reflects the investment (unit of measurement: 10,000 yuan) in scientific and technological innovation. The number of authorized patents and the turnover in the technology market reflect the output of scientific and technological innovation.
(3) ***Social Betterment.*** Registered urban unemployment rate (%), the number of basic pension insurance participants (unit of measurement: 10,000 persons), and investment (unit of measurement: 10,000 yuan) in social security and employment guarantee reflect the social development level.
(4) ***Environmental Protection.*** Total wastewater discharge (unit of measurement: 10,000 metric tons), total sulfur dioxide discharge (unit of measurement: 10,000 metric tons), and the total amount (unit of measurement: 10,000 metric tons) of industrial solid waste reflect environmental pollution status.
(5) ***Ecological Conservation.*** The urban green space coverage rate (%), per capita green space (square meters), total afforestation area (hectares), and the coverage (%)of nature reserves reflect the ecological status.

Analyzing from these five aspects, inter-provincial development disparities, waste of resources, or industrial structures that are problematic or not eco-friendly enough can restrict the development of the YRB to some extent.

As shown in Table 5, the nine YBR provinces delivered unstable performances in pursuing overall high-quality development from 2011 to 2019. During this period, Qinghai and Gansu did better than the other seven YRB provinces, and Shandong did the worst. After excluding the three provinces, there was not much difference in the performances of the remaining six provinces. After examining the nine provinces' changes in the five dimensions of high-quality development from 2011 to 2019, we find that in this indicator system that highlights environmental protection and ecological conservation, the nine provinces get relatively low economic development indexes and show only slight differences in the other four indexes.

Table 5 Composite development indexes for high-quality development in the YRB

Year province	2011	2012	2013	2014	2015	2016	2017	2018	2019
Shanxi	0.180 0	0.178 0	0.170 8	0.153 7	0.165 6	0.147 4	0.151 1	0.150 0	0.130 2
Inner Mongolia	0.156 0	0.161 8	0.149 1	0.135 3	0.125 0	0.121 9	0.123 7	0.126 3	0.123 6
Shandong	0.232 6	0.259 9	0.242 5	0.232 4	0.219 4	0.222 7	0.226 6	0.235 2	0.236 6
Henan	0.203 0	0.225 8	0.205 3	0.174 6	0.180 9	0.189 3	0.233 2	0.221 8	0.209 5
Sichuan	0.193 3	0.245 9	0.225 7	0.244 1	0.229 3	0.239 3	0.191 7	0.235 5	0.249 4
Shaanxi	0.164 6	0.183 0	0.172 3	0.140 5	0.138 4	0.128 0	0.173 6	0.183 8	0.162 2
Gansu	0.187 3	0.163 2	0.162 2	0.113 1	0.136 8	0.109 7	0.185 9	0.179 7	0.147 5
Qinghai	0.146 2	0.131 2	0.129 3	0.091 4	0.116 5	0.106 8	0.164 1	0.157 6	0.138 0
Ningxia	0.105 1	0.093 8	0.098 8	0.118 0	0.113 4	0.104 4	0.162 1	0.112 4	0.112 1

3.3 Spatial Pattern and Evolution of High-Quality Development in the YRB

The evolution of the YRB provinces' high-quality development status from 2011 to 2019 revealed that Sichuan surpassed Shandong to top the list in 2019, delivering the best performance in pursuing high-quality development. Sichuan made progress in all dimensions: Its indexes increased in a balanced manner, reflecting its success in pursuing high-quality development. Shandong ranked second only to Sichuan: This ranking was mainly attributed to its strong economic performance. Shandong had the highest economic development index among the nine YRB provinces. Henan's ranking was also good for its relatively high indexes of social betterment, innovation-driven progress, environmental protection, and ecological conservation. Among these nine provinces, Shandong and Henan respectively ranked first and second in the ranking of regional GDPs and that of authorized patents. Nevertheless, since they had the largest and second largest provincial populations among the nine province, they did not differ much from the other seven provinces in terms of per capita numbers. In addition, Shandong and Henan ranked high in the emission of "the three wastes" (i.e., waste gas, wastewater, and solid waste) and low in area of nature reserves, which prevented them from getting high rankings in an evaluation emphasizing environmental protection and ecological conservation. There was not much difference in the high-quality development status in Inner Mongolia, Shanxi, Shaanxi, and Gansu Province: The performances of these provinces were not stable. The low rankings of Qinghai and Ningxia, which got high environmental protection and ecological conservation indexes, were caused by their weak economic performance and inadequate social security benefits. With a small population, Qinghai faced low environmental pressure and enjoyed a well-preserved ecological environment. It had high ecological conservation index, and its emission date of "the three wastes" (i.e., waste gas, wastewater, and solid waste) remained low by the end of 2020.

1. **The Temporal Evolution of the High-Quality Development Status in the YRB Provinces.** A timeline-based evaluation of the high-quality development in the nine YRB provinces from 2011 to 2019 found that all the investigated provinces' performances got better as time went by, except for some slight fluctuations.

In 2013, the high-quality development status worsened in the YRB provinces, except Qinghai, Ningxia, and Inner Mongolia. In 2017, the high-quality development status improved in the YRB provinces, except Sichuan. This was probably related to the provinces' efforts to boost development after the promulgation of the 13th Five-Year Plan in 2016.

The high-quality development indexes for Qinghai and Gansu, which had no serious environmental pollution problems or ecological challenges, increased with slight fluctuations from 2011 to 2019: Their biggest increases occurred in 2017. Sichuan's high-quality development index followed a fluctuating ascending trend from 2011 to 2014 and then a fluctuating descending trend from 2015 to 2017. Ningxia followed a fluctuated upward trend: Its high-quality development index declined in 2015, surged in 2017, and began a slight decline in 2018. After a decline in 2014, the high-quality development index for Inner Mongolia increased year by year but decreased slightly in 2018. Shaanxi's index declined briefly in 2014 and then increased steadily. Henan's index remained basically stable: after the decline between 2012 and 2014, it kept rising until 2017; it began to decline slightly in 2018. The index of Shanxi remained basically stable between 2011 and 2019, following a slightly upward trend in general. That of Shandong followed a fluctuated upward trend: there was slight declines in 2013, 2018, and 2019, and a big increase in 2017. In 2019, Shandong and Sichuan were the only two YRB provinces that achieved growths in high-quality development index.

2. **The Overall and Provincial Analyses of High-Quality Economic Development Status.** Figure 2 shows the YRB provinces' differences in high-quality development status, and Table 6 gives out their economic development indexes. In terms of economic development, the nine YRB provinces experienced fluctuations fluctuated after 2011, following upward or downward trends in their development. In this period, the main factors affecting the provincial economies was probably the recovery from the 2008 Financial Crisis and the implementation of China's 12th Five-Year Plan that had an impact on the economic restructuring and development mode transformation of the YRB provinces. Shandong had the strongest provincial economy in the YRB. Because of its advantageous ports, Shandong had the highest proportion of foreign investment in regional GDP. Its main export destinations included: the ASEAN, the EU, South Korea, the United States, and Japan. As the second most populous YRB province, Henan had high regional GDP. Its low economic development index was mainly a result of its low per capital GDP. Besides, its insufficient economic activity was also an important reason. In comparison with Shandong, Henan did not have any geographical advantages to engage in foreign trade and thus lagged behind it in economic development index. Qinghai and Gansu had relatively weak provincial economies. Their low rankings were caused by the lack of regional economic vitality and the

low added values of the secondary and tertiary industries, which suggested that these two provinces should continue focusing on their economic development. In 2020, the nine YRB provinces maintained their economic and social stability by coordinating their epidemic-prevention-and-control efforts with their economic development.

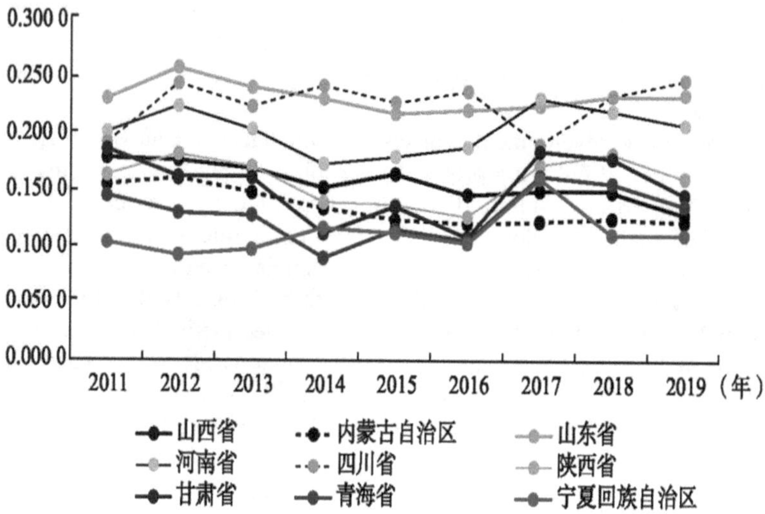

Fig. 2 High-quality development of provinces in the YRB

山西省	Shanxi
内蒙古自治区	Inner Mongolia

(continued)

Table 6 Economic development indexes for provinces in the YRB

Year province	2011	2012	2013	2014	2015	2016	2017	2018	2019
Shanxi	0.040 6	0.044 9	0.043 1	0.055 2	0.062 7	0.051 3	0.028 9	0.028 5	0.023 7
Inner Mongolia	0.041 6	0.040 0	0.038 8	0.042 1	0.030 5	0.033 6	0.025 9	0.022 0	0.021 0
Shandong	0.052 2	0.061 5	0.059 2	0.065 6	0.049 3	0.044 3	0.043 3	0.038 9	0.045 4
Henan	0.017 1	0.017 4	0.016 9	0.016 2	0.014 4	0.015 9	0.012 4	0.014 3	0.013 7
Sichuan	0.028 8	0.035 7	0.033 2	0.029 5	0.028 1	0.029 3	0.022 3	0.020 4	0.032 2
Shaanxi	0.024 3	0.034 9	0.035 0	0.029 0	0.024 6	0.024 0	0.029 5	0.035 6	0.025 8
Gansu	0.010 5	0.011 1	0.014 6	0.029 2	0.034 4	0.028 1	0.016 4	0.015 4	0.014 3
Qinghai	0.024 6	0.020 4	0.020 0	0.016 2	0.024 9	0.022 4	0.020 1	0.017 6	0.014 6
Ningxia	0.032 3	0.024 6	0.030 4	0.044 9	0.036 3	0.029 9	0.086 8	0.035 5	0.035 4

(continued)

山东省	Shandong
河南省	Henan
四川省	Sichuan
陕西省	Shaanxi
甘肃省	Gansu
青海省	Qinghai
宁夏回族自治区	Ningxia
(年)	(year)

Table 7 gives out some economic indicators for the YRB provinces. The following is a detailed analysis.

In 2020, Shanxi's industries remained generally stable under the impact of the COVID-19 epidemic. Industrial firms above the designated size (industrial enterprises, each with annual main business revenue of at least 20 million yuan) in the province achieved a total profit of 96.38 billion yuan, a year-on-year decline of 20.2%: Its mining sector fell by 33.2% year-on-year; Its manufacturing sector fell by 2.3%; Its production sector and supplies of electricity, heat, gas and water increased by 7.7%. Different changes in profits were found in its five pillar industries. The profits of its equipment manufacturing industry increased; those of its energy, material, chemical, consumer goods, and other industries declined. Its provincial GDP reached 1765.193 billion yuan, an increase of 4.07% year-on-year: the added value of its secondary industry was 767.544 billion yuan, an increase of 2.80% year-on-year; that of its tertiary industry was 902.981 billion yuan, an increase of 4.15% year-on-year; its total imports and exports reached 150.58 billion yuan, an increase of 4.0% year-on-year. Its economy recovered rapidly and investment growth remained stable. Industrial investment constituted the main driving force for the rapid growth of investment in the province.

In 2020, the industrial added value of industrial firms above the designated size in Inner Mongolia achieved a year-on-year increase of 0.7%: the added value of its mining sector decreased by 7.1% year-on-year; that of its manufacturing sector increased by 8.4%; that of its production sector and supplies of electricity, heat, gas and water increased by 3.7%. As to its industries, most increases were found in its chemical raw materials and chemical products manufacturing industries, electricity and heat production and supply industries, non-metallic mineral product industries, tobacco products industries, and ferrous and non-ferrous metal smelting and rolling processing industries. Its raw materials manufacturing industries realized accelerated growths, and its consumer goods manufacturing industries improved significantly. In 2020, the total volume of the province's imports and exports was 104.33 billion yuan, down 4.9% from the previous year. Its total exports fell by 7.4% and total imports fell by 3.7%. Its total imports from and exports to BRI countries reached 62.87 billion yuan, down 12.0% from the previous year. In 2020, its provincial GDP was 1735.982 billion yuan, an increase of 0.86% year-on-year: The added value of its

Table 7 Some economic development indicators for provinces in the YRB

Indicator year province	GDP (yuan)		Value added to the secondary sector (yuan)		Value added to the tertiary sector (yuan)	
	2019	2020	2019	2020	2019	2020
Shanxi	1,696,160,000,000	1,765,193,000,000	746,630,000,000	767,544,000,000	867,000,000,000	902,981,000,000
Inner Mongolia	1,721,253,000,000	1,735,982,000,000	676,310,000,000	686,800,000,000	858,610,000,000	846,700,000,000
Shandong	7,054,050,000,000	7,312,900,000,000	2,817,180,000,000	2,861,219,000,000	3,725,170,000,000	3,915,305,000,000
Henan	5,371,780,000,000	5,499,707,000,000	2,303,560,000,000	2,287,533,000,000	2,604,650,000,000	2,676,801,000,000
Sichuan	4,636,380,000,000	4,859,876,000,000	1,718,790,000,000	1,757,110,000,000	2,436,830,000,000	2,547,110,000,000
Shaanxi	2,579,317,000,000	2,618,186,000,000	1,177,950,000,000	1,136,258,000,000	1,202,260,000,000	1,255,174,000,000
Gansu	871,830,000,000	901,670,000,000	286,242,000,000	285,200,000,000	479,660,000,000	496,650,000,000
Qinghai	294,110,000,000	300,592,000,000	115,390,000,000	114,355,000,000	148,530,000,000	152,807,000,000
Ningxia	374,848,000,000	392,055,000,000	158,710,000,000	160,896,000,000	188,140,000,000	197,358, 000,000

Data source The CEInet Statistics Database

secondary sector was 686.8 billion yuan, up 1.55% year-on-year; that of the tertiary sector was 846.7 billion yuan, a year-on-year decrease of 1.39%.

In 2020, the industrial development in Shandong Province remained stable, actualizing a total industrial added value of 2311.1 billion yuan, up by of 3.6% from the previous year. The industrial added value of its industrial firms above the designated size increased by 5%: Its equipment manufacturing industry increased by 12.6%, and high-tech manufacturing industry increased by 9.8%. The operating revenue of its service firms above the designated size (service firms, each with annual main business revenue of at least 10 million yuan) increased by 3.5%. Their operating profits changed from negative to positive, increasing by 0.7%. The added value of industrial firms above the designated size was mainly contributed by the mining industries, such as coal mining and washing industry and oil and gas mining industry, and manufacturing industries. The growth rates for the output of their major industrial products are listed as follows: − 8.6% for raw coal, − 32.0% for coke, − 15.9% for natural gas, − 0.3% for crude oil, 9.2% for machine-made paper and paperboard, 4.1% for cement, and 1.6% for flat glass, etc. The biggest drop is found in the output of coke. In terms of foreign trade, the province's main international trade partners in 2020 were the ASEAN, the United States, the EU, South Korea, and Japan. Its total imports and exports increased by 7.5% over the previous year: Its total exports grew by 17.3%; Its imports fell by 4.1%. Its total imports from and exports to BRI countries reached 660.82 billion yuan, an increase of 9.1% year-on-year. In 2020, its provincial GDP was 7312.9 billion yuan, an increase of 3.67% year-on-year: The added value of its secondary sector was 2861.219 billion yuan, an increase of 1.56% year-on-year, and that its tertiary industry was 3915.305 billion yuan, up 5.1% from the previous year.

In 2020, the added value of industrial firms above the designated size in Henan Province increased by 0.4% compared with the previous year. For the industrial firms, mining, manufacturing, and electricity-heat-gas-and-water-production-and-supply sectors grew by 4.2%, − 0.3%, and 4.8%, respectively. As to the output of their products, knowledge-intensive and high-added-value new products of intelligent manufacturing, electronic information, and new energy industries grew rapidly. The output of optoelectronic devices, sensors, engines, and lithium-ion batteries increased by 89.2%, 73.3%, 40.3% and 32.7%, respectively. The provincial GDP was 5499.707 billion yuan, an increase of 2.38% year-on-year: The added value of its secondary sector was 2287.533 billion yuan, a year-on-year decrease of 0.7% and that of its tertiary industry was 2676.801 billion yuan, an increase of 2.77% year-on-year.

In 2020, the total profits of industrial firms above the designated size in Sichuan Province reach 319.77 billion yuan, up by 13.4% over the previous year. Their total operating revenue was 4525.01 billion yuan, up by 5.5% from the previous year. For these industrial firms, mining, manufacturing, electricity- heat-gas-and-water-production-an-supply sectors grew by 22.5%, 12.4%, and 14.5%, respectively. In terms of industries, the added values provided by oil and gas mining industries; ferrous metal smelting and rolling processing industries; wine, beverage and refined tea manufacturing industries; computer, communication and other electronic equipment manufacturing industries; and pharmaceutical manufacturing industry increased respectively by 31.8%, 55.2%, 19.4%, 43.1% and 7.5%. In 2019, the province received foreign investments amounting to $92.3 billion, a year-on-year increase of 2.3%. As of the end of 2019, Sichuan attracted investments from 352 of the global top 500 enterprises, of which 247 are overseas enterprises. In 2020, its provincial GDP reached 4859.876 billion yuan, an increase of 4.82% year-on-year: The added value of its secondary sector was 1757.11 billion yuan, up by 2.23%, and that of its tertiary industry was 2547.11 billion yuan, up by 4.53%.

In 2019, the added value of industrial firms above the designated size in Shaanxi Province increased by 5.7% year-on-year: Its mining sector grew by 5.2%; its manufacturing sector grew by 4.7%; its electricity, heat, gas, and water production and supply sector grew by 9.5%. In 2020, the industrial added value of its industrial firms above the designated size increased by 1% year-on-year. In 2019, there were 323 new foreign-invested projects in the province, utilizing direct foreign investments that totaled $7.729, an increase of 12.9% year-on-year. The amount of foreign direct investments confirmed through signed contracts was $2.981 billion, a year-on-year decline of 44.6%. In 2020, the total imports and exports of the province increased by 1.8% year-on-year. In 2020, its provincial GDP reached 2618.186 billion yuan, an increase of 1.51% year-on-year: the added value of its secondary industry reached 1136. 258 billion yuan, a year-on-year decrease of 3.54%; the added value of its tertiary industry was 1255.174 billion yuan, up 4.4% year-on-year.

In 2020, the added value of industrial firms above the designated size in Gansu Province increased by 6.5% year-on-year: Its mining sector increased by 3.3%; its manufacturing sector increased by 6%; its power, heat, gas, and water production and supply sector increased by 12.5%. For the industrial firms above the designated size, the added value of industrial strategic emerging industries, high-tech industries, and equipment manufacturing industries increased by 14.9%, 22%, and 22%, respectively. In 2019, the province confirmed 17 foreign-direct-invested projects through contracts, utilizing direct foreign investments that amounted to $82.05 million, an increase of 62.8% year-on-year. The turnover of its contracted projects reached $351.69 million, up 33% year-on-year. Its total imports from and exports to BRI countries amounted 20.09 billion yuan, an increase of 2.8% year-on-year. In 2020, its total import and export volume was 37.28 billion yuan, down 2% from the previous year. Its provincial GDP in 2020 was 901.67 billion yuan, an increase of 3.42% year-on-year: The added value of its secondary industry was 285.2 billion yuan, a year-on-year decrease of 0.36%; The added value of the tertiary industry was 496.65 billion yuan, an increase of 3.54% year-on-year.

In 2020, the total industrial added value of Qinghai decreased by 0.2% year-on-year. The added value of its electricity and heat production and supply sector, non-ferrous metal smelting and rolling processing sector, coal mining and washing sector, oil and natural gas mining sector, and non-ferrous metal mining and dressing sector increased respectively by 11.1%, 14.1%, 4.5%, 1.5% and 10.1%. The above-mentioned five industrial sectors contributed 57.3% of the provincial industrial added value. Its total imports from and exports to BRI countries reached 670 million yuan, down 54.5% year-on-year. The total imports and exports, according to the statistics of its customs, reached 2.28 billion yuan, down 39.2% year-on-year. Affected by the COVID-19 epidemic, its total imports and exports decreased significantly. It provincial GDP was 300.592 billion yuan, an increase of 2.2% year-on-year: The added value of its secondary industry was 114.355 billion yuan, a year-on-year decrease of 0.9%; The added value of its tertiary industry was 152.807 billion yuan, up 2.88% year-on-year.

In 2020, the total industrial profits of Ningxia was 20.39 billion yuan, a year-on-year decline of 9.0%. Its once-profitable mining sector suffered a loss of 3.9 billion yuan, an increase of 16.9% year-on-year. Its electricity, heat, gas, and water production and supply sector fell by 5.3%. For its industrial firms above the designated size, the oil, coal, and fuel processing sector fell by 23.4%; the power and heat production and supply sector fell by 10.5%; the coal mining and washing sector suffered a loss of 3.99 billion yuan; the ferrous metal smelting and rolling processing sector suffered a loss of 3.13 billion yuan. Among the 36 industrial sectors of the province, 24 saw increases in their profits, and the remaining 12 suffered profit losses. In 2020, the provincial total imports and exports reached 12.317 billion yuan, down 48.8% from the previous year, under the impact of the COVID-19 epidemic. Its provincial GDP was 392.055 billion yuan, an increase of 4.59% year-on-year: The added value of its secondary industry was 160.896 billion yuan, a year-on-year increase of 1.38%; The added value of its tertiary industry was 197.358 billion yuan, an increase of 4.9% year-on-year.

3. **The Overall and Province-Specific Analyses of Innovation-Driven High-Quality Development Status in the YRB.** According to the innovation-driven progress indexes, all the nine YRB provinces saw sharp increases in the number of their authorized patents and the total turnover of their technology markets (excluding that of Inner Mongolia) from 2011 to 2019. Their innovation ability was greatly improved. During this period, the professional and technical personnel in state-owned enterprises and institutions increased in number in all YRB provinces except Inner Mongolia and Shandong: In Shandong, the professional and technical workforce could find better employment opportunities in non-state-owned enterprises; As to Inner Mongolia, this province did not have much appeal to such workforce, and the turnover of its technology market declined from 2011 to 2019. It is time that Inner Mongolia does something to change this situation. In 2020, when the nine provinces had to deal with public health challenges brought by the COVID-19 epidemic, they still took multiple measures to maintain economic stability, boost development, and encourage scientific innovation.

In 2019, the number of patent applications in Shanxi was 31,705, an increase of 17%. Among them, the number of invention patent applications was 8,424, and the number of authorized patents was 16,598. In 2020, high-tech manufacturing investments in Shanxi grew rapidly, up by 37%. By the end of 2020, the Shanxi-Great Bay Area Innovation Center was put into operation in Shenzhen, facilitating 50-plus cooperation projects for 20 resource platforms and a number of venture capital institutions.

In 2020, Shanxi increased its investment in scientific and technological research and R&D activities, expanding its fiscal expenditure on science and technology.

In 2020, the number of newly-authorized patents in Inner Mongolia was 17,985, an increase of 62.6% year-on-year. The number of invention patents per 10,000 people was 2.7, an increase of 0.4 year-on-year. The contract transaction prices totaled 25.45 billion yuan, of which the technology transaction prices was 3.26 billion yuan, up by 92.6%. The number of product quality inspection agencies reached eight, including five national-level testing centers. The operating revenues of R&D institutes under the firms above the designated level increased by 17.2% compared with the previous year. The growth rate of investments in high-tech manufacturing industry changed from negative to positive, with significant increase in investments to a number of new infrastructure projects such as 5G network facilities and data centers.

In 2020, Shandong boasted about 87,000 invention patent applications, about 27,000 newly-authorized invention patents, and 198 national-level enterprise technology centers, of which nine were newly established. The number of its national-level technology-based SMEs reached 18,203, up by 91.3% from the previous year. A total of 225 technology business incubators (TBIs) run in the province, including 98 national-level and 127 provincial-level TBIs, and 30 provincial-level "government-industry-university-research-finance-service-practice" innovation and entrepreneurship communities were established.

In 2019, there were 206 provincial-level critical laboratories and 16 national-level critical laboratories in Henan, running 35 provincial-level major science and technology projects. A total of 144,010 patent applications were submitted, and 86,247 patents were authorized. As of the end of the year, the number of its valid invention patents reached 37,311; the number of invention patents per 10,000 population was 3.88. The technology contract transaction prices totaled 23.407 billion yuan. In 2020, the provincial investments in high-tech manufacturing industries increased by 24.3%, and the added value of its high-tech manufacturing sector increased by 8.9%.

The R&D activities in Sichuan was hindered by the COVID-19 epidemic, but the adverse influence was within control. In 2020, when the epidemic situation impacted the R&D output of enterprises, Sichuan's enterprises above the designated size still managed to achieve a 11.7% increase in the added value of the high-tech manufacturing sector, which was 7.2 percentage points higher than the industrial added value created by industrial firms above the designated size. Communication and other electronic equipment manufacturing industries became important drivers for the provincial economy.

In 2020, Shaanxi had 6,198 high-tech enterprises that held valid qualification certificates. A total of 49,928 technology contracts were signed throughout the year, with a transaction amount of 153.368 billion yuan. Among them, there were 16,234 technology development contracts, 556 technology transfer contracts, 3,951 technology consulting contracts, and 29,187 technology service contracts.

In 2019, there were 27,637 patent applications from and 14,894 patent authorizations (including 1,154 invention patents) given to Gansu. The province had 7,432 valid invention patents, and the number of invention patents per 10,000 population was 2.82. A total of 5,921 technical contracts were signed, with a total transaction amount of 19.642 billion yuan. There were 5 national-level engineering technology research centers and 25 national-level enterprise technology centers running in the province. A total of 1,479 provincial-level scientific and technological achievements were registered throughout the year. In 2020, for industrial firms above the designated size, the added values in strategic emerging industries. high-tech industries, and equipment manufacturing industries increased by 14.9%, 22.0%, and 22.0% respectively compared with the previous year.

In 2020, Qinghai achieved 580 scientific and technological achievements at and above the provincial and ministerial levels, an increase of 35 from the previous year. Among them, there were 149 basic theoretical achievements, 415 applied technical achievements, and 16 soft scientific achievements. It had 7,227 patent applications (including 1,742 invention patent applications) filed, an increase of 2,217 from the previous year, and 4,693 patents (including 333 invention patents) authorized, an increase of 1,650 over the previous year. The relevant turnover reached 1.056 billion yuan, up by 16% from the previous year.

In 2019, Ningxia had three national-level engineering technology research centers, 50 provincial-level engineering technology research centers, three national-level key laboratories, and 32 provincial-level key laboratories (including two laboratories jointly run by the province and relevant ministries). It registered 233 scientific and technological achievements at the provincial level, and the annual number of its patent applications was 9,268, of which 2,524 were invention patent applications. A total of 1,930 technology contracts were signed, with a transaction amount of 1.674 billion yuan. In 2020, Ningxia increased its investment in high-tech manufacturing sector, with a year-on-year growth of 60.4%. The output of its high-tech products such as industrial robots, CNC machine tools, and voltage changers increased remarkably.

4. **The Overall and Province-Specific Analyses of Social Betterment in the High-Quality Development of the YRB.** Regarding social betterment, the YRB provinces have worked unremittingly to improve people's livelihoods and have continuously increased its investments in employment, social security, and other social aspects, as shown in Table 8. Shandong, Sichuan and Henan rank high in the list, and Ningxia ranks first from the bottom. In general, all the nine YRB provinces succeeded in continuously improving people's livelihoods from 2011 to 2019. During this period, their registered urban unemployment rates declined with some fluctuations, and the participation rates of basic pension insurance kept rising. During this period, the provinces' spending on social security and employment promotion continuously increased, and so did the average number of their college/university students. Located in the central and western regions of China, the nine YRB provinces have relatively backward provincial economies and large poverty-stricken populations. To build a well-off society in an all-round way, the YRB has made solid progress in stabilizing employment, ensuring people's livelihoods, and lifting the poor people out of poverty. All the nine provinces have expanded their fiscal expenditures on social betterment and the improvement of basic public services.

Table 8 Some social development indicators for provinces in the YRB

Indicator year province	Registered urban unemployment rate (%)		Urban residents living on subsistence allowances (approximate number)		Unemployment benefit payments (yuan)	
	2018	2019	2018	2019	2018	2019
Shanxi	3.26	2.71	358,000	280,400	1,177,000,000	1,439,000,000
Inner Mongolia	3.58	3.70	388,000	338,900	991,000,000	1,228,000,000
Shandong	3.35	3.29	159,000	132,800	6,446,000,000	7,769,000,000
Henan	3.02	3.17	500,900	440,700	2,072,000,000	4,848,000,000
Sichuan	3.47	3.31	937,000	768,400	8,078,000,000	10,287,000,000
Shaanxi	3.21	3.23	259,000	215,700	1,418,000,000	4,214,000,000
Gansu	2.78	3.00	496,700	396,700	565,000,000	804,000,000
Qinghai	2.97	2.24	76,600	64,700	356,000,000	504,000,000
Ningxia	3.89	3.74	97,000	93,200	302,000,000	480,000,000

Data source The CEInet Statistics Database

In 2018, the registered urban unemployment rate in Shanxi Province decreased to 3.26%, down 0.17% from the previous year. The provincial expenditure on social security and employment was 6.716 billion yuan, an increase of 3.72%. The number of people participating in basic pension insurance reached 8.376 million, an increase of 4.87%. In 2019, the registered urban unemployment rate in Shanxi was further reduced to 2.17%, down by 0.55 percentage points from the previous year. In 2020, while handling challenges brought by the COVID-19 epidemic, Shanxi achieved basic stability in labor remuneration and employment by increasing its expenditures on social security and employment promotion by 13.8% year-on-year. About 488,000 people were newly-employed in its urban areas, and about 377,000 rural laborers were successfully-transferred, both numbers exceeding the province's annual targets. At the end of December, 2020, the registered unemployment rate in urban areas was 3.1%, achieving the provincial goal of keeping the rate under 4.5%. Unremitting efforts were exerted to safeguard and improve people's livelihoods.

In 2018, the registered urban unemployment rate in Inner Mongolia was reduced to 3.58%, down by 0.05 percentage points from the previous year. Its expenditures on social security and employment promotion promotionreached 70.720 billion yuan, an increase of 0.43% year-on-year. The number of people participating in the basic pension insurance was 7.3353 million, an increase of 5.35% over the previous year. The coverage rate of social pensions reached 100%. In 2019, the registered urban unemployment rate in the province was 3.7%, an increase of 0.15 percentage points over the previous year. In 2020, while doing a good job in epidemic prevention and control, the province vigorously and orderly enhanced its social security systems and increased the employment rate. In the first three quarters of the year, the growth rate of per capita disposable income for urban residents (according to permanent residence registration records) changed from negative to positive. In 2020, about 232,000 people were newly-employed in its urban areas, 31,000 fewer than the previous year. At the end of the year, the registered urban unemployment rate was 3.8%, an increase of 0.1 percentage points year-on-year.

In 2018, the registered urban unemployment rate in Shandong was 3.35%, which was 0.05 percentage points lower than the previous year. The provincial expenditure on social security and employment promotion was 125.399 billion yuan, an increase of 9.73% year-on-year. The number of people participating in the basic pension insurance was 27.6267 million, an increase of 3.68% over the previous year. In 2019, the registered urban unemployment rate in Shandong was reduced to 3.29%, a decrease of 0.06 percentage points from the previous year. In the first three quarters of 2020, the registered urban unemployment rate was 3.1%, down 0.22 percentage points year-on-year. In 2020, the per capita disposable income of residents was 32,886 yuan, an increase of 4.1% over the previous year. Shandong has increased its investments in various social fields such as people's livelihoods, food industry, pharmaceutical industry, and health and social work.

In 2018, the registered urban unemployment rate in Henan was 3.02%, an increase of 0.26 percentage points over the previous year. The province's spending on social security and employment promotion amounted to 129.845 billion yuan, an increase of 10.64% year-on-year. The number of people participating in the basic pension insurance was 20.0654 million, an increase of 5.43% over the previous year. In 2019, the registered urban unemployment rate increased to 3.17%, up by 0.15 percentage points over the previous year. In 2020, to implement the strategy of rural revitalization and stabilize the overall situation of employment, this province with a large population and a large rural labor force focused on supporting start-up businesses run by returning rural workers, trying to boost employment through entrepreneurship.

In 2018, the registered urban unemployment rate in Sichuan was 3.47%, down by 0.54 percentage points from the previous year. The provincial expenditure on social security and employment promotion was 164.417 million yuan, an increase of 8.69% year-on-year. The number of people participating in basic pension insurance was 25.4371 million, an increase of 8.20% over the previous year. In 2019, the registered urban unemployment rate in Sichuan was reduced to 3.31%, down by 0.16 percentage points from the previous year. In 2020, the province continuously improved its social security systems and generally completed its poverty-alleviation tasks.

In 2018, the registered urban unemployment rate in Shaanxi was 3.21%, which was 0.07 percentage points lower than the previous year. The provincial spending on social security and employment promotion promotion reached 79.386 billion yuan, an increase of 9.53% year-on-year. About 9.9199 million people participated in the basic pension insurance, up 30.90% year-on-year. In 2019, its registered urban unemployment rate was 3.23%, an increase of 0.02 percentage points over the previous year. In 2020, 432,500 people were newly employed in its urban areas. The registered unemployment rate in urban areas was 3.63%, and the number of people participating in the basic pension insurance reached 11.5695 million at the end of the year.

In 2018, the registered urban unemployment rate in Gansu was 2.78%, an increase of 0.07 percentage points over the previous year. The province spent about 50.477 billion yuan on social security and employment promotion, up 7.25% year-on-year, and saw the number of people participating in basic pension insurance increase to 4.5471 million, up by 5.47% over the previous year. At the end of 2019, there were 7,829 community-level elderly care institutions and facilities in the province, including 955 community-powered care institutions and facilities, and 3,631 mutual-aid facilities. In 2019, a total of 5.185 million surplus laborers in urban and rural areas were transferred to areas that required labor force. Among them, 1.911 million people were transferred to areas outside the province, and 3.274 million people, within the province. In 2019, the registered urban unemployment rate in Gansu was 3%, 0.02 percentage points higher than the previous year. In 2020, the province increased it expenditure on the improvement of people's livelihoods by 4.3% year-on-year. About 356,800 people were newly-hired for urban jobs, and the employment situation was stable.

In 2018, the registered urban unemployment rate in Qinghai was 3.89%, an increase of 0.02 percentage points over the previous year. About 20.379 billion yuan was invested into efforts to enhance social security and boost employment, an increase of 9.2% year-on-year. The number of people participating in basic pension insurance was 1.4509 million, an increase of 4.65% over the previous year. In 2020, the registered urban unemployment rate in Qinghai was reduced to 2.1%, with about 61,900 people newly-employed for jobs in urban areas and around 1.1158 million laborers transferred out of the agricultural and pastoral areas. In that year, the growth rate of the per capita disposable income for all residents in the province was 6.3%, and people's livelihoods were continuously improved under the impact of the COVIV-19 epidemic. At the end of 2020, the number of basic pension insurance participants reached about 4.2047 million in the province, an increase of about 65,500 from the previous year.

In 2018, the registered urban unemployment rate in Ningxia was 3.89%, an increase of 0.02 percentage points over the previous year. The provincial expenditure on social security and employment promotion grew to 175.99 billion yuan, an increase of 7.77%. The number of people participating in basic pension insurance increased to 2.1613 million, up by 5.05% over the previous year. In 2019, the registered unemployment rate in Ningxia decreased to 3.74%, down by 0.15 percentage points from the previous year. In 2020, the employment situation in the province was generally stable. About 73,000 people were newly-employed for urban jobs, and the registered urban unemployment rate was 3.92%.

In 2020, the YRB provinces strove to promote high-quality development while fighting against the COIVD-19 epidemic. They prioritized issues regarding people's livelihoods and took various measures to solve problems regarding employment and living standards, thus stabilizing the employment situation and advancing the poverty-alleviation campaign.

5. **The Overall and Province-Specific Analyses of Environmental Protection Work in the High-Quality Development of the YRB.** We evaluate the nine provinces' environmental protection work from the following aspects: wastewater discharge volume, industrial solid waste generation, and sulfur dioxide emissions. Ningxia, Qinghai, and Gansu are ranked in the top three in the environmental performance index list. Their economic backwardness explains the small environmental damage they have caused. Shandong ranks last because its rapid economic development has caused a large emission of "the three wastes" and environmental deterioration. The environmental performance indexes from 2011 to 2019 for the nine YRB provinces are shown in Table 9. It is obvious that most areas of the YRB still faces grave environmental challenges. In response to the country's green development policy, all YRB provinces have stepped up their efforts to curb pollution. They have greatly reduced the emission of sulfur dioxide, but their wastewater discharge volume and industrial solid waste generation have still been increasing since 2011, which is an inevitable consequences of their economic development. To achieve high-quality development, the YRB provinces must continuously step up their efforts to control environmental pollution and enhance environmental protection.

Beginning from 2011, Shandong remained the largest wastewater discharger among the nine YRB provinces, followed by Henan and Sichuan; Qinghai, Ningxia, and Gansu were ranked first, second and third from the bottom in the list. The reason Shandong discharged so much wastewater was that it had the strongest provincial economy, especially the strongest industrial economy, among the nine YRB provinces. Located in the western region of China, Qinghai, Ningxia, and Gansu had relatively backward economies. Therefore, their industrial sector did not generate much wastewater in comparison with other YRB provinces. Among the nine provinces, Shanxi was the largest industrial solid waste generator, followed by Inner Mongolia and Shandong. The problem should be attributed to their rich

Table 9 Some environmental status indicators for provinces in the YRB

Year province	2011	2012	2013	2014	2015	2016	2017	2018	2019
Shanxi	0.011 3	0.010 5	0.010 9	0.011 1	0.010 7	0.010 2	0.009 0	0.010 0	0.008 7
Inner Mongolia	0.013 6	0.012 4	0.013 6	0.013 1	0.012 0	0.013 3	0.011 7	0.012 3	0.015 0
Shandong	0.003 9	0.004 2	0.003 7	0.003 9	0.004 3	0.003 1	0.003 0	0.003 5	0.003 5
Henan	0.011 7	0.010 9	0.009 8	0.010 9	0.012 0	0.014 1	0.013 2	0.011 3	0.016 0
Sichuan	0.019 0	0.017 5	0.016 9	0.017 4	0.018 6	0.015 8	0.013 2	0.015 4	0.014 8
Shaanxi	0.025 3	0.023 1	0.022 4	0.023 1	0.022 3	0.022 1	0.019 2	0.020 6	0.022 4
Gansu	0.029 5	0.026 8	0.026 6	0.027 7	0.026 8	0.026 2	0.022 9	0.024 6	0.027 4
Qinghai	0.031 8	0.028 9	0.029 6	0.030 6	0.028 4	0.023 2	0.023 6	0.025 5	0.025 5
Ningxia	0.033 4	0.030 2	0.029 6	0.031 2	0.030 4	0.027 7	0.024 3	0.027 0	0.028 8

mineral resources, and this was especially true for Shanxi, a leading coal producer of China. As to the emission of sulfur dioxide, Shandong's emission was much higher than those of the other eight YRB provinces. The large emission was a by-product of its rapid economic development. From 2015 to 2019, the sulfur dioxide emissions of the YRB provinces declined remarkably year by year because of the green-development-related energy saving and emission reduction measures. At the end of 2019, Shandong did not differ much from the other eight YRB provinces in terms of sulfur dioxide emissions. Its progress in the environmental performance was obvious.

Due to the long-term high-intensity exploitation of mineral resources, Shanxi faced problems such as destroyed groundwater resources and increasingly grave constraints of resources and environment. Therefore, the province attached great importance to environmental protection, and expanded its expenditure on energy conservation and environmental protection by 14.4% in 2020.

In 2019, the installed renewable electricity capacity accounted for 31.3% of the total installed capacity in Inner Mongolia, with 72.8 billion kWh of green power generated from solar energy and wind power. The use of new energy continuously reduced the environmental pressure brought by economic development in 2020, when the energy consumption of the industrial firms above the designated size in the province increased by 4.5% year-on-year.

Shandong achieved green economic development by developing new energy sources and adopting energy-conservation, consumption-reduction and emission-reduction measures. In the first three quarters of 2020, the new electricity generation reached 61.04 billion kWh, an increase of 10.6%, and the six high-energy-consuming industries contributed 93.8% of the provincial coal use reduction.

In 2019, the energy consumption per unit of value added for industrial firms above the designated size in Henan decreased by 14.13% year-on-year, completing the country's energy consumption goal. The water-quality monitoring system in the province found out that the proportion of water that meet the I to III standards was 68.1%. Besides, a total of 19 new large and medium-sized mineral deposits were discovered in the province that year. In 2020, Henan introduced more comprehensive measures for pollution prevention and control, upgrading its performance in environmental protection.

In 2019, 85 among the 87 nationally-selected water-quality monitoring sections in Sichuan were qualified as having excellent water quality, the excellent rate as high as 97.7%. That year, the province arranged a total of 2.36 billion yuan as its environmental protection special fund for the implementation of its 285 critical emission-reduction projects, and issued mandatory clean-production requirements to 221 production enterprises, achieving remarkable progress in environmental protection. In 2019, the proportion of days with good air quality in the province was 89.1%, which was 0.7 percentage points higher than that of the previous year.

In 2019, the energy consumption per unit product of calcium carbide for key energy-consuming industrial enterprises in Shaanxi decreased by 2.1%. Among the 1,257 seawater quality monitoring points in China's coastal waters, 76.6% reached the national I- or II-grade seawater quality standards; among its cities, 46.6% met the air quality standards.

In 2019, the industrial energy consumption of industrial firms above the designated size in Gansu decreased by 2.7% year-on-year. Among the 38 surface water sections covered by the monitoring system in the province, 94.7% reached and exceeded the grade-III standards. The proportion of days with good air quality in the province reached 93.1%.

In 2019, the proportion of days with good air quality in cities at and above the municipal (prefectural) level in Qinghai was 96.1%. The comprehensive energy consumption of industrial enterprises above the designated size in the province decreased by 2.6% year-on-year. The comprehensive energy consumption per ton of steel of major energy-consuming industrial enterprises decreased by 10.1%. In 2020, the energy consumption per unit of value added for industrial firms above the designated size in the province decreased by 4.4% year-on-year.

In 2019, the water quality of the entire Ningxia section of Yellow River was rated as grade-II excellent. The average number of days with good air quality for Ningxia's cities and towns was 321, accounting for 87.9% of the total number of days in the whole year.

It is necessary to be aware that unreasonable industrial structures cause environmental pollution and that such environmental damage restricts economic development. The unreasonable exploitation of resources that causes not only wastes but also ecological crises. The extensive economic growth model leads to pollution, which hinders the growth of traditional industries as well as that of emerging industries. Such issues increase the difficulties in the industrial restructuring when the region tries to transform its old economic growth drivers into new ones. When these problems regarding resources, environment, industrial structure, and economic growth model are closely intertwined, they restrict the high-quality development of the region. To achieve high-quality development, all the YRB provinces shall pay attention to environmental pollution when pursuing economic growth. They must prioritize the environmental protection endeavor and resort to science and technology, trying their best to protect their environment while actualizing economic growths.

6. **The Overall and Province-Specific Analyses of Ecological Conservation in the High-Quality Development of the YRB.** When evaluating the ecological conservation work of the nine YRB provinces, we consider the following aspects: urban green space coverage rate, per capita green space, afforestation area, and nature reserve coverage. Generally speaking, their ecological performance indexes fluctuate: some provinces follow an upward trend, while others, a downward trend, as shown in Tables 10 and 11.

The green space coverage rate and per capita green space area in the nine YRB provinces continued to increase from 2011 to 2019, showing the YRB's progress in green development. In terms of afforestation areas, Inner Mongolia, Shandong, and Henan saw slight decreases in their afforestation areas, and the other six YRB provinces saw continuous increases in this respect. The YRB provinces must realize

Table 10 Some ecological status indicators for provinces in the YRB

Year Province	2011	2012	2013	2014	2015	2016	2017	2018	2019
Shanxi	0.027 2	0.025 3	0.025 4	0.025 0	0.024 8	0.028 8	0.027 0	0.020 5	0.020 7
Inner Mongolia	0.019 3	0.015 4	0.015 1	0.009 6	0.011 1	0.012 8	0.010 1	0.012 2	0.017 9
Shandong	0.010 7	0.010 1	0.011 6	0.013 4	0.014 0	0.018 3	0.017 6	0.016 4	0.018 3
Henan	0.033 8	0.032 3	0.033 0	0.031 3	0.033 5	0.037 4	0.033 3	0.029 8	0.030 3
Sichuan	0.023 7	0.023 9	0.026 1	0.029 5	0.023 5	0.021 4	0.019 9	0.019 2	0.019 0
Shaanxi	0.024 0	0.021 2	0.023 9	0.022 5	0.022 0	0.029 2	0.028 2	0.033 6	0.036 7
Gansu	0.045 8	0.040 2	0.034 4	0.037 4	0.042 8	0.050 2	0.041 9	0.053 8	0.052 9
Qinghai	0.034 6	0.032 7	0.036 6	0.036 9	0.044 9	0.056 0	0.049 4	0.054 2	0.060 8
Ningxia	0.016 4	0.016 7	0.015 7	0.021 0	0.022 6	0.024 0	0.021 2	0.019 7	0.018 7

Table 11 Ecological status indicators for provinces in the YRB in 2019

Indicator province	Urban green space coverage (%)	Afforestation area (hectare)	National nature reserve (whole number)
Shanxi	42.29	347,355	8
Inner Mongolia	40.52	720,285	29
Shandong	41.8	168,297	7
Henan	41.03	196,493	13
Sichuan	41.85	400,370	32
Shaanxi	39.32	333,451	26
Gansu	36.03	369,085	21
Qinghai	35.21	220,576	7
Ningxia	41.34	92 254	9

Data source The CEInet Statistics Database

that afforestation and expansion of vegetation regions are of great significance for their success in solving their ecological problems such as soil erosion.

As to the provincial nature reserve coverage rates, most YRB provinces made progress: Shandong, Shaanxi, Gansu, Qinghai and Ningxia increased their rates, but slight decreases were seen in the other four YRB province. Among the nine YRB provinces, Qinghai had the highest nature reserve coverage rate, followed by Gansu, Sichuan, and Inner Mongolia; Henan and Shandong had the lowest and second lowest coverage rates.

By the end of 2019, Shanxi had 3.572 million hectares of forest area, with its forest coverage rate reaching 22.8%. That year, natural disasters caused the province direct economic losses of 12.08 billion yuan, affecting 1.422 million hectares of crops, an increase of 68.8%. Among them, 289,000 hectares of the affected areas resulted in complete yield loss, an increase of 53.2%.

Inner Mongolia has large, concentrated areas of deserts and sandy lands. In 2020, it increased its afforestation area by 915,000 hectares and returned 45,000 hectares of farmland to forest. It added 46,000 hectares of vegetation areas for its natural forest resource protection project and 79,000 hectares, for its sandstorm source control project that benefited Beijing and Tianjin. A total of 182 nature reserves were established in the province, including 29 national-level nature reserves and 60 provincial-level ones. By stepping up its governance efforts, it continuously reversed desertification in 2020.

In 2019, Shandong increased its urban green space coverage rate to 41.8%, its total afforestation area to 168,297 hectares, and its number of national nature reserves to seven. In 2020, Shandong's forest land area was 3.564 million hectares, and its forest coverage rate was 18.25%. Its ecological environment and air quality were strikingly improved: The average PM 2.5 concentrations continued to decrease; The excellent and good rates of its surface water increased significantly.

In 2019, Henan expanded its total afforestation area by 196,493 hectares, of which 165,000 hectares were artificially planted. In 2020, the province continuously stepped up its afforestation and ecological protection work.

In 2019, Sichuan had 166 nature reserves, with their total areas covering about 83,000 square kilometers, accounting for 17.1% of the provincial land area. Nine counties in the province were rated as national-level ecological-civilization demonstration counties, and three practice-and-innovation bases were built in the province to highlight the idea that lucid waters and lush mountains should be regarded invaluable assets.

In 2019, Shaanxi Province boasted 333,500 hectares of newly-added afforestation area and 26 national-level nature reserves. In 2020, the afforestation area of the province was further increased by 324,500 thousand hectares; the number of its national-level nature reserves remained 26; the forest coverage rate of the province reached 43.06%.

In 2019, there were 56 nature reserves in Gansu, which included 21 national-level ones, 12 national-level geo-parks, and 24 provincial-level geo-parks. Due to the impact of natural disasters, a 10%-reduction in production occurred in about 161,400 hectares of its cultivated area and a 30%-reduction, in about 99,200 hectares of cultivated area.

In 2019, there were 11 nature reserves in Qinghai, covering about 217,800 square kilometers of land. Among them, there were seven national-level nature reserves that covered an area of about 207,400 square kilometers. Over the period, the province finished its soil-erosion-control work for a total of 1,746.05 square kilometers of land, and increased its afforestation area by about 289,000 hectares.

In 2019, Ningxia boasted an afforestation area of 92,254,000 hectares. It added about 293 square kilometers of artificial forest and about 306 square kilometers of tending area. It had 14 nature reserves, including nine national-level nature reserves and five provincial-level nature reserves.

Because of the socio-economic impact of the COVID-19 pandemic in 2020, people across the globe feel more awe about nature and increasingly realize the importance of ecological protection and environmental governance. The ecological conservation of the YRB is an importance part of China's efforts in environmental protection. The nine YRB provinces shall keep in mind the severity of their ecological problems and the importance of their ecological conservation work and actively strive for progress in this regard. They shall adhere to ecological priority and green development policy and coordinately push forward their ecological protection campaign and high-quality development endeavor so that the Yellow River can play its role in benefiting the people. They shall carry forward the fine cooperation practices formed during their joint fight against the COVID-19 epidemic and build an inter-provincial ecological-protection-and-environment-governance mechanism, continuously enhancing the ecological conservation measures in the YRB.

3.4 Conclusion

In the evaluation of high-quality development status from 2011 to 2019 for the nine YRB province, Qinghai gets the highest evaluation. Possible reasons include its small population, low environmental and ecological pressure, and high environmental and ecological performance indexes. As to the high-quality economic performance, Shandong outshines its peers in the YRB, and Qinghai, Gansu, and Inner Mongolia are ranked low in this respect. Over the period, most YRB provinces greatly improve their innovation abilities and their people's livelihoods. Sichuan and Henan are ranked high in the evaluation of social betterment, and Ningxia is ranked first from the bottom in this respect. Qinghai, Gansu, and Ningxia are ranked in the top three in the environmental performance evaluation. The nine provinces' ecological environment indexes fluctuate, with some provinces following an upward trend and some provinces, a downward trend. As we adopt the entropy weight method (EWM) in determining the weighting of different indexes in this chapter, which causes some deficiencies in weight determination, we will adopt the expert scoring method and select more comprehensive assessment index systems in the follow-up in-depth studies.

In 2020, while fighting against the COVID-19 epidemic, the nine YRB provinces managed to improve their performance in economic development, social betterment, innovation-driven progress, environmental protection, and ecological conservation. While maintaining economic and social stability, they comprehensively consolidated foundations for high-quality. Despite these achievements, it is a must to realize the complexity of our external environment and the current challenges for the high-quality development in the YRB. In the future, the YRB provinces shall carry forward the fine cooperation practices formed during their joint fight against the epidemic and build an inter-provincial ecological-protection-and-environment-governance mechanism. They shall actively participate in the "dual circulations" of China and strive to realize their structural transformation.

4 New Paradigm for the High-Quality Development of the YRB

The Yellow River flows through several national ecological function areas, and most of them are ecologically fragile. The Loess Plateau has severe soil-erosion problem problem, and its ecosystem and economic foundations are weak. The lower reaches of the Yellow River has problems such as floods and insufficient water supplies. From 2011 to 2019, the high-quality development indexes for the nine YRB provinces fluctuated. In 2020, while fighting against the COVID-19 epidemic, the YRB provinces managed to enhance their performance in economic development, social betterment, innovation-driven progress, environmental protection, and ecological conservation. In the future, the YRB needs an all-round development plan for strengthened connections between its central cities and their surrounding small- and medium-sized cities, improved division of labor and cooperation among metropolitan areas, accelerated construction of high-speed transportation networks and modern information networks. By enhancing the inter-provincial and intra-provincial economic ties, the coordinated economic development of YRB will be achieved. For high-quality development of the YRB, relevant local governments shall guide different localities to give full play to their comparative advantages and find their development drivers. They shall uphold the new development philosophy as a guide to improve the sustainable development capacity of their region.

4.1 New Spatial Paradigm

For the high-quality development of the YRB, it is a must to implement the new concept of green development. Innovation must be taken as the main development driver. A new spatial paradigm featuring high efficiency must be developed. Ecological conservation work and industrial development must be coordinated. In detail, it is essential do well in the following aspects.

First, it is necessary to cultivate regional growth poles that can boost the development of their surrounding areas. Regional-level central cities should be developed intensively, and their economic and population-carrying capacities must be enhanced. Despite the weak economic performance of the YRB, the relevant provincial capitals and core cities still have the potential to become regional growth poles. There are a few city clusters worth looking forward to such as the Central Plains City Cluster and the Guanzhong Plain City Cluster.

Second, it is essential to make reasonable spatial planning. Under the constraint of rational distribution of water resources in the YRB, function zones of the region must be reasonably divided,, such as free trade pilot zones and new national-level ecological function areas. The ecological conservation work and the economic development must be coordinated, and the upper, middle, and lower reaches of the Yellow River need to be considered as a whole.

Third, it is a must to promote the development of metropolitan areas. China is determined to boost the development of the YRB metropolitan areas during the 14th Five-Year Period, including metropolitan areas with provincial cities as their centers and those in the Yellow River Ji-Shaped Bend. In view of the differences of the metropolitan areas, it is necessary to formulate appropriate development policies according to their different characteristics. Connections between central cities and their surrounding small- and medium-sized cities must be strengthened to boost the development of the surrounding cities. Industrial division of labor and cooperation within a metropolitan area need to be strengthened for stronger economic ties and coordinated economic development.

4.2 New Economic Development Paradigm

The pursuit of high-quality economic development entails not only growth speed but also growth quality, i.e. the stability and sustainability of economic growth, the rationality of economic structure, and the efficiency of economic development. Different YRB areas have different resource endowments: some are suitable for developing agriculture; some are rich in mineral resources. This requires them to utilize their comparative advantages created by their different local conditions when pursuing economic development. The YRB should accelerate the transformation of its growth drivers and cultivate new drivers. It shall encourage technological innovation and realize industrial upgrading. Under the current situation, the YRB can try to realize the upgrading by developing digital economy, especially in its remote areas. It shall enhance its ties with other economic regions of China, and take an active part in the BRI development to boost its opening-up and give full play to its economic potential, It should be noted that Qinghai, Gansu, and Inner Mongolia still have problems such as unreasonable economic structure, imperfect market mechanism, and backward infrastructure, which will cause certain constraints on economic development. Unremitting efforts are required for the settlement of these issues.

4.3 New Population Paradigm

To pursue high-quality development, it is a must to embrace the people-centered philosophy and safeguard and improve people's livelihoods. While propelling people to gather in the central cities, metropolitan areas, and city clusters, it is necessary to see that residents in ecologically fragile areas move out of these regions to facilitate the ecological restoration and environmental protection work in the YRB. In this way, the efficiency of the economy will be improved through the gathering of population, and the dual interests of high-quality development and ecological conservation will be guaranteed. It should be noted that in this process, to ensure and improve people's livelihoods, it is necessary to do well in the following aspects: changing the existing urban–rural dual structure (a series of concepts to maintain the dual economic form of the modern urban sectors and the rural traditional ones and the separated dual social systems between the rural and urban societies): coordinating the urban and rural developments; provide sufficient employment opportunities for the population entering cities; increasing investment in education and improving the education quality; enhancing the social security system; improving medical conditions and infrastructure facilities. In poverty-stricken areas, the people-centered philosophy of development should be thoroughly implemented to ensure improvement in infrastructure and public services.

4.4 Coordination Between Ecological Protection and Industrial Development

Most of the ecological function areas through which the Yellow River flows are ecologically fragile: the Loess Plateau has a serious soil erosion problem, a weak eco-system, and staggering economic foundations; the lower reaches of the Yellow River faces a high risk of flooding and insufficient water resources. These problems restrict the economic and social development of provinces along the River. Some YRB provinces including Shandong realize economic development but caused environmental pollution such as the emission of "the three wastes" (i.e., waste gas, wastewater, and solid waste). This serves as a reminder that while developing industries, it is a must to control pollutants and strengthen environmental governance: environmental protection underpins high-quality development, the two aspects enhancing each other. To conclude, it is necessary to adopt a systematic thinking approach when considering the YRB's development issues. It is essential to regard the YRB as a whole while taking into consideration the particularities of its different parts and divide the region into different function areas and ecological protection areas based on their different characteristics. For instance, the Loess Plateau should be regarded as an ecological protection area, targeting at its soil erosion problem, and provinces in the western region shall develop industries with their distinct characteristic such as eco-tourism.

Reference

Li, R. (2019). Identification and classification of metropolitan circle in China. *City, 2019*(02), 61–72.

Chapter 4
Strategy for the Development of the Chengdu-Chongqing Economic Circle

Given the development laws proven in China and the new era's requirements for regional economic development in the country, the strategy of developing the Chengdu-Chongqing Economic Circle (CCEC) was put forward, which highlighted high-quality development of the Chengdu-Chongqing region and a new paradigm for the development of the entire western region. The strategy entails increasing the influence of existing regional systems for coordinated development, strengthening and implementing existing regional development plans, and providing a solid strategic foundation for China's high-quality economic development and domestic circulation.

The chapter analyzes the value, significance, and main content of the development strategy for the CCEC and the current development status of the region.

1 Background of the CCEC Development Strategy

On January 3, 2020, General Secretary Xi Jinping proposed at the 6th Meeting of the Central Committee for Financial and Economic Affairs the development plan for the CCEC, requiring to build it into a growth pole for the high-quality development of the western region. Since being publicly announced, this development strategy has attracted attention from all sectors.

In fact, the concept of CCEC was created on the basis of the formerly-proposed ideas about the Chengdu-Chongqing region and the Chengdu-Chongqing City Cluster. Its creation respects the objective laws and targets problems caused by inequality and inadequate needs. It gives a different mission to the Chengdu-Chongqing region given the new facts in the new era. Table 1 collates information about policies (promulgated in the period from 2003 to 2020) related to the CCEC development. Among them, the "Regional Planning of the CCEC" issued by the PRC State Council in 2011 first recognizes the CCEC's strategic value and defines its as an

J. Sun, *Regional Economic Development in China in 2020*,
https://doi.org/10.1007/978-981-95-3604-7_4

essential support for the accelerated development of the western region, the enhanced inland opening-up, and the improved comprehensive strength of the country. The "Development Plan for Chengdu-Chongqing City Cluster", announced by the PRC State Council in 2016, first requires the CCEC to explore a new path for the development of city clusters in the western region by doing well in the following aspects: respecting the urban development laws; giving full play to comparative advantages; making up for shortcomings; removing bottlenecks; strengthening coordination; optimizing structure. According to General Secretary Xi Jinping's requirements proposed in 2020, the development of the CCEC entails the following: embodying objective development rules; leveraging comparative advantages; pursing synergistic development; promoting the rational flow and efficient aggregation of industries, population, and various production factors; strengthening the influence of the two central cities, namely Chongqing and Chengdu; building the Chengdu-Chongqing region into a national-level economic center and scientific and technological innovation center, a new frontier of reform and opening-up, a livable place for high-quality life, and a strong driver for high-quality development. In a word, the development strategy of the CCEC has been endowed with great historical and practical significance. It means enormous economic opportunities for the Chengdu-Chongqing region, planning to turn it into the top economic hub and growth driver for the development of the western region. It is of great significance to the high-quality economic and social development of China.

To begin with, the CCEC development strategy is proposed as an essential part of China's coordinated regional development plan. The coordinated regional development strategy includes continuously advancing the development campaign of the western region, revitalizing the northeastern region in an all-round way, promoting the rise of the central region, and encouraging the trail-blazing development of the eastern region. At a more detailed level, it entails the coordinated development of the BTH region, the GBA, the YRD, the YREB, the YRB, and the CCEC. Based on the development campaign of the western region, the CCED development strategy is formulated to boost the high-quality development of the western region and realize full coverage of detailed regional development policies. It enriches the theoretical connotation and practical significance of China's regional coordinated development strategy.[1]

Furthermore, the CCEC is proposed against such a background: the western region pursues development and opening-up and strives for a new development paradigm. The CCEC is projected to become the growth pole of this region. Chengdu and Chongqing are the central cities of the western region in terms of economic performance, population size, and public-service quality. They play a pivotal role in the economic development of the western region. With the promulgation of the CCEC development strategy, their status in the region has been further enhanced. This is of great significance for the in-land opening-up and high-quality development in the

[1] Please refer to the relevant statements in the PRC state council's "2020 Report on the Work of the Government.".

Table 1 Policies for the development of the CCEC

Year	Title	Content point(s)	Issuing authority (authorities)
2003	A Preliminary Study on the Principal Areas of the Western Region Development Campaign	This policy document first proposed the development of the CCEC	The Chinese Academy of Sciences
2004	Research on Key Economic Belts of the Western Region Development Campaign	This policy document named the CCEC as an economic center of the YREB in the upper reaches of the Yangtze River	The Western Development Office
2005	Research Project for the 11th Five-Year Plan: Joint Prosperity for the CCEC—Seven Strategies and Action Plans for Its Future	This is the first thematic study on the CCEC	The National Development and Reform Commission
2006	The 11th Five-Year Plan for the Development Campaign of the Western Region	This policy document explicitly proposed the CCEC development strategy	The National Development and Reform Commission and the Western Development Office
2007	Agreement on Promoting Sichuan-Chongqing Cooperation for the CCEC Development	The policy document proposed the goal of turning the CCEC into the fourth economic growth pole in China	Sichuan provincial people's government
2008	Framework Agreement on Deepening Sichuan-Chongqing Economic Cooperation	The policy document called for deepened Sichuan-Chongqing cooperation for the CCEC development	Sichuan Provincial People's Government and Chongqing Municipal People's Government
2009	Post-Earthquake (Wenchuan) Restart of the Development Plan for the CCEC	The policy document determined the geographical scope of the CCEC and the establishment of a unified work and coordination mechanism for the CCEC development	The National Development and Reform Commission
2011	Regional Planning of the CCEC	The CCEC was defined as an important support for accelerating the development of the western region, enhancing China's inland opening-up, and increasing the comprehensive strength of the country	The State Council
2014	National New-Type Urbanization Plan (2014–2020)	The policy document called for accelerated development of the Chengdu-Chongqing City Cluster	The State Council

(continued)

Table 1 (continued)

Year	Title	Content point(s)	Issuing authority (authorities)
2016	Development Plan for the Chengdu-Chongqing City Cluster	This was a guiding and binding document for the development of the Chengdu-Chongqing City Cluster	The National Development and Reform Commission and the Ministry of Housing and Urban–Rural Development
2019	Key Tasks for New-Type Urbanization in 2019	This document evaluated the implementation of the "Development Plan for the Chengdu-Chongqing City Cluster" and proposed further policy support	The National Development and Reform Commission
2019	Overall Plan for the New International Land-Sea Trade Corridor in the Western Region	The policy document proposed to let Chengdu and Chongqing play essential radiative roles as key regional hubs	The National Development and Reform Commission
2020	Key Tasks for New-Type Urbanization and Integrated Urban–Rural Development in 2020	An outline of the development plan for the CCEC was compiled	The National Development and Reform Commission
2020	Guiding Opinions on Promoting the Development Campaign of the Western Region and Starting a New Paradigm	This policy document emphasized the necessity of promoting the coordinated development of the Chengdu-Chongqing City Cluster and that of steadily carrying out comprehensive regional governance	The State Council

Source of information Jiang and Li (2020) and information available on the Internet

western region. With the implementation and enhancement of the CCEC development strategy, the western region can realize synergistic development together with the other economic regions and give full play to its own resource endowment, industrial structure, and favorable policies. It will further increase the flow speed of its resources, developing a new strategic paradigm.

Last but not least, the CCEC development strategy is proposed as a response to the region's development status and development trends. Different from other city clusters where one central city exerts a strong influence on its surrounding areas and promotes their development, the Chengdu-Chongqing City Cluster has two central cities, namely Chengdu and Chongqing. Given this fact, the ideal development mode shall give full play to the advantages of the two cities and allow them to attract more people for their economic progress. The continuous economic growth of the two cities will drive the rapid development of their surrounding urban communities and economic circles and realize cooperation with other western cities such as Xi'an and

Kunming. In this way, these two cities will jointly become the economic centers of the western region and promote the high-quality development and opening-up of the region.

2 The Strategic Value and Significance of the CCEC Development Strategy

The "Development Plan for Chengdu-Chongqing City Cluster" issued by the PRC State Council in 2016 clearly describes the Chengdu-Chongqing City Cluster as a primary hub for the development of the western region, a strategic support for the YREB, and an important demonstration area for the promotion of new-type urbanization. In General Secretary Xi Jinping's remarks, there are new arrangements for the development of the Chengdu-Chongqing region and city cluster from a higher strategic perspective.

From the perspective of pursuing coordinated regional development, the CCEC development strategy is a supplement and improvement to China's regional coordinated development plan. The regional coordinated development plan originally entails five development strategies for four major economic regions, and among them, only one strategy is formulated for the western region. Given the need to foster new development paradigm and realize coordinated regional development, the strategic importance of the CCEC becomes widely recognized: The space spillover benefits generated by the Chengdu-Chongqing region can facilitate the high-quality development in the western region and the coordinated regional development across China.[2]

From the perspective of pursuing new-era development, the CECC is an exploration of new regional development ideas such as a dual-center economic circle based on the thorough implementation of China's new development philosophy. Apart from the CECC, there are three dual-center economic circles in China, i.e. the Harbin-Changsha city cluster, the Beijing-Tianjin city cluster, and the Lanzhou-Xining city cluster. Such dual-center economic circles often face resource waste problems caused by internal conflicts of interest and the unreasonable distribution of benefits. If the Chengdu-Chongqing dual-center economic circle can successfully solve such problems when boosting the development of the western region, it will play, as Yi (2021) has pointed out, an inspirational and exemplary role for the other dual-center economic circles in China and help them to develop in a better way.

As Zhang and Xiong (2020) suggested, from the perspective of fostering a new development paradigm in the western region, the CCEC constitutes an economic engine for high-quality development. The establishment of the CCEC is conducive

[2] The five development strategies originally included in China's regional coordinated development plan are as follows: the coordinated development strategy for the BTH region, the integrated development strategy for the YRD, the development strategy for the YREB, the development strategy for the GAB, and the ecological protection and high-quality development strategy for the YRB.

to a new development paradigm for the development campaign of the western region: The CCEC helps to highlight the environmental priorities, boost the progress of the campaign, and promote high-quality development. Firstly, the CCEC, through joint efforts of the two cities, pursues a green economy while enhancing its environmental protection efforts. It strives to build a mechanism that protects its ecological interests, namely a good ecological environment in the upper reaches of the Yangtze River, for its belief in the concept of ecological civilization. Secondly, the CCEC, situated at the junction of the BRI economic belt and the YREB, has location advantages for inland opening-up. By consolidating its trade ties at home and abroad, it can facilitate cooperation that benefit all and realize a new development paradigm featuring joint governance, shared interest, and mutual benefits. Thirdly, with further development, the CCEC will be able to optimize the existing industrial structures and improve the flow speed of production factors. It will become a growth pole that boosts the development of the entire western region and foster a new development paradigm where different localities to realize joint development by giving full play to their comparative advantages and upholding the concept of mutual benefit.

3 Main Contents of the CCEC Development Strategy

At the 6th Meeting of the Central Committee for Financial and Economic Affairs, General Secretary Xi Jinping remarked that the development of the CCEC should be a systematic project that required strong top-level design and overall coordination efforts. He emphasized the leading role of central cities, the allocation function of the factor market, and the concept of integrated development, requesting Chengdu and Chongqing to plan together, deploy together, cooperate with each other and implement together in this economic endeavor. To be more specific, the CCEC shall do well in the following aspects: strengthening transportation infrastructure, accelerating the establishment of a modernized industrial system, enhancing collaborative innovation and development, optimizing the territory spatial planning, stepping up ecological environment protection, pushing forward institutional innovation, and exerting join efforts to improve public services that benefit all. In a word, from a higher strategic point of view, there are new characteristics and interpretations regarding the basic thinking, strategic objectives, and main tasks of the CCEC development strategy.

3.1 Basic Thinking

The CECC shall make use of objective laws and its own comparative advantages to boost its regional economic development, giving full play to the strategic value of its dual-center structure. First, the economic performance of Chengdu and Chongqing shall be further enhanced, and the relationship between the two cities should be handled reasonably. With increased economic influence and positive spillover effects, these two cities will become important economic centers of the country. Second, these two cities shall be built into top-notch scientific and technological innovation centers of the western region: Relevant investments need to be increased; Relevant professional personnel need to be trained; The R&D abilities of scientific research institutions must be enhanced; The transfer of superior technologies shall be actively received. Third, the CCEC shall develop itself into a new frontier of reform and opening-up in the western region by vigorously reforming and innovating its systems and mechanisms, stepping up its opening-up, and leveraging its proximity to inland countries bordering China for strong trade ties. Fourth, the CCEC shall become a livable place for high-quality life: It must highlight ecological civilization and environmental protection, ensure equalization of basic public services, and build smart cities and park-like cities.

To sum up, the basic thinking regarding the development of the CCEC is based on a double-center development logic: Enhance the central cities for the improvement of the economic circle, which is conducive to the overall progress of the entire economic region. It provides the Chengdu-Chongqing region with a feasible blueprint and guides the two cities to do well in the following aspects: jointly building a new inland frontier of reform and opening-up, a high-standard market system, and a first-class business environment; constructing a new land-sea transportation system in western China for the BRI cooperation; actively participating in the dual circulations of China. It requires reasonable handling of the relationship between the two central cities and their relationship with their surrounding areas: Chengdu and the main urban districts of Chongqing shall be enhanced in terms of development ability and comprehensive competitiveness; The urban development shall include connotation enhancement besides spatial extension; An integrated metropolitan area shall be developed by leveraging the influence of the central cities and adopting balanced development methods. The future development of the Chengdu-Chongqing region, which is based on its regional advantages and endowments, calls for more coordinated efforts at the national and regional levels and in various sectors such as industries, innovation, ecological protection, and public services.

3.2 Strategic Objectives

From a higher strategic perspective, the development of the CCEC shall serve the western region's need for high-quality development; the new development pattern of China, in which domestic and foreign markets reinforce each other, with the domestic market as the mainstay; and the socialist modernization process. It is a must to adhere to the strategic direction of the supply-side structural reform and insist on expanding domestic demand: Production, distribution, circulation, and consumption shall rely more on the domestic market, and the supply system shall change to meet domestic demand, forming a high-level dynamic balance of demand and supply. From the supply side, it is necessary to improve product quality, ensure flexible flow of production factors, and maintain the market supply environment. This calls for further development of regional industries and closer intra-regional cooperation. By upgrading industrial integration, tapping the potential of industrial development, the CCEC will be able to create high-quality, high-efficiency, and high-value industrial and innovative chains to meet the needs of its people and its regional economic development. From the demand side, it is essential to carry out the demand-side reform, remove barriers, and make up for shortcomings. After smoothly connecting all links in production, distribution, and circulation, consumption sectors, the demand will help to drive high-quality improvement in supply side and overall efficiency. With strengthened connections in sectors such as infrastructure, public services and products, the development needs of individuals will be stimulated. This is conducive to the establishment of a coordinated information network for regional demand. With accurate information about demand, the CCEC's development goals will be clarified, and its high-quality development will be realized on the basis of constantly meeting the needs in production activities and daily lives.

3.3 Main Tasks

The CCEC shall strengthen its transportation infrastructure, accelerate the establishment of a modernized industrial system, and enhance its capacity for collaborative innovation and development. It shall optimize its territory spatial planning, step up its ecological conservation work, and advance its institutional innovation. The two cities shall exert concerted efforts to improve their public services and ensure equal accessibility to such services.

1. **The Strategic Value of Strengthening the CCEC Development.** Situated in the hinterland of China, the Chengdu-Chongqing region connects the country's southern, central, southwestern, and northwestern regions and pools resources from these regions. With the help of the Yangtze River, Chengdu and Chongqing can cooperate with a large number of cities and utilize this advantage as a strategic support for its participation in the BRI cooperation, the western region development campaign, and inland opening-up. Therefore, it is necessary to admit the region's advantages in inland location, inland opening-up, and domestic demand potential, just as Yan and Yao (2020) has pointed out. With enhanced economic connections between Chengdu and Chongqing, the CCEC's influence will be expanded to Yunnan, Guizhou, Tibet, Qinghai, and other provinces, and its interaction with them will be boosted. The development of the CCEC will facilitate China's economic interaction with countries in the Southeast Asia such as India, Myanmar, Laos, Vietnam, Thailand, etc. Targeted economic and trade policies need to be formulated for the full utilization of the resources and factors of the western region.
2. **Consolidation of the Transportation Infrastructure in the CCEC.** Critical issues that need current and future attention are increasing investment in

infrastructure and speeding up the flow of factors, which decides the Chengdu-Chongqing region's ability to conduct inter-provincial and inter-national interactions and achieve its strategic goals. Within the Sichuan Basin, where the CCEC is located, it is relatively easy to build transportation facilities. Within this region, the transportation hubs shall be further strengthened and a modernized comprehensive transportation system shall be established, to increase mobility and accessibility. The economic and trade channels connecting the western region of China and bordering countries shall be established by leveraging the CCEC's location advantages and utilizing the Yangtze River waterway, China–Europe railways, and other important traffic routes. When developing transportation infrastructure surrounding the CCEC, it is a must to make scientific, rational plans based on specific terrain features and build a modernized transportation network featuring a hub-and-spoke structure with the CCEC as the center. This is conducive to improved overall traffic accessibility in the western region, the surrounding areas' wider participation in the development of the CCEC, and the new paradigm for the development campaign of the western region.

3. **Accelerated Efforts to Establish a Modernized Industrial System in the CCEC.** The overall goal of a modernized industrial system is to coordinate real economy development, scientific and technological innovation, modern financial, and human resources, which is in line with the purpose of the CCEC development. According to its own actual situation, the Chengdu-Chongqing region strives to modernize its agriculture sector, develop its equipment manufacturing industry and modern service industry, and optimize its industrial structure. It plans to improve the core competitiveness of its industries to meet the requirements of high-quality development and explore a better and faster economic development path. When building a modernized industrial system, it is necessary to strengthen the guiding role of policies, effectively leverage the positive role of the market in the adjustment of regional industrial structure, and formulate industrial policies based on comparative advantages. It is essential to take into consideration the basic conditions of the Chengdu-Chongqing region and its neighboring provinces when choosing key industrial sectors for the CCEC. To be more specific, the regional industrial division of labor system should be considered for better inter-provincial cooperation, and a platform for cooperation and exchange should be built. In this way, the CCEC will have production chains with its unique competitiveness, which will contribute to the increase of the overall strength of the region.
4. **Collaborative innovation and Development Ability of the CCEC.** It is a must to give equal prominence to Chengdu and Chongqing and break through the constraints of administrative divisions, when establishing scientific and research network and promoting the flow of scientific research personnel and resource sharing. In this way, a mechanism of scientific and technological innovation, which features a clear division of labor, a mutual-enhancement effect, and an open attitude, will be built. This mechanism will constantly enhance the scientific and technological innovation ability of the Chengdu-Chongqing region, turning them into top-notch innovation centers of the western region. Corresponding

innovative development plans need to be formulated to clarify short-term and long-term innovation goals and further stimulate the innovation vitality of the CCEC. Market demands, laws, and regulations shall be clarified for the protection of the R&D interests and output. Exchanges and cooperation with countries and regions with high innovation abilities should be strengthened to facilitate the technological innovation, management innovation, and industrial innovation in the CECC and Provide long-term support for the growth of its innovation capability. All these measures are aimed at promoting the formation of a virtuous circle of innovation and development in the CCEC, which helps the CCEC to make breakthroughs in industries, public services, infrastructure, and other aspects, and promote the development of its surrounding areas.

5. **Territory Spatial Planning of the CCEC.** According to the development objectives of the CCEC and based on the coordination among population, resources, and environment, reasonable spatial plannings shall be formulated for ecological function, and production and living activities. In this process, it is essential to do well in the following aspects: ensuring the organic unity of economic, social, and ecological benefits; controlling the development intensity; adjusting the spatial layout structure; creating a production space featuring high efficiency, a living space boasting a livable environment, and an ecological space with picturesque scenery. Simply put, the core purpose of spatial planning is to coordinate production, living, and ecological function spaces: Increase urban space for better population and economic carrying capacities while strictly implementing the ecological protection red line, permanent basic farmland system, and urban development boundaries. (*Qiwen qida*) Facing the increasingly high requirements for economic and population carrying capacities in the CECC, it is a must to respect objective laws, put ecology and safety in a more prominent position, and rationally control the scale of big cities. When enhancing the two central cities, it is essential to develop other big cities to boost economic development. It is necessary to improve the efficiency of resource utilization and the quality of development and implement development modes according to local conditions. In this way, the new-type, people-orientated urbanization can be promoted.
6. **Enhanced Ecological Conservation Efforts of the CCEC.** In view of its location in the upper reaches of the Yangtze River, the CCEC must always pay attention to the influence of its own production and living activities, preventing its economic endeavors from adversely impacting other provinces in the YRB. The CCEC shall bear in mind that green mountains and lucid water are invaluable assets and constantly enhance its ecological conservation and environmental protection efforts. Chengdu and Chongqing shall jointly establish a new mechanism of ecological protection that benefit all, to coordinate their ecological conservation efforts, jointly solve environmental pollution problems, and implement regional collaboration for joint prevention and control. They shall restore their ecology through joint research, exploration, and implementation, and strengthen their cooperation on the management of cross-border rivers such as the Jialing River and the Fujiang River. They shall adopt more flexible methods

to solve pollution problems such as differentiated management methods for enterprises that cause different levels of pollution and the establishment of a pollutant market similar to carbon market, which helps to alleviate the emission-reduction pressure of enterprises at different development stages. They shall drastically improve their energy utilization efficiency and increase their investment in relevant R&D activities, promote green development in different industries across the region.

7. **Advanced Institutional Innovations in the CCEC.** The CCED needs systematic, regionally applicable, coordinated, and shared institutions and mechanisms. It needs a better top-level design to create and cultivate a good business environment, so as to enhance the central roles of Chengdu and Chongqing and develop them into regional economic centers. A coordinated city cluster management mode shall be explored to realize the integrated development of the Chengdu-Chongqing City Cluster. Responsibilities shall be further clarified, and more power shall be shifted to grass-roots levels, seeing that local governments have the capacities and power to actively participate in the CCEC development and provide better services. The relationship between the government and the market should be reasonably handled for the establishment of a more effective local market mechanism, the active participation of micro subjects, and the decisive role of the market in resource allocation. In this process, the government shall give full play to their supervision and regulation functions to ensure high-quality economic development. High-standard, internationally-oriented regional policies, laws, and regulations should be formulated for international communication and production cooperation tasks given by the state. In a word, the CCEC must explore extensively and try its best to fully utilize the advantages of the region.
8. **A Jointly-Built Public-Service System That Benefits All.** Promote Chengdu-Chongqing in-depth cooperation in the fields such as education, health care, sports, and government affairs, and build a number of service projects that achieve comprehensive coverage. Unremitting efforts must be exerted to promote the integration of their public-service systems, implement an inter-city policy coordination mechanism, and realize mutually-beneficial inter-regional cooperation. In the field of education, educational alliances should be established at all levels such as preschool education, primary education, vocational education, and higher education; the Western Science City and the Yangtze River Education Innovation Belt shall be built. In the field of health care, Chongqing and Chengdu shall build a joint prevention-and-control system for epidemic outbreaks and public health emergencies, which includes a command center, medical supply warehouses, drill training centers, and an information-sharing mechanism for health emergencies and infectious diseases. In the field of sports, the two cities shall strengthen cooperation in public sports services, sports events, sports facilities, sports tourism, talents, and jointly hold high-quality outdoor sports competitions such as mountaineering or mountain marathon competitions. In the field of transportation, they shall strengthen cooperation on road transportation, waterway transportation, ports, multimodal transportation, and industry management and coordination. In the field of government affairs, they shall establish a joint tax

management system to mutually recognize taxpayers' basic information, qualifications, and credit evaluation results: Such mutual recognition facilitates the flow of market factors. As to the joint management of housing provident fund, fund transfers between Chengdu and Chongqing need a one-stop solution that ensures easy withdrawal instead of requiring complicated application procedures in both cities.

4 The Status Quo of the CCEC

As mentioned above, the CCEC is a new term for the Chengdu-Chongqing region and its predecessor is the Chengdu-Chongqing City Cluster. A comparative study of 19 city clusters in 2018 (see Table 2) reveals that the Chengdu-Chongqing City Cluster has the following characteristics: large population and medium urbanization level and per-capita income level. This shows that there is still much room for development in the Chengdu-Chongqing City Cluster and that the economic vitality of its micro subjects shall be further stimulated to explore a distinctive development path. It is worth noting that compared with other city clusters in the western region, the Chengdu-Chongqing City Cluster ranks first in terms of economic scale and development level. This suggests that it has the potential to become a growth pole that drives the regional development. Therefore, the CCEC shall aim at high-quality economic development and become a livable place for high-quality life.

To further understand the development status of the CCEC in terms of population, industries, and public services, this chapter compares the performance of the CCEC with that of the western region and that of the nation from 2009 to 2018 and studies its changes in population size, economy, industries, and public services in this period. Through the above comparison and study, this chapter analyzes the development status of the CECC and makes a summary. It should be noted that this chapter's definition of the Chengdu-Chongqing City Cluster is in line with that of the "Development Plan for Chengdu-Chongqing City Cluster", issued by the PRC State Council in 2016.[3,4]

[3] The Chengdu-Chongqing City Cluster includes 27 districts and counties of Chongqing, namely Yuzhong, Wanzhou, Qianjiang, Fuling, Dadukou, Jiangbei, Shapingba, Jiulongpo, Nan'an, Beibei, Qijiang, Dazu, Yubei, Banan, Changshou, Jiangjin, Hechuan, Yongchuan, Nanchuan, Tongnan, Tongliang, Rongchang, Bishan, Liangping, Fengdu, Dianjiang, Zhongxian; and 15 cities of Sichuan, including Chengdu, Zigong, Luzhou, Deyang, Mianyang (excluding Beichuan County and Pingwu County), Suining, Neijiang, Leshan, Nanchong, Meishan, Yibin, Guang'an, Dazhou (excluding Wanyuan City), Ya'an (excluding Tianquan County and Baoxing County), and Ziyang.

[4] The latest data included in the comparative study is those collected in 2018 because the latest statistical yearbook published by Sichuan Province is the one for the year 2018.

Table 2 Basic information about 19 city clusters of China in 2018

City cluster Degree of integration	City cluster name	Permanent population (approximate number)	Proportion in the national population (%)	Urbanization rate (%)	GDP (trillion yuan)	Proportion in the national GDP (%)	Per-capita GDP (yuan)
Super mega city clusters Grade-one integration	The BTH City Cluster	112,700,000	8.1	65.4	8.4	9.3	74,373
	The YRD City Cluster	154,010,000	11.0	68.6	17.7	19.7	115,997
	The PRD City Cluster	63,010,000	4.5	85.3	8.1	9.0	128,625
	The Chengdu-Chongqing City Cluster	100,150,000	7.2	53.8	5.8	6.3	57,426
Maga city clusters Grade-two integration	The Mid-Yangtze River City Cluster	126,770,000	9.1	60.9	8.3	9.2	65,630
	The Central Plains City Cluster	69,050,000	4.9	49.7	3.8	4.2	54,946
	The Harbin-Changchun City Cluster	46,250,000	3.3	58.8	2.7	3.0	59,072
	The Beibu Gulf City Cluster	42,110,000	3.0	50.4	2.1	2.3	50,532
	The Guanzhong Plain City Cluster	40,380,000	2.9	58.5	1.8	2.0	44,577
	The Hohhot-Baotou-Ordos-Yulin City Cluster	11,510,000	0.8	69.9	1.5	1.7	130,321
	The Lanzhou-Xining City Cluster	15,260,000	1.1	59.3	0.6	0.7	38,008
Large city cluster Grade-three integration	The Shandong Peninsula City Cluster	100,470,000	7.2	56.7	7.9	8.8	78,508
	The Cross-Strait City Cluster	59,510,000	4.3	64.9	4.2	4.7	70,735

(continued)

Table 2 (continued)

City cluster Degree of integration	City cluster name	Permanent population (approximate number)	Proportion in the national population (%)	Urbanization rate (%)	GDP (trillion yuan)	Proportion in the national GDP (%)	Per-capita GDP (yuan)
	The Mid-Southern Liaoning City Cluster	30,540,000	2.2	58.6	2.2	2.4	72,082
	The Central Shanxi City Cluster	16,270,000	1.2	60.4	0.9	1.0	52,243
	the Central Guizhou City Cluster	27,020,000	1.9	51.7	1.2	1.3	44,412
	The Central Yunnan City Cluster	22,700,000	1.6	47.9	1.1	1.2	48,458
	The Ningxia Riverside (the Yellow River) City Cluster	5,640,000	0.4	61.4	0.4	0.4	62,057
	The Tianshan North Slope City Cluster	6,920,000	0.5	50.6	0.6	0.7	82,370

Note The information in the table comes from Evergrande Research Institute's "*Zhongguo Chengshi Fazhan Qianli Paiming:2019* (China's Cities of Potential Rankings 2019)", which is sorted out by Zhang and Xiong (2020)

4.1 The CCEC in Comparison with the Western Region

According to the Changes in regional GDPs and incomes (see Fig. 1), the CCEC's share in the western region's aggregate output kept increasing from 2009 to 2019: At the end of 2019, the proportion reached 30.4%; The strong growth momentum suggested further growth. This shows that the Chengdu-Chongqing region has gradually become the main economic driving force and guiding energy of the western region, determining whether the western region can get rid of its backwardness in the future. The per-capita GDP and per-capita disposable income of urban residents in the CCEC were 2.48 and 3.22 times higher than the average data of the western region. This shows that the living standards in Chengdu-Chongqing region is higher than its western pears, and that there is a huge income gap in the western region. In the future, the western region shall rely on the Chengdu-Chongqing region for higher economic growth and better living standards for all residents.

In accordance with Changes in public services (see Fig. 2), either the number of books in public libraries or the number of beds in health institutions in the Chengdu-Chongqing region accounted for more than 20% of the western total number. This indicates that the CCEC outshines its western peers in public services. Its advantage in the medical resources was particularly prominent, which accounted for nearly 28% of the total amount of the western region. This reflects that there are great differences in medical and health services within the western region. With better public welfare and better urban-service quality, the Chengdu-Chongqing region keeps attracting population inflows, which is conducive to income growth and economic development.

Changes in population, finance, and industrial level (see Fig. 3) reveal the following facts. The CCEC boasted 26.56% of the total population of the western region in 2019; Either its fiscal revenue or fiscal expenditure took up 15% of the

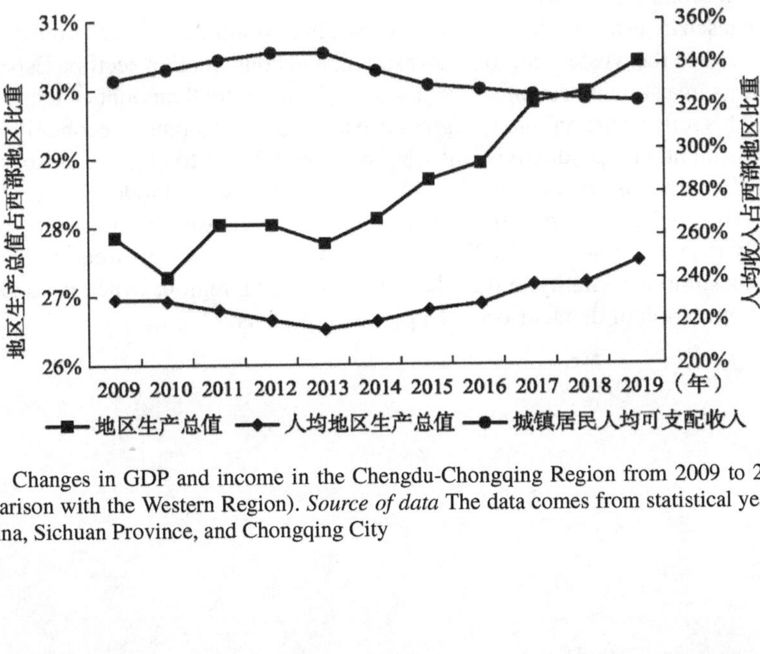

Fig. 1 Changes in GDP and income in the Chengdu-Chongqing Region from 2009 to 2019 (in Comparison with the Western Region). *Source of data* The data comes from statistical yearbooks of China, Sichuan Province, and Chongqing City

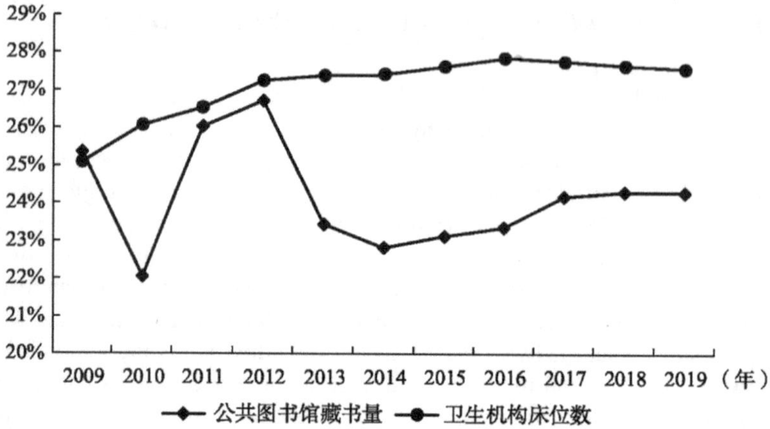

Fig. 2 Changes in public service work in the Chengdu-Chongqing Region from 2009 to 2019 (in Comparison with the Western Region). *Source of data* The data comes from statistical yearbooks of China, Sichuan Province, and Chongqing City

total amount of the western region, and its fiscal revenue was obviously higher than its fiscal expenditure; Its proportion in the western region's total industrial output exceeded 30%. Different comparisons, especially those about population and industries, confirm the pivotal position of the CCEC in the western region. In the future, it is necessary to further strengthen the CCEC's attraction to population and its abilities to develop industries and conduct industrial structure adjustments, ensuring sufficient labor supply for its economic growth and its lasting economic influence in the western region. Besides, given the CECC's good financial situation, remarkable improvements in its municipal administration, people's livelihoods, living environment are also worth looking forward.

Changes in investment and consumption (see Fig. 4) show the CCEC's outstanding contribution to the western region's investment and consumption sectors. Especially in 2029, its investment accounted for nearly 35% of the total amount of the western region. This reflects the region's progress in economic development, economic structure adjustment, and productivity distribution, which helps to enhance material and cultural foundations for high-quality life and facilitates the modernization of the region. The CCEC's contribution to consumption decreased before 2017, but its gradually rose and reached 25.02% in 2019. To some extent, this trend reflects the steady consumption vitality in the Chengdu-Chongqing region, which is conducive to the formulation of demand-oriented production plans.

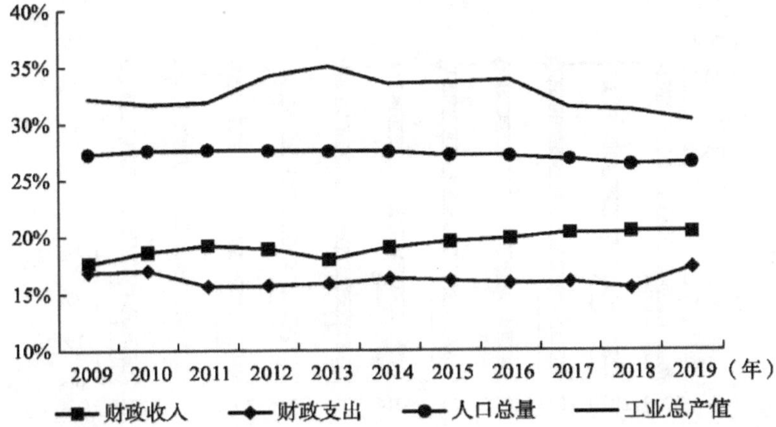

Fig. 3 Changes in population, finance, and industrial levels in the Chengdu-Chongqing Region from 2009 to 2019 (in Comparison with the Western Region). *Source of data* The data comes from statistical yearbooks of China, Sichuan Province, and Chongqing City

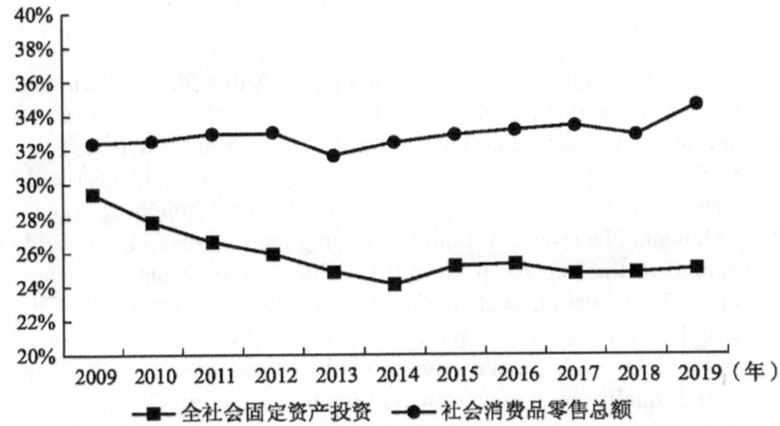

Fig. 4 Changes in investment and consumption in the Chengdu-Chongqing Region from 2009 to 2019 (in Comparison with the Western Region). *Source of data* The data comes from statistical yearbooks of China, Sichuan Province, and Chongqing City

4.2 The CECC in Comparison with the Nation

Regarding changes in population size (see Fig. 5), the Chengdu-Chongqing region's population accounts for about 7% of the national total, but this proportion is in a slight downward trend that lasts until 2019, when the proportion starts to pick up again. The localities contributing the most to the population reduction in the region include Chongqing's Liangping and Fengdu and Sichuan's Dazhou, which are located far away from the main urban areas of Chengdu and Chongqing. This

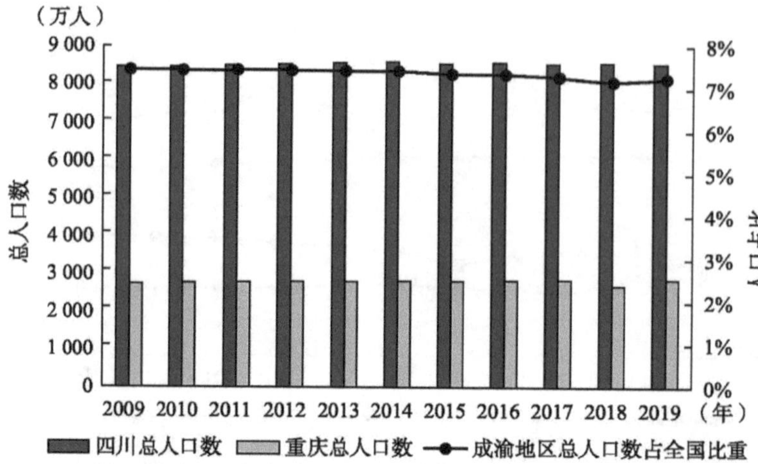

Fig. 5 Changes in population in the Chengdu-Chongqing Region from 2009 to 2019 (in Comparison with the National total Numbers). *Data source* The data comes from statistical yearbooks of China, Sichuan Province, and Chongqing City; In this statistic, Sichuan and Chongqing refer to the regions under the jurisdiction of these two provinces in the Chengdu-Chongqing region

reveals the population siphoning effect in the region. Although this phenomenon is almost inevitable in the process of urban development, it still serves as a reminder that the relationship between central and peripheral cities shall be handled carefully.

As to the changes in per-capita disposable income of urban residents (see Fig. 6), the CCEC realized continuous growth in this regard, with Chongqing doing a little better than Sichuan. Nevertheless, both Chongqing and Sichuan lagged behind the national average in this regard. It is worth noting that such lagging-behind is a recent situation. There was not such a difference in 2009. This requires the Chengdu-Chongqing region to enhance the quality of its regional economic development as soon as possible, improve its people's living standards, and strive to become a growth pole for the high-quality economic growth of the western region.

According to the changes in the three categories of industries (see Fig. 7), the CCEC's contribution to the national GDP remained around 6.3% for a long time. In the CCEC, industries developed rapidly and the industrial structure changed constantly. From 2018 to 2019, its industrial structure where the secondary industry accounted for the largest proportion and the tertiary industry, the second largest proportion was changed, and the tertiary sector began occupying the largest part. This shows that the modern service industry in the CCEC has developed rapidly and gradually become the main industrial sector supporting the regional economic development. This point can also be confirmed by regional policy documents, which point out that the dominant industries of Chengdu are electronic information industry, aerospace industry, food processing industry, tourism, and cultural and creative industry and that Chongqing's advantageous industries are automobile manufacturing industry, electronic information industry, equipment manufacturing industry,

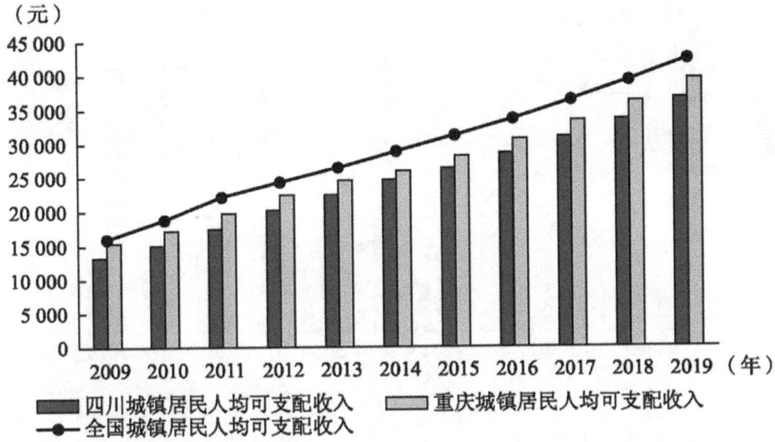

Fig. 6 Changes in per capita disposable income of urban residents in the Chengdu-Chongqing Region From 2009 to 2019 (in Comparison with the National Numbers). *Data source* The data comes from statistical yearbooks of China, Sichuan Province, and Chongqing City; In this statistic, Sichuan and Chongqing refer to the regions under the jurisdiction of these two provinces in the Chengdu-Chongqing region

chemical industry, material processing industry, food processing industry, and textile industry. Coordination among these industries is highly possible one of the key coordination tasks of the CCEC in the future, as it will help the two central cities to form a positive synergy and jointly promote the development of their surrounding areas. In the future, the CCEC also needs to further consolidate its industrial structure and promote its rationalization and upgrading for the high-quality development of the regional economy.

Regarding changes in industrial output value (see Fig. 8), the CCEC's share in the national industrial output value increased year by year from 2017 and reached 5.66% in 2019, but its regional industrial output value basically stopped growing in 2016. These two seemingly contradicting facts suggest that the reduction in the total industrial output value of the CCEC is a result of the changes in domestic and international environments, such as its supply-side structural reform and industrial restructuring. With the further adjustment of its industrial structure and the industrial optimization brought by the supply-side reform, industries in the CCEC will flourish again. For instance, in 2019, the total industrial output value of Sichuan Province rebounded to 4,018.314 billion yuan, and the added value of industrial enterprises above the designated size in Chongqing increased by 6.2% year-on-year.

As to changes in consumption (see Fig. 9), the CCEC's contribution to the national retail sales of consumer goods remained around 6%. Sichuan's consumption capacity increased year by year but Chongqing's consumption capacity decreased slightly in 2018. Given its advantages in tourism, urbanization, Internet celebrity economy, and night-time economy, the CCEC will make a greater contribution to the consumption sector against the backdrop of the "dual-circulation" development paradigm. In 2019,

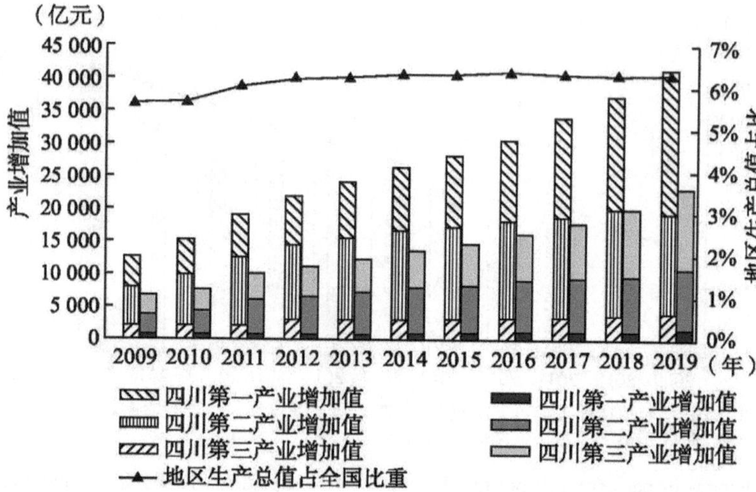

Fig. 7 Changes in the three categories of industries in the Chengdu-Chongqing Region from 2009 to 2019 (in Comparison with the National Numbers). *Data source* The data comes from statistical yearbooks of China, Sichuan Province, and Chongqing City; In this statistic, Sichuan and Chongqing refer to the regions under the jurisdiction of these two provinces in the Chengdu-Chongqing region

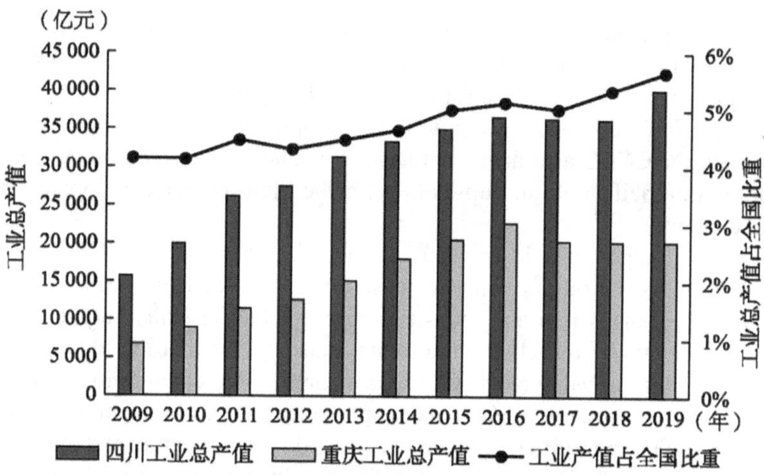

Fig. 8 Changes in industrial output value in the Chengdu-Chongqing Region from 2009 to 2019 (in Comparison with the National Numbers). *Data source* The data comes from statistical yearbooks of China, Sichuan Province, and Chongqing City; In this statistic, Sichuan and Chongqing refer to the regions under the jurisdiction of these two provinces in the Chengdu-Chongqing region

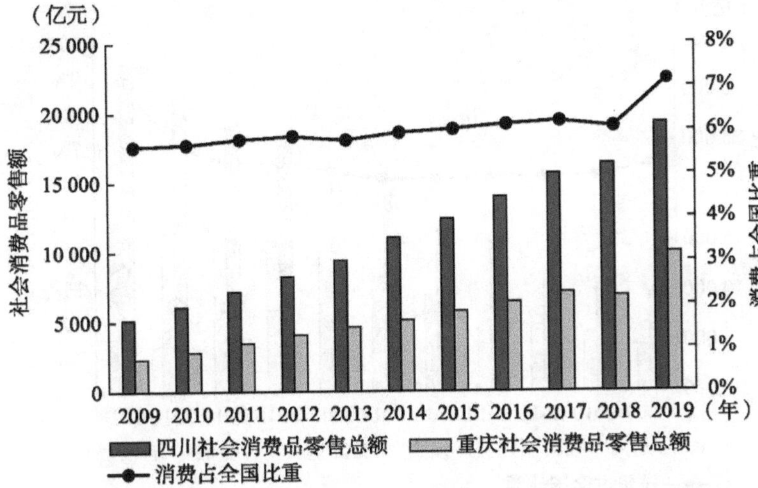

Fig. 9 Changes in consumption in the Chengdu-Chongqing Region from 2009 to 2019 (in Comparison with the National Numbers). *Data source* The data comes from statistical yearbooks of China, Sichuan Province, and Chongqing City; In this statistic, Sichuan and Chongqing refer to the regions under the jurisdiction of these two provinces in the Chengdu-Chongqing region

the CCEC saw significant growth in consumption, contributing more than 7% of the national consumption. Despite that, the local governments in the CCEC still need to continuously improve its people's living standards and satisfy their needs as a way of further stimulating their consumption and improving their consumption quality.

As for changes in investment (see Fig. 10), the CCEC's share in the national fixed assets investment was close to 9% In 2019. The investment of Sichuan Province rose constantly, but that of Chongqing decreased slightly in 2017 and then continued to increase. This shows that the investment environment in the CCEC is constantly improving, and the total amount of investment keeps increasing. With optimized economic structure and productivity distribution, the endogenous development of the region will be further boosted.

Regarding public services (see Fig. 11), the CCEC provided relatively high-quality public services and continuously made progress in medical and health services and education. The Chengdu-Chongqing region shall pay more attention to the equalization of public services when enhancing the quality of its public services. In this way, the two cities can jointly offer high-quality public services that benefit all and foster a long-term mechanism to promote regional high-quality development.

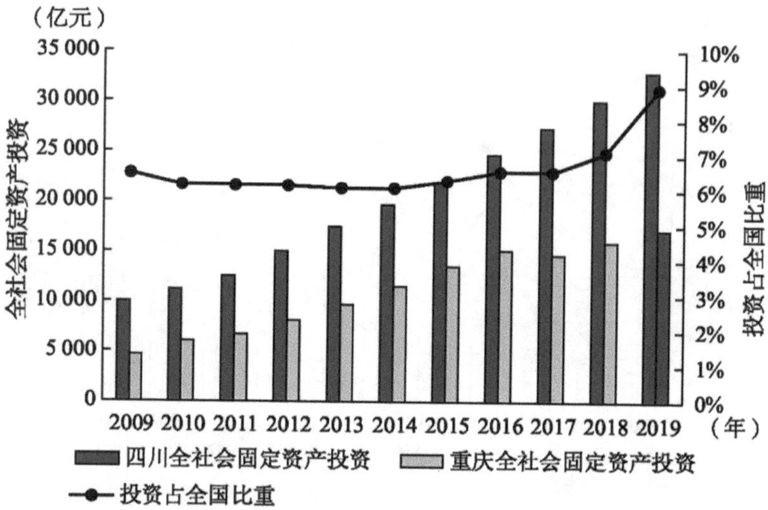

Fig. 10 Changes in investment in the Chengdu-Chongqing Region from 2009 to 2019 (in Comparison with the National Numbers). *Data source* The data comes from statistical yearbooks of China, Sichuan Province, and Chongqing City; In this statistic, Sichuan and Chongqing refer to the regions under the jurisdiction of these two provinces in the Chengdu-Chongqing region

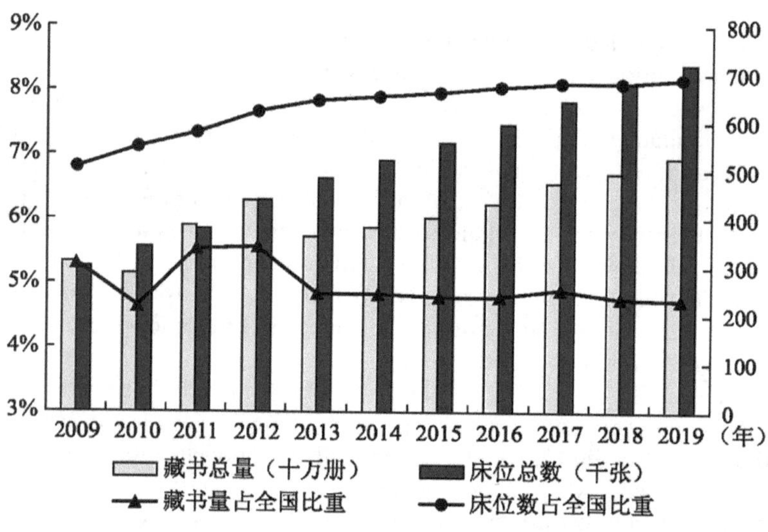

Fig. 11 Changes in public services in the Chengdu-Chongqing Region from 2009 to 2019 (in Comparison with National Numbers). *Data source* The data comes from statistical yearbooks of China, Sichuan Province, and Chongqing City; In this statistic, Sichuan and Chongqing refer to the regions under the jurisdiction of these two provinces in the Chengdu-Chongqing region

4.3 Summary of Development Status

Two main points can be summed up based on the various comparisons above. First, the Chengdu-Chongqing region has almost absolute advantages in the western region. Given this, the formulation of the national-level development strategy for the CCEC is of great significance for the overall development of the western region. Second, the CCEC's proportions in various national total numbers remained basically unchanged over the years. With such stable economic performance, the CCEC has the potential to become the fourth growth pole of our nation and the top economic center of the western region. The comparison between the CCEC and the western region serves as a reminder that the CCEC development calls for the proper handling of the relationship and development gap between Chengdu and Chongqing, especially contradictions and competition for development opportunities between the two cities. To be more specific, it is necessary to discover and give full play to the comparative advantages of the two cities and establish a long-term cooperation mechanism for their common prosperity and progress and their coordinated development. To overcome the problems regarding innovation, cooperation, and benefit distribution, the CCEC needs to do well in the following aspects.

1. **Further-Boosted Innovation Vitality of the CCEC.** In 2019, the proportion of science and technology investment in the Chengdu-Chongqing region was below 2%, lower than the national average of 2.23%. It was middle-ranking in terms of innovation output, without any innovation advantage at the national level. Given that it currently pursues high-quality economic development, and that the 5th Plenary Session of the 19th Central Committee highlights innovation as the most important support for our modernization endeavor, the CCEC has to increase its investment in innovation for regional high-quality development, reform its innovation-related institutions for a continuous inflow of innovative talents, create a fine innovations environment safeguarding the interests of innovation subjects, and actively receive industrial technology transfers. In a word, it needs to turn itself into a top-notch innovation center of the western region and earnestly improve its regional competitiveness, which is conducive to the overall progress of the western region. To achieve this goal, the CCEC also needs to create a good innovative atmosphere, actively introduce and foster high-tech enterprises, and increase its appeal to innovation talents. This calls for increased financial investment in scientific and technological innovation and the promulgation of corresponding incentive programs for scientific and technological innovation at the county level.
2. **Strengthened Coordination Between the Two Central Cities for Stronger Resultant Force.** At present, there is still competition between Chengdu and Chongqing. These two central cities have not yet reached a consensus on their cooperation and urgently need an institution or mechanism that ensures their win–win cooperation. Jiang and Li (2020) has pointed out that it is necessary to strengthen the complementarity between the two cities because there are homogenization problems regarding their leading industries, especially their electronic

information industries and automobile manufacturing industries. Such problems can lead to competition for resources between the two cities, which cause unnecessary factor losses. For a stronger resultant force of the CCEC, it is a must to address the homogenization problems and pursue differentiated and mutually-complementary development. The two central cities shall decide on their own development focuses through thorough communication and flexibly adjust their industrial relations through the market. Through differentiated development, they will gain their complementary advantages and realize common progress as an integrated whole.

3. **Balanced Relations Between the Central and Peripheral Cities in the CCEC.** Efforts must be exerted to avoid excessive siphon effects. It is essential to see that the development of the two central cities benefits their surrounding cities, counties, and towns as well as the entire western region. To achieve this goal, it is a must to speed up the economic development at the county level and even the township level while boosting the development of Chengdu and Chongqing. By promoting the integration of urban and rural areas and the development of new-type urbanization, the non-central areas' ability to receive industrial transfer will be strengthened. Regarding the coordinated development within the CCEC, the two central cities shall shoulder the responsibility of guiding the non-central areas, formulating appropriate assessment standards for them, and boosting their development.

5 Suggestions for the CCEC Development

The CCEC has been given a high expectation: its historical mission is to create a new development paradigm for the western region. The above analysis reveals that although the Chengdu-Chongqing region has the potential to become the fourth economic growth pole of China, it is still troubled by multiple internal development gaps regarding its industrial structure, industrial development, consumption, investment, and public services. The relationship between the central cities and their surrounding areas still needs to be optimized. In a word, the CCEC development shall develop its central cities into economic drivers for the western region and the entire country. In this regard, we put forward the following five suggestions.

First, a mechanism for the integrated development of the CCEC shall be established. Chengdu and Chongqing should set up corresponding leading groups under the leadership of the Central Government: they shall jointly build relevant systems and institutions and formulate feasible plans and policies; they shall exerted concerted efforts to turned the Chengdu-Chongqing region into a national-level economic hub, science and technology innovation center, opening-up frontier, and livable place for high-quality living, realizing mutually-beneficial development. The traditional administrative barriers shall be removed for the creation of an environment that encourages differentiated development and the pursuit of comparative advantage. The integration of public service systems, social-governance and ecological-conservation efforts shall be promoted and a regional compensation system shall be established. A variety of explorations on the integrated development will facilitate the timely communication and sharing of information, which helps the CCEC to become an integrated whole when participating in the "dual-circulation" paradigm of China. The CCEC needs an all-inclusive cooperative network and a regional industrial cooperation platform. On the basis of equality, all involved localities shall abide by the jointly-formulated work plans and performance appraisal lists and promote the integrated development of the CCED within the scope of supervision.

Second, public services need to be given more prominence, and more investment in infrastructure needs to be made. The Chengdu-Chongqing region shall regard the equalization of public services as one of its long-term development goals. It shall formulate policies to coordinate the relationship between public services and green economy, guaranteeing that its residents can enjoy equal opportunities in education, employment, social security, health care, housing security, cultural and sports, and environmental improvement. Such equality helps to enhance the quality of public services and increase the CCEC's appeal to population and industries. Located in the Sichuan Basin, the Chengdu-Chongqing region boasts relatively flat terrain, large population, and high population density. Such location advantages are conducive to high-quality public services, especially high-quality infrastructure. Increased investment in infrastructure help to satisfy the development needs of the CCEC, the ever-increasing commute needs of its residents, and the surging demand for lower-cost logistics and transportation. The CCEC shall also leverage the golden waterway of the Yangtze River, the river-side railways and expressways, the Chongqing-Xinjiang-Europe International Railway, the Chengdu-Europe Express Rail, and China–Europe freight trains for stronger connections to the Silk Road Economic Belt, participating in the global industrial cooperation and competition.

Third, a modernized industrial system should be established. For its future development, the CCEC needs a modernized industrial system, progress in industrial services, and stronger economic clusters. As to the adjustment methods for establishing a modernized industrial system, they shall be based on the regional comparative advantages and development goals and modified according to actual needs: The low-end technologies, products, and services shall be reduced; Overcapacity and low-end links should be addressed; New development space shall be created; Emerging industries such as big data, new energy, new materials, intelligent equipment, and loT shall be included as an dispensable part in the industrial system. The CCEC shall give full play to its factor endowments in its modernized industrial system and attach importance to the development of factors such as science and technology, talents and capital, which is conducive to a strategic support system and a strong defense system against the impact of the declining real economy. In this way, all-round development at micro, meso, and macro level become possible. The CCEC must build a policy system to support the development of its modern industries, which safeguards the interests of all parties and promote coordinated development. While ensuring economic and social development, it can relax relevant regulatory requirements for emerging industries to promote the vigorous development of modern service industries, modern agriculture, big data industry, and other industries. Regarding the administrative procedures related to the modernized industrial system such as administrative appraisals, bidding processes, and qualifications, steps should be simplified, and the administrative power should be shifted to lower levels. Internet-assisted administrative measures should be adopted to improve the overall work efficiency.

Fourth, ecological civilization shall be highlighted, and the natural and urban environments shall be improved. For its location in the upper reaches of the Yangtze River, the CCEC has to pay close attention to ecological and environment issues while pursuing high-quality development. It shall urge multiple departments to participate in the environmental protection work in the YRB through policy deployment. To be more specific, feasible policies and measures shall be formulated to save resources and protect the environment, and the responsible person system should be established, turning the CCEC into a development model featuring the organic combination of ecological civilization and industrial civilization. The CCEC shall uphold the green development philosophy and promote the ecological civilization consciousness that features respect for, adaptations to, and protection of nature. It shall promote an ecological culture, actively advocating green life and vigorously developing circular economy. It shall foster and promote an ecological culture and develop ecological industries and products, striving to integrate ecological civilization into its economic, political, cultural endeavors and achieve environmental friendliness and green economic development. The CCEC must work hard to achieve the goal of building park-like cities. For this end, it shall improve its urban environment and refined management ability and always consider the ecological needs when pursuing urban development. Different parts such as forest land, park land, and water conservancy land shall be integrated to form a system for better management. Urban residents shall be encouraged to actively participate in activities aimed to maintain community hygiene and optimize urban environment. In this way, the life quality and happiness index of the people will be significantly improved.

Fifth, innovation shall be integrated into all development links of the CCEC. The innovation we discuss here not only refers to technological or scientific innovation but also includes ideological, institutional, and management contents that play an important role in the CCEC development. Ideological innovation can help the CCEC explore a distinctive development path and flexibly handle its relation with the central authorities for simplified administrative procedures, the power shift to lower levels, and the thorough implementation of strategic plans. This can meet the actual needs of residents in the new era and bring them benefits. Institutional innovation is conducive to the vitality of the market economy and the market's decisive role in resource allocation. With effective market mechanism, the market and relevant factors will help the regional economy discover its unique advantages and mobilize its micro subjects. The innovation of management systems helps to eliminate institutional barriers, improve relevant reward and punishment systems and cadre evaluation standards, and renew working philosophy and methods. Through such innovation activities, the government can streamline its procedures, and optimize its staff training systems. Technological innovation can effectively improve the innovation strength of the region. It encourages the region to actively absorb advanced foreign technologies and improve its technology market and technology transfer management, which is conducive to the transformation of scientific and technological achievements and smooth connections between its economic and sci-tech sectors.

References

Chenyu diqu shuangcheng jingjiquan quanmian tuijin gonggong fuwu yitihua (Comprehensively promote the integration of public service sectors in the Chengdu-Chongqing Economic Circle). *Xinhua News Agency*. http://www.gov.cn/xinwen/2020-04/29/content_5507491.htm.

Jiang, Y., & Li, X. (2020). Research on economic integration of Sichuan, Chongqing and Guizhou to boost the construction of Chengdu Chongqing Economic Circle. *West Forum, 30*(05), 43–56.

Qiwen qida! Sichuan youhua guotu kongjian buju, zhexie difang jiang sheji tiaozheng (Seven questions and seven answers: Sichuan's efforts to optimize its territory spatial planning will influence the following localities). *Sichuan Fabu*. http://dnr.sc.gov.cn/scdnr/ywbb/2020/8/21/25757fbcd8ee4e3689b8f982a1fc2d13.shtml.

Yan, X., & Yao Y. (2020, April 01). *Gao zhiliang jianshe chengyu diqu shuancheng jingji quan* (High-quality development of the Chengdu-Chongqing Economic Circle). *Chengdu Daily*, 2020, A7.

Yi, M. (2021). An analysis of promoting the construction of Chengdu-Chongqing twin-city economic circle in the new era: Historical review and reality research. *West Forum, 2021*(3), 72–81.

Zhang, Z., & Xiong, Y. (2020). Strategic thoughts and suggestions of promoting the integrated development of Chengdu-Chongqing twin cities economic circle. *Western China, 2020*(02), 1–12.

Chapter 5
Improvements to China's New Urban System

China's urbanization rate of permanence residence has been on the rise for a long time. It increased by about 4.5% during the 13th Five-Year-Plan period and hit more than 60% at the end of 2019. Given the brand-new development environment at home and abroad in the new era, urbanization has been given more significance and expectation in China. Industrial space layout in urban area is positioned as a way to improve the efficiency of factor allocation, promote the differentiated development of emerging industries, and stimulate new development drivers. The upgrading of urban facilities and public services is thought of as a reasonable way to increase public consumption, expand domestic demand, and foster a strong domestic market. The benefits that urbanization bring to rural areas help to tap the full potential of counties and towns, remove barriers in the flow of barriers, and promote the integrated development of urban and rural areas. Aggregation and radiation effects of cities can improve the current urban system and city clusters, promoting coordinated regional development.

The development of urbanization can be divided into different stages. According to this rule, China's urbanization is in the critical period of moving from the fast development stage to the maturity stage. During this transition period, the advanced growth of big cities is remarkable. This phenomenon conforms to the objective laws of urbanization but may lead to development imbalances within city clusters or economic regions and over-sized cities. The population density is generally high in China's mega cities (mega city: a city with a resident population of more than five million) and super mega cities (super mega city: a city with a resident population of more than 10 million). The "one-city dominance" phenomenon is still obvious in some regions, and the small and medium-sized cities are insufficient in quantity. Generally speaking, the quality and resilience of its cities are inadequate, and some of its cities has already begun shrinking.

The essence of urban development is the aggregation of various factors in a certain spatial area. The externalities it brings about constitute an important driving force of

J. Sun, *Regional Economic Development in China in 2020*,
https://doi.org/10.1007/978-981-95-3604-7_5

urbanization. The positive externalities include easy market accessibility and cost-effectiveness of daily life, which are attributed by scholars to sharing, matching, and learning mechanisms. As to the negative externalities, they are market congestion, insufficient supplies of public goods, urban environment deterioration, etc. Based on this, we presume that there is an inverted 'U' relationship between city scale and city efficiency: city efficiency increases with the expansion of city scale, and after reaching a certain scale, it decreases with the scale expansion. At present, there are a large number of studies explaining this situation from an empirical perspective. Wang and Xia (1999) has estimated aggregation benefits and costs of different cities and concluded that the aggregation benefit of a city reaches its peak when its population size is about one to four million and that further population growth will diminish the benefit. Foreign researchers have come to similar conclusions on Chinese cities: They generally hold that that the peak appears when the population size is between 2.9 million and 3.8 million.

To promote the healthy, high-quality urbanization and accelerate modernization in China, it is of strategic importance to study the scales of our cities and the characteristics of our urban systems in different regions.

1 Spatial Characteristics of Chinese Cities

There are obvious gaps in urban space among cities in China. Such gaps are almost inevitable in the process of urbanization, and they constitute a typical feature of our current urbanization stage. Given that our economic regions differ in geographical conditions, factor endowment, and policy environment, cities across our country have different development speeds and scales. From the perspective of urbanization factors, scales mentioned here entail economic aggregates, population, and land areas. There are inter-regional differences among cities of the same type and intra-regional differences among cities of different types. Therefore, cities in different regions of China are often in different stages of urbanization and face different urbanization problems. Generally speaking, the spatial characteristics of China's cities can be concluded as follows: the shrinking cities in the northeastern region, the excessively high urban density in the eastern coastal areas, and the "one-city dominance" in the central and western regions.

1.1 Shrinking Cities in the Northeastern Region

Haeussermann and Siebel (1988) first introduced the concept of city shrinking or shrinking cities as a metaphor for the demographic and economic decline of cities due to de-industrialization. At present, there is no unified conclusion on the definition of shrinking city or the reasons for city shrinking. Vishman defines city shrinking as the loss of urban residents: a city enters a state of shrinking when it loses more than 10,000 residents for two consecutive years. Fernandez et al. described it as population loss and economic recession and attributed it to the overly dependence on a single industry and the poor ability to cope with the impact exerted by economic transformation, resource depletion, or changes in domestic and foreign market environments. There are many available research papers on shrinking cities in the United States, Germany, Britain, Japan, etc. but those on city shrinking in China is still insufficient. In China, the term "shrinking small and medium-sized cities" first appeared in "Key Tasks for New-Type Urbanization in 2019," a document issued by the National Development and Reform Commission (NDRC), and relevant research shows that from 2007 to 2016, 84 cities started shrinking with some of them continuing their spatial expansion.

Most existing studies take population loss as an important criterion to judge city shrinking. Therefore, this chapter extracts the population data (2015–2019) of cities at the prefecture level in the three northeastern provinces from official statistics and counts shrinking cities according to changes in their populations in the five year. Those with negative population growth are counted as shrinking cities. The data processing results are shown in Table 1.

Generally speaking, most cities in the three northeastern provinces were in a state of shrinking except central cities of city clusters, and the shrinking cities were mostly concentrated in the marginal areas of the Harbin-Changchun City Cluster and the Mid-Southern Liaoning City Cluster. Among the three provinces, Liaoning had the least shrinking cities, and among all the cities, small and medium-sized cities were more prone to city shrinking.

Surprisingly, when their small and medium-sized cities shrank, the three provinces realized higher overall urbanization rates. The urbanization rates of Liaoning, Jilin, and Heilongjiang increased from 67%, 54.8% and 58% in 2014 to 67.5%, 56.6%, 59.4% in 2018. These increases seem to contradict the fact that most cities in the region are shrinking. Therefore, we study relevant calculation methods and compare the shrinking situations at the city level and the municipal district level for a more comprehensive analysis of the population changes and a test to determine whether there is a trend of urban hollowing. The classified statistical results of shrinking cities are shown in Table 2.

Table 2 shows that cities that experienced both overall and partial shrinking constituted the largest group and existed most commonly in the northeastern region. Most prefecture-level cities in the region were in a shrinking state, but these three northeastern provinces also had some cities with increasing populations. About 1/4 of the overall-shrinking cities experienced population loss across their territories, and only three cities in Liaoning, namely Shenyang, Dalian, and Panjin, saw overall population growths: population increases in all their municipal districts. According to the above findings, we made the following judgments: Most small and medium-sized cities in the northeastern region were in a state of shrinking but its big cities, especially central cities such as Dalian, Shenyang, Harbin, and Changchun, were in the state of overall expansion. It is worth noting that the expansion speed of municipal districts was much higher than the speed of the cities' overall expansion or shrinking. This suggests that there was no urban hollowing trend in the cities and that their population agglomeration effects were getting increasingly strong.

Table 1 Shrinking cities in the northeastern region

Province	Total		Liaoning		Jilin		Heilongjiang	
City type	All cities	Small- and medium-sized cities	All cities	Small- and medium-sized cities	All cities	Small- and medium-sized cities	All cities	Small- and medium-sized cities
Total number of cities	34	2.4	14	9	8	6	1.2	9
Shrinking cities	25	2.1	9	7	6	5	1.0	9
Proportion of shrinking cities (%)	65.30	87.50	64.29	77.78	75.00	83.33	83.33	100.00

Data source The data in this table are extracted from China Statistical Yearbooks.

Table 2 Classification and Quantity of Shrinking Cities

	Overall-shrinking cities	Overall-expanding cities	Total
Partially-shrinking cities	25	0	25
Partially-expanding cities	6	3	9
Total	31	3	34

Data source The data in this table are extracted from China Statistical Yearbooks.

From the perspective of population migration, the northeastern region's big and mega cities, especially its central cities, still absorbed an increasing number of new comers. From 2014 to 2018, the population growth rates of Dalian, Shenyang, Changchun, and Harbin were 33%, 13%, 20%, and 16%, respectively. Nevertheless, beneath the exciting population increases, there is an underlying apprehension: big cities in the region also face the risk of shrinking. On one hand, the northeastern region, especially its small and medium-sized cities, has been troubled by population outflow for years, and big cities in this region rely only on intra-regional migration to achieve population growth. On the other hand, because of various development restrictions, big cities in the region have lagged far behind their peers in other parts of China. It is fair to say that although they do not shrink in terms of absolute numbers, big cities in the northeastern region have already entered a relatively shrinking state in comparison with their counterparts in other economic regions. These cities need to get prepared in advance to realize high-quality urban population growth.

To intuitively reflect the relatively shrinking state of these big cities, this chapter compares them with their counterparts in the eastern and central regions: these cities are similar in city function, scale, population, and economic strength. We summarize their development differences during the period from 2009 to 2018 in Table 3. As shown in the Table, cities whose development status was similar to that of Shenyang, Dalian, Changchun, or Harbin in 2009 achieved much higher growth than these four cities did in the following 10 years: They outshone the four cities in terms of scale expansion speed and economic strength increase. Besides, there were big gaps among the three northeastern cities—Shenyang, Dalian, and Changchun—in terms of urban expansion and population growth. Such intra-regional differences among big cities were rarely seen in the other regions.

It can be seen that the relatively shrinking state of big cities in the northeastern region is mainly manifested in two aspects. First, their development speed lags behind that of their counterparts in the other regions, which reduces their appeal to the people. Secondly, their urban population growth rate lags behind their urban space expansion rate.

Table 3 Comparison Between Northeastern Cities and Their Counterparts in Other Economic Regions

Item / Region		2009			Growth rate		
		Urban long-term residents (million people)	Urban built-up area (km^2)	GDP (billion yuan)	Urban population-long-term residence (%)	Urban built-up area (%)	GDP (%)
The northeastern region	Shenyang	5.1223	395	366.7	17.3	41.8	53.2
	Dalian	3.0201	258	302	32.4	56.6	96.6
	Changchun	3.6232	365	205.3	22.0	48.5	178.0
	Harbin	4.747	345	227.2	16.1	28.1	109.4
The other economic regions	Wuhan	5.1497	475	388.9	71.7	52.4	281.8
	Qingdao	2.7547	273	278.9	88.0	161.9	233.6
	Jinan	3.4824	336	251.3	59.1	56.0	189.4
	Hangzhou	4.2944	393	407	47.9	56.5	207.3
	Zhengzhou	2.8501	337	143.2	34.7	61.4	320.0

Data source The data in this table are extracted from China Statistical Yearbooks.

1.2 The Excessively High Urban Density in the Eastern Coastal Areas

Cities are growth poles and supporting nodes of regional economic development. Nevertheless, densely-populated big cities in the eastern coastal areas, especially the overcrowded super mega and mega cities there, cannot perfectly fulfill their normal functions and even suffer the loss of urban development efficiency. This is because the local governments put too much emphasis on exerting the agglomeration effect of cities to achieve economies of scale when conducting urban development.

The government-led urban development pursues the agglomeration effect of cities and the economies of scale but neglects the estimation of and preparation for negative externalities. Such a mindset often leads to excessive resources concentration and high urban density in a city, which is harmful to the efficiency of urban development. The sustainable driving force of urban development comes from the labor transfer from the primary industry to the secondary and tertiary industries. The concentration of enterprises and population brings not only technology, information, infrastructure, and human capital spillovers but also negative externalities such as increased costs of living, production, and urban governance. There is an inverted U shape impact of urban scale on urban governance efficiency. The marginal effect increases with the expansion of a small city, but when the city expands to a certain scale, the marginal effect starts decreasing. In China, nearly 30% of cities have gone beyond this scale and faced problems brought by the decreased efficiency.

In accordance with "China City Statistical Yearbook 2019," the overly crowded cities in 2018 were mainly concentrated in the PRD, the YRD, and the Beijng-Tianjin region. Among them, Shenzhen was the most crowded: Its urban density reached 5,963 people per square kilometer, which was 41 times the national average. Shanghai had the second highest population density, with 3,816 people per square kilometer. In addition to its economic attributes, a city also serves as a comprehensive platform that accommodates various factors and individuals and thus needs to provide sufficient public services. It achieves its best efficiency only when its factor aggregation scale matches its public services supply capacity. In reality, however, these two aspects often cannot be promoted simultaneously. The potential of urban development will be harmed when the urban density is too high and when the supply capacity of urban public services and functions cannot satisfy the demand of urban development.

A city's ability to resist risks will decrease with its unlimited expansion. This is proven by what has happened under the influence of the COVID-19 epidemic. Because of its impact on our urban system and regional economy, China now attaches great importance to the pressure-bearing and recovery capacity of our cities and economic regions. This issue is also very noteworthy from the perspective of urban development. First, with continuous expansion, a city will have to face increasingly complicated problems and the growing pressure of preventing external shocks (e.g., extreme weather, environmental pollution, water crisis, public health events, etc.) from undermining its governance system and regional support capacity. Second, to guarantee its ability to resist risks, a city needs improvements regrading its resilience and density made simultaneously with the its scale expansion, which is almost impossible to achieve in reality. As such improvements usually lag behind the scale expansion, a more effective method is to limit the urban scale and appropriately reduce the urban population density. We feel obliged to mention the potential risks of two sites: the Xinfadi Market in Beijing and the world's largest centralized storage yard in Shenzhen. These two sites were constructed under the influence of traditional urban development thinking and traditional preference for the centralized control and economies of scale.

A city with excessive urban density may lead to polarization that will harm the inter-city ties, the urban system, and the resilience of the whole region. The normal functioning of an urban system is based on the connections and division of labor among cities at different levels. Without such a foundation, a group of cities cannot be referred to as an urban system. With the continuous improvement in economy and city quality, especially in the context of high-quality development, our division of labor will be increasingly specialized, and the specialized industrial sectors established based on strengthened inter-city economic ties will become long-term growth drivers of our regional economic development. Existing studies have fully proved the positive correlation between such economic ties and economic growth. To ensure the establishment of such ties, it is essential to prevent excessive urban density and wide urban gaps.

1.3 The "One-City Dominance" in the Central and Western Regions

The "one-city dominance" phenomenon is attributed to the widening differences in city size, which is an inevitable thing in the process of urbanization. Nevertheless, it seems to become a typical characteristic of the urbanization in the central and western regions. The scales of cities are determined by their development speeds, which are influenced by geographical conditions, factor endowment, and policy environment. From the perspective of urbanization factors, the term city scale entails economic aggregate, population, and land area; scale gaps can be divided into two categories, namely, the inter-regional scale gaps among cities of the same type and intra-regional scale gaps among cities of different types. From the perspective of regional high-quality development, this chapter takes provincial capitals as the research objects and explores how the economic and population gaps within an urban system leads to the phenomenon of one-city dominance.

First, we need to determine whether the scales of these cities are within a reasonable range. According to the GDP data and population data in "China Urban Statistical Yearbook 2019", we find the largest city of each province and calculate its primacy ratio and proportion in the province. Our research excludes Xinjiang, Tibet, Qinghai, Hainan, and Taiwan for incomplete urban statistics. We also exclude the four municipalities directly under the Central Government, i.e. Beijing, Shanghai, Tianjin, and Chongqing and the two special administrative regions, namely Hong Kong and Macao. Our statistical results are shown in Tables 4 and 5.

It is generally believed that a primacy ratio lower than two suggests a reasonable structure and balanced development and a ratio higher than two reflects excessive concentration. In more than half of the provinces involved in the statistics, the primacy ratio is greater than two. The highest ratio appears in Sichuan, which is 9.61: The economy of its largest city (Chengdu) is 9.61 times that of its second largest city (Mianyang). The proportions of these cities in their provinces are high, generally exceeding 20%. From the perspective of population, nearly a half of the provinces involved see a primacy ratio higher than two. The highest ratio is 6.13, which is for Xi'an, the most populous city of Shaanxi. Generally speaking, the primacy ratios regarding economic concentration are higher than those about population concentration. This suggests that within a province, the economic concentration effect of its largest city or central city is much stronger than its population concentration effect.

To sum up, the scale differences among cities in China can be described from the following four aspects.

First, the "one-city dominance" phenomenon is prominent in the central and western regions. In these regions, a province's primary city or largest city (by economy and population) is always its provincial capital. These cities' primacy ratios (by economy and population) are much higher than those of their counterparts in the other regions.

Table 4 The Provinces' Largest Cities by Economy and Their Primacy Ratios in 2018

Item ranking	Province	The largest city in the province	Primacy ratio	Proportion in the province (%)
1	Sichuan	Chengdu	9.61	30.93
2	Shaanxi	Xi'an	7.09	33.50
3	Hubei	Wuhan	6.90	39.26
4	Guizhou	Guiyang	6.11	31.39
5	Shanxi	Taiyuan	5.66	21.52
6	Gansu	Lanzhou	5.55	28.83
7	Yunnan	Kunming	4.51	34.43
8	Jilin	Changchun	4.18	39.58
9	Hunan	Changsha	4.02	19.10
10	Jiangxi	Nanchang	3.48	18.09
11	Henan	Zhengzhou	3.37	12.60
12	Ningxia	Yinchuan	2.96	33.51
13	Anhui	Hefei	2.60	17.30
14	Guangdong	Guangzhou	2.30	29.76
15	Heilongjiang	Harbin	1.93	28.88
16	Zhejiang	Hangzhou	1.85	22.16
17	Jiangsu	Nanjing	1.46	13.59
18	Guangxi	Nanning	1.32	16.14
19	Shandong	Qingdao	1.28	11.95
20	Inner Mongolia	Baotou	1.14	16.46
21	Liaoning	Dalian	1.06	23.07
22	Hebei	Tangshan	1.02	9.94
23	Fujian	Xiamen	1.00	13.39

Data source The data in this table are extracted from China Statistical Yearbooks.

Second, the primacy cities that have established "one-city dominance" do better in economic concentration than in population concentration. In other words, the "one-city dominance" is more an economic concentration phenomenon than a population concentration phenomenon.

Third, primary cities' proportions in their provincial economies are high, generally surpassing 1/3. The proportions of Changchun and Wuhan are even close to 40%.

Fourth, in some eastern provinces, there is a multi-center urban development trend. In provinces such as Liaoniang, Shandong, and Fujian, the primacy ratios of the primary cities are low, and their provincial capitals are no longer their primary cities in terms of economy.

When our urbanization reached the stage of rapid development, large-scale reconstruction and expansion projects led to the revaluation of land factors, which worked

Table 5 The Provinces' Largest Cities by Population and Their Primacy Ratios in 2018

Item Ranking	Province	The largest city in the province	Primacy ratio	Proportion in the province (%)
1	Shaanxi	Xi'an	6.13	21.27
2	Heilongjiang	Harbin	4.02	15.61
3	Hubei	Wuhan	3.88	16.53
4	Sichuan	Chengdu	3.70	10.14
5	Ningxia	Yinchuan	2.55	17.54
6	Hunan	Changsha	2.52	5.06
7	Jilin	Changchun	2.46	18.42
8	Yunnan	Kunming	2.27	10.80
9	Zhejiang	Hangzhou	2.15	12.70
10	Jiangsu	Nanjing	1.91	8.90
11	Guangxi	Nanning	1.90	6.84
12	Henan	Zhengzhou	1.86	3.38
13	Shanxi	Taiyuan	-1.66	8.30
14	Guangdong	Guangzhou	1.65	10.31
15	Gansu	Lanzhou	1.58	8.49
16	Guizhou	Guiyang	1.55	10.59
17	Liaoning	Shenyang	1.50	14.34
18	Jiangxi	Nanchang	1.36	6.19
19	Anhui	Hefei	1.22	3.97
20	Fujian	Fuzhou	1.17	7.38
21	Hebei	Shijiazhuang	1.16	5.50
22	Inner Mongolia	Baotou	1.13	7.25
23	Shandong	Jinan	1.07	5.49

together with the Tax Sharing System in creating the land finance of local governments. As a result, local governments expanded their investments in public infrastructure, which promoted the accelerated urbanization of land, and stepped up their effort to attract enterprises with cheap land prices and preferential policies. With the urbanization of industries, i.e. industries gathering in cities, relevant workers including those from rural areas moved to cities for high-paying jobs, realizing the urbanization of people. In the best case, the urbanization of land, that of industries, and that of people shall be promoted simultaneously. Nevertheless, in reality, provincial capitals tended to gather too much political and economic resources and face great economic pressure and overly-accelerated urbanization of land. Cities where the urbanization of land led that of industries and that of people usually had endless demand for population and industries. They were eager to absorb resources instead of sharing its resources to others, so their aggregation effect is far greater than their radiation effect. It is worth mentioning that urbanization will inevitably lead to rising land prices, house price, and costs of living, which will hinder the urbanization of industries and population, and even cause de-industrialization trends/

As to the drivers of our urbanization, they can be divided into three categories: the government, the market, and the population. The government, often a local government, intervenes in the process of urban development through administrative means and policy making, actively attracting factors to gather in cities. The market allocates certain free-circulating resources to certain cities according to changes in supply and demand. The population influences the urbanization through its movement, i.e. the migration from the rural areas to the urban areas, which is carried out for better development opportunities and living conditions. It is generally believed that these three forces are intertwined and work together, but an analysis from a historical perspective reveals that our urbanization is mainly influenced by government efforts and population movement. First of all, let us take a look at the government efforts. China's national strategies played a decisive role in our urbanization. Once a place was included in such a strategy, it would see huge progress in its urban construction. The main-stream regional policies of local governments in the 1990s highlighted the efforts to accelerate urbanization. Provincial capitals with centralized political resources enjoyed policy advantages and thus sped up their urbanization processes. Next, let us analyze the influence of population movement, which remains a key factor in all stages of our urban development. People tend to migrate to places offering better opportunities, payments, public services, and infrastructure conditions when the future benefits are greater than the cost of moving. Because of their advantages in urban functions and services, big cities represented by provincial capitals expanded faster than other cities over the past years.

2 Multi-Center Urban Development Mode

The establishment of a multi-center urban development mode is probably the most effective and feasible means to limit the excessive expansion of cities. First, this means does not put excessive restrictions on the expansion of large cities, which may lead to stagnation of urban development and greater efficiency losses. Second, it serves the core purpose of urban expansion restriction: forcing large cities to play their radiating and driving roles. This will promotes the development of hinterland cities in China.

According to the sizes and functions of their central cities, urban systems can be summarized into three modes, namely the mono-center, bi-center, and multi-center mode. It is worth noting that the center we discussed here is a relative concept of space and scale. That is, a multi-center urban system may be regarded as a single center at a more macro level. This chapter focuses on non-mono-center urban systems at the provincial level. For convenience of expression, these provincial urban systems are referred to as multi-center urban systems hereinafter. The newly-fostered central cities discussed hereinafter are mainly prefecture-level cities.

In the existing research and practice regarding urban systems, we find an increasing interest in multi-center urban systems. This provides us with sufficient research cases to explore the multi-center urban systems in China: e.g. the city cluster along the Atlantic coast of the United States with New York as its center and Boston, Philadelphia, Washington, and Baltimore as its sub-centers; the Great Lakes city cluster with Chicago as its center and Detroit, Cleveland, Rochester, Pace, and Buffalo as its sub-centers; the multi-center Japanese city cluster taking Tokyo as its center and involving Osaka-Kobe metropolitan area and Nagoya metropolitan area; the city cluster in the southeast of the UK, with London as its primary city and Birmingham, Liverpool, and Manchester as its central cities. It can be predicted that in the future urban system planning, this multi-polarization idea will be further implemented. The EU has already adopted the multi-center urban development mode as a theoretical tool and development goal. The United States has also listed the multi-center urban development mode as a major issue in its long-term plans. According to the scales and spatial distribution of central cities in China, we find two suitable urban development modes: the satellite-cities mode and the urban belt mode.

2.1 The Satellite-Cities Mode

Satellite cities are smaller cities that are near to a large city that is the center of a city cluster. The satellite-cities mode is widely applicable to the urban system development in the central and western provinces whose primary cities have exceptionally high primacy ratios. This mode will not change the position of a primary city. It will enable the city to smooth its factor circulation channels connecting to other cities in the hinterland and give full play to its radiating role. Under such circumstance, other cities in the city cluster or a relatively remote district of the primary city will an opportunity to develop into a growth pole of the cluster.

Within a region or province, a three-level urban system consisting of central cities, satellite cities, and counties shall be established through the combination of central cities and satellite cities at all levels. In this way, towns or townships will benefit from the radiating effect of cities, and central cities in the hinterland will get dual identities: while serving as satellite cities for primary cities, they will become the central cities for counties. That is, the task of driving regional development through the radiating effect is gradually shifted from primary cities to its satellite cities. Leveraging the relatively scattered distribution of satellite cities, the influence of cities in their region or provinces will be significantly enhanced.

The establishment of a three-level urban system through the development of satellite cities entails two aspects. First, the factor carrying capacity of satellite cities needs to be improved. This requires improvement in infrastructure, public services, and social securities systems. To be more specific, these cities have to strengthen their transportation facilities, underground pipe networks, and transportation hubs; they must offer better education, medical care, cultural, and health-care services; they need to attract more economic, legal, advertising service agencies to settle in;

they have to see improvement in utilization efficiency of factors. Second, the factor circulation channels of satellite cities need to be smoother. Traffic and logistics channels between satellite cities and central cities need to be upgraded. Overlapping industries should be integrated. Information exchanges and the application, promotion, and popularization of emerging technologies should be encouraged. An urban culture highlighting harmony in diversity shall be fostered. Among the primary cities in the central and western regions of China, Zhengzhou has already achieve initial results in promoting a multi-center urban development mode: With its rapid development in recent years, it has established railway and urban expressway networks that connect its satellite cities such as Fuyang, Anyang, Nanyang, and Changzhi.

The satellite-cities model has its inherent development logic, which is embodied in the following three aspects regarding the process in which a factor is transferred step by step from a central city to a county through a satellite city.

1. **Urban Development** As mentioned above, the prerequisite for the transfer of resources from a central city to its satellite city is the improved factor carrying capacity and factor utilization efficiency of the satellite city. The transfers or redistribution of factors cannot directly boost urban construction in the satellite city. Therefore, when building our ideal urban system through the development of satellite cities, the primary task is to strengthen the urban facilities and functions of satellite cities. It should be noted that the facilities mentioned here refer to not only infrastructure facilities but also cultural and educational services, social services, medical and health services, etc. Possible methods include the following: building local universities on the basis of locally-based scientific research institutes; enhancing knowledge and technology interactions with other localities; preserving unique local cultures and highlighting local characteristics through city images, featured colors, or city landmarks; improving urban services and urban efficiency to ensure the quality of life for urban residents.
2. **Intra-Regional Positioning** After meeting the basic requirements for urban construction, satellite cities have to participate in inter-regional or inter-provincial division of labor to benefit from the spillover effect of central cities and increase their own influence. In order to ensure that satellite cities participate in the division of labor in an orderly manner and avoid redundant construction and local protectionism, it is essential to determine the role of each satellite city in the industrial chain based on its proximity to its central city in the early stage of its development. In the meantime, it is necessary to create an environment encouraging the division of labor and inter-city cooperation and foster regional alliances that boast strong economic and cultural ties among alliance members.
3. **Industrial Transfer** The development of satellite cities aims to improve the urban systems and alleviate the "one-city dominance" phenomenon in the central and western provinces. Given this, a satellite city's success is determined by whether it has the ability to receive the transfer of industries and relevant resources from its central city. To receive such a transfer, the satellite city needs to develop related supporting industrial systems and grasp the industrial upgrading trend. They

also need to see the development prospects for industries, especially for high-tech industries, and strive to receive a whole-industrial-system transfer, which involves all links in design, R&D, manufacturing, and sales. In this process, the satellite city should also pay attention to the development of its local small and medium-sized enterprises and provide economic and policy supports based on their needs. In this way, a geographical economic network can be established.

2.2 The Urban Belt Mode

In the urban belt mode, growth poles are connected into a development axis and cities of different sizes and functions are distributed along both sides of the axis. Like the satellite-cities mode, it is also a multi-polar development mode. Nevertheless, there are obvious differences between the two: The former has only one central city as its primary growth pole and many satellite cities as sub-level development growth poles, which are weaker than the central city in all aspects; the latter features more than two equally-influential central cities and consists of areas they affect.

Compared with the process of influence transmission in the satellite-cities model, the urban belt mode breaks through the center-city-and-hinterland geographical restrictions. In the latter, there is no longer a positive correlation between the scales and development levels of non-central cities and their proximity to their central cities. In other words, these non-central cities enjoy equal status. It should be noted that the prerequisites for breaking through such geographical restrictions include the following: remarkable improvement in transportation facilities and inter-city traffic flow efficiency;,the popularization of modern information and communication technologies. That is, the urban belt mode puts forward higher requirements for the quality and balanced development of non-central cities.

The process of urban system development in this mode is as follows: A primary city and central cities that have similar economic strengths form the development axis and then they develop together with non-central cities to form the urban belt. In this mode, the boundary of the urban belt is not fixed. With the growth of the central cities, more non-central cities will be brought into their spheres of influence, and the urban belt will be further expanded. Compare with the satellite-cities model that is still influenced by administrative divisions, the urban belt mode breaking through such restrictions and has stronger expansion and radiation capabilities and higher requirements for inter-city connections and the subjective initiatives of market subjects. The CCEC features a typical urban belt development mode. Its main development axis consisting of its two primary cities, namely Chengdu and Chongqing, and three central cities, i.e. Wanzhou, Qianjiang, and Nanchong. On its both wing, there are tow sub-axes: One takes Chengdu as its primary city and Mianyang and Leshan as central cities; the other with Chongqing as the primary city involves central cities such as Yibin and Luzhou. Such a dual-primary-city urban system, to a certain extent, mitigates the negative impact of "one-city dominance". Besides, it offers a possible explanation to the contradiction between Chengdu's super high primacy ratio and its mediocre performance in nearly all aspects in a nationwide comparison.

As a multi-pole urban development mode, the urban belt mode shares many similarities with the satellite-cities mode but also has its own characteristics. It puts more emphasis on the spatial layout, division of labor, and inter-city cooperation in urban and industrial development instead of industrial transfer. Relevant key points can be concluded in the following three aspects.

1. **Infrastructure Construction** The formation of an urban belt often depends on certain geographical conditions, such as rivers and coastlines, or traffic trunk lines. The factor flow channels constructed on the basis of these conditions constitute the prerequisite for the formation of the urban belt mode and its urban system. Given this, it is essential to strengthen the connections between central and non-central cities by building inter-city rapid transit rail and expressway systems, enhancing transportation hub facilities, and constructing an information-sharing network with high coverage and convenient service. It is a good choice to leverage, ports, airports, traffic trunk lines to improve the construction of supporting facilities.
2. **Industrial Layout** On the basis of smooth factor circulation channels, central cities, non-central cities. and small and medium-sized towns can establish a system for their division of labor and cooperation. The core of this system is industrial layout. Based on the advantageous, leading industries of central cities and their own characteristics, non-central cities shall develop advanced manufacturing and modern service industries, trying to develop themselves into regional manufacturing centers or trade and logistics centers. In this way, the urban development space will be expanded and the utilization rate of space will be improved, which will strengthen the cities's appeal to population and resources.
3. **Inter-City Cooperation** The urban belt mode calls for frequent and close cooperation among non-central cities. To this end, they shall remove administrative barriers, carry out inter-city cooperation on the basis of free factor circulation and industrial coordination. To be more specific, they shall promote the region integration of the technology market, capital market, and administrative norms; ensure intra-region coordination in public services, public culture, and ecological governance; and normalize the coordination and compensation mechanisms to guarantee regional interests. The non-central cities' role of overall coordinating shall be enhanced, and the urban belt or city cluster shall pursue common development.

3 Conclusion and Policy Recommendations

Since 1978, China has adopted a series of urban development plans and rapidly increased its urbanization rate, creating a new path of urbanization with Chinese characteristics and our current urban systems. As early as 1980, China put forward the policy of controlling the scale of big cities, rationally developing medium-sized cities, and actively developing small and medium-sized cities in the document entitled "Summary of National Conference on Urban Planning." At present, this policy is still in line with the development status of our country but it is not fully implemented. To achieve their primary goal of economic growth, local governments tend to concentrate the superior resources of their own regions in their primary cities with the help of their administrative power. This explains why the "one-city dominance" phenomenon exists extensively in the western and central provinces and why their small and medium-sized cities are insufficient in quantity and low in quality. Generally speaking, the scale and number of our small and medium-sized cities determine that they can hardly play the role of enriching our current urban systems.

To solve this difficulty regarding our urban development, the Central Economic Work Conference has proposed to promote the development of city clusters and multi-center, multi-level, and multi-node urban networks. On the basis of conforming to the objective laws of urban development, efforts shall be made to explore multi-polar urban development modes. Such efforts underpins our endeavors to comprehensively improve the quality of our cities, promote our people-oriented urbanization, and turn our cities into fulcrums for our regional development.

To pursue multi-polar urban development, it is essential to rationally control the scales of the primary or central cities. The development of new-type infrastructure and the application and popularization of new technologies create great opportunities to re-plan the development path for our non-central cities, comprehensively improve the quality of our cities, and narrow the development gaps among them. The inter-city links need to be strengthened. The factor carrying capacity and utilization efficiency of non-central cities shall be increased, and so shall their political and economic status. These cities shall be integrate into urban systems and their industrial division of labor. In this way, new mono-center or multi-center urban systems with multiple levels will be established.

3.1 Limited Central-City Expansion

China's urbanization rate of permanent residence has exceeded 60%, and a batch of super mega cities and mega cities have emerged. In the urban development of its central and western regions, the polarization phenomenon is obvious. A series of problems caused by the excessive urban expansion have attracted attention. Nevertheless, the super mega cities in the eastern region and the primary cities in the central and western regions are projected to continue their expansion in the foreseeable future. Under such circumstances, it is very important to limit the expansion of central cities and improve the efficiency of our urban development.

It should be noted that the size limitations on central cities do not mean the abandonment of these cities or a halt in their economic development. Such limitations are aimed to enable to cities to shift their focus from economic goals to more comprehensive development goals that involve the production, living, and ecological needs of their enterprises and residents. Their scale expansion speeds are reduced in exchange for the overall progress in urban quality, layout, and smartness; the improved ability to ensuring equality in their quality public service systems; and the increased city influence.

1. **Optimized Urban Layout** To optimize the urban layout, it is necessary to coordinate new-urban-area construction activities to the renovation of old urban areas. Old urban areas boast a large amount of mature public service resources but face limited development space. In the new urban areas, the situation is just the opposite. Because of the competition for development space and high-quality resources, new and old urban areas can hardly follow the same development path. This requires the clear positioning of different urban areas, which involves different development modes and a clear order of development. In detail, when old urban areas have better development conditions than the new areas do, resources should be concentrated to complete the upgrading of the old areas; when new areas have greater development potential in comparison with old ones, key urban functions and high-quality urban resources shall be transferred to the new areas. The development of Quanzhou, a big city in Fujian Province, is a typical example. To well preserve its ancient buildings, it has its old urban area called Licheng District kept only residential, cultural, and tourism functions and related supporting services and facilities. In its new urban area, which is situated near the old one, it advocates economic efficiency as the priority and builds a modernized urban area.
2. **Increased Urban Efficiency**To improve urban governance capacity and operational efficiency, it is a must to build smart cities through the application and popularization of new technologies. Smart cities are one of the development goal of our future urban development, and central cities that boast rich scientific and technological resources are better prepared to achieve this goal. First, smart cities mean higher urban carrying capacities and urban development efficiency because the application of information technology, eespecially the loT, transform and empowers cities' natural, economic, and social systems. Second, smart cities mean new opportunities. New high-tech industrial clusters will be developed when enterprises, scientific research institutes, and colleges and universities work together to push forward smart-city related projects that can be applied in cities to solve the traffic, medical, and health needs of residents.
3. **Improved Urban Environment** Urban environment improvement is one of the most important methods used to enhance the comprehensive strength of cities in the next stage of our urbanization, which highlights high-quality development. Its core part is to promote the people-oriented urbanization, i.e. making a city safer, healthier, and more livable, and turning it into a place for high-quality living. In addition to the efforts devoted to the ecological protection, it is necessary to guarantee basic public services for all city residents including the new comers. These services include those regarding household registration, basic social securities, and all services that satisfy people's basic demands and enhance their living experience, which should be taken into account in the urban planning

process. Such services include the expansion of the transportation and underground pipeline networks; the construction of highly-accessible convenience facilities and full-fledged districts and communities; the expanded urban ecological space and public space; the improvement in the total amount and quality of public services; and the basic elimination of inequalities among different districts and those among different groups of people.

3.2 *Enhanced Quality of Non-Central Cities*

The purposes of limiting the expansion of primary cities include the following: transferring the limited resources to cities and towns at the lower levels; boosting the overall urban development at the lower levels; narrowing the development gap between primary cities and other cities; and enabling cities at the lower levels to absorb industrial transfers from cities at the higher levels and become an integral part of their urban systems. A non-central city is usually in a position of being dominated by the primary city of its urban system and relies on the resource transfers from the primary city. Nevertheless, this does not mean that a non-central city has to be completely attached to its primary city. It had better explore its own development path, make full use of the opportunities brought about by the new technological revolution, and rely on itself to realize leap-forward development.

1. **Consolidated City Economic Strength.**The primary task of a non-central city is to enhance it economic strength, which relies on the healthy development of its leading industries and determines whether it will become a new growth pole in its economic region. The emergence of new technologies and their integration with traditional industries provide our cities with more choices. While adapting to its regional industrial framework, a non-central city shall actively develop strategic high-end industries such as information technologies, bio-pharmaceuticals, high-end equipment manufacturing, new energy vehicles, new energies, and new materials. It shall focus on the technical requirements of its urban construction, optimize its environment for technological innovation and development, and promote the application of new technical achievements in the traditional industries of its primary city. In this way, its products will be circulated in the urban system it belongs to, which can be referred to as a "miniature internal circulation."

   The construction of industrial parks shall be regarded as important as industrial development and urban construction because they offer space for the development of a city's leading industries and constitute a an integral part of a city. When planning and designing industrial parks, it is a must to stick to the same standards for urban development: production areas shall be separated from living areas; in each living area, there shall be a high-level service complex; industrial parks shall be included in urban landscape designs to eliminate the sense of separation between industrial parks and urban areas.
2. **Increased Soft Power of City Culture.**Compared with central cities, non-central cities have much weaker cultural soft power. They shall increase their popularity

by enhancing their city images and increase their influence by promoting their city culture.

First, the non-central cities must strengthen their cultural service systems. This does not just entail relevant facility improvements but also systematic efforts that ensure the facilities' compatibility with relevant urban development plans and urban environment, culture, and customs. Second, regarding cultural inheritance and accumulation, it is necessary to avoid short-sighted thinking, e.g. paying too much attention to short-term profits. It is essential to proceed from cultural inheritance, develop cultural industries or products that boast rich connotations and unique experiences, and establish a complete cultural industry chain with unique characteristics. Third, regarding city images and city landmarks, they are designed to embody different city cultures and used as the most convenient means to promote cultural development. When designing such images and landmarks, it is essential to avoid formalism problems such as overstatements and over-packaging. The quality of landmark buildings shall be guaranteed, and their quantity shall be under control. To sum up, attention shall be paid to the creation and extension of cultural cores, and excessive waste shall be avoided. In addition, it is necessary to strengthen supervision over the construction of cultural facilities, preventing them from becoming real estate speculation opportunities.

3. **Enhanced City Influence.**A non-central city's influence and it position in its urban system can be improved through major projects and pilot programs of major regional strategies, regional projects, and policies. Such projects and programs will bring it resources and preferential policies and boost development in this city and its city cluster, which is conducive to the establishment of an urban system.

   Examples in practice are as follows. The establishment of areas for the exploration into relevant reform policies, such as national-level new areas, independent innovation demonstration areas, free trade pilot zones, intelligent manufacturing demonstration areas, cultural inheritance demonstration areas, and ecological civilization demonstration areas, turn some non-central cities into regional leading cities and give them a first-mover advantage and a chance to comprehensively improve their strengths. The construction of large-scale regional facilities such as airports, ports, high-speed railway networks, inter-city railway networks, and communication hubs open up factor circulation channels for cities at all levels and increase their factor utilization efficiency. With resources gathered and distributed for major regional strategies and major projects, more corporate headquarters, R&D centers, and financial centers were built. This propel more non-central cities to participate in regional division of labor and cooperation, and in this process, their influences increase.

3.3 Consolidated Intra-Links of Urban Systems

A perfect urban system needs not only well-developed central and non-central cities but also strong ties among all its cities and towns. With such ties, the spontaneous adjustment of by the market will be realized: The size of the central city will naturally be limited within a reasonable range; The non-central cities will achieve accelerated development; A virtuous circle within the system can be established. The intra-links of an urban system are formed when its cities and towns are well developed and closely linked to one another in the horizontal or vertical division of labor. To this end, it is a must to clarify the functions of and connections among cities and towns at all levels in this urban system.

1. **Determine the Functions of Cities and Towns at All Levels.** At present, urban systems have been basically established across China, but within each system, the division of functions among cities and towns is still unclear. Therefore, there are problems such as overlapping and functions, loss of functions, and the lack of intra-system cooperation and coordination. To solve such problems, it is necessary to establish and constantly optimize our urban-system development mechanism and clarify the specific responsibilities and functions of cities and towns at different levels. The ultimate purpose is to build a multi-level network for our cities and towns, which consists of national-level central cities, regional-level central cities, small and medium-sized cities, counties, and central towns.

   (1) ***National-level central cities.*** These cities are at the top levels of our urban systems. According to the different development characteristics of our economic regions, they bear different national strategic responsibilities and play their special roles. The development focus of these cities is improving their openness and international influence. They shall strive to gain certain control over the international economic activities and scientific and technological innovation activities while strengthening their own radiating and leading roles in regional economic development and serving as the bases of national scientific and technological innovation. They must strengthen their functions as gateways and hubs in traffic and information networks and play an exemplary role in promoting cultural development and green development. They shall play important supporting roles for our national strategic arrangement.

   (2) ***Regional-level central cities.*** Second only to national-level central cities, regional-level central cities are key fulcrums of our economic regions and city clusters. Compared with cities at the lower levels, they gather more urban resources and do better in cultural development, social securities, and public services. Different national-level central cities, they are not required to assume the responsibility of internationalized development but they has to shoulder the responsibility of boosting regional development through its radiating function. They have gathering, radiating, guiding, and coordinating functions, but in specific situations, they major functions may be different. With these four functions, they boost the development of regional urban systems and create new growth poles that drive regional development and the opening-up endeavor, forming metropolitan areas (circles) in different regions.

(3) ***Small and Medium-Sized Cities.*** This term refer to non-central cities with specific functions, including cities with featured industries, featured cities, and inland border port cities (towns). The development of these cities and towns is aimed to solve practical problems that areas outside the city clusters lack sustainable growth driver and face great pressure brought by their natural and cultural protection tasks. It is also intended to satisfy the huge demand of the growing middle-income groups for leisure vacations and experience consumption. By innovating the development path, featured cities are developed and the diversified development of urbanization is realized. For their future development, these cities shall continuously adhere to their differentiated development paths, formulate appropriate policies, and reasonably choose their leading industries. They must ensure simultaneous quality improvements in their industrial and urban development and pursue intensive development and smart growth.

(4) ***Counties.*** Counties are the most fundamental units of our urban systems, covering the widest area and occupy the largest proportion. Their progresses underpin the success of our urban–rural integrated development, the sustainability of our urban systems, and our urbanization process. Given their differences in resource endowment and development conditions, they follow different development paths and have different development status. Therefore, they need targeted guidance on their development modes and paths. Generally speaking, it is necessary to promote the industrial development and population growth of counties and encourage qualified counties to upgrade themselves into cities for better social governance and urban construction abilities. As to featured counties, they shall give full play to the characteristics and advantages in ecologic, scenic, historical resources, protect these resources, and pursue characteristic development and moderately-concentrated construction.

2. **Strengthened the Ties Among Cities and Counties Within an urban System.** Although China has constructed the general framework of its urban-town system, many of its cities and towns have not yet established stable connections. In many cases, an urban system can be described as consisting of a city and its surrounding towns: it is more an agglomeration than a cluster. As to the intra-links of an urban system, they are often based on administrative instructions and influenced by the administrative hierarchy. This leads to blind competition within the system, resulting in redundant construction and waste in sectors such as infrastructure construction,

   In view of the above, it is necessary to strengthen intra-links of our each urban system and establish their coordination mechanisms. For the coordinated development of an urban system, it is a must to do well in the following aspects: formulating feasible, effective plans; eliminating administrative barriers within an urban system; promoting the unification of markets and administrative institutions; ensuring basically equal acceptability to infrastructure facilities; giving full play to the radiating and leading roles of central cities; advancing the formation of a network structure with an urban system; strengthening regional integration. The governments should take advantage of its vertical management advantages to guide development, horizontally clarifying the division of labor and complementary cooperation and preventing problems such as overlapping urban industrial structures and malignant competition.

References

An, H. (2018). *Quyu jingji xue (Regional economics)*. Higher Education Press.

Au, C., & Henderson, J. V. (2006). Are Chinese cities too small? *The Review of Economic Studies, 73*(3), 549–576.

Chen, C., Zhang, P., Liu, X., & Zhang, Z. (2009). Urbanization, industrial efficiency and economic growth. *Economic Research Journal, 44*(10), 4–21.

Gill, I., & Kharas, H. (2007). *An East Asian renaissance: Ideas for economic growth*. The World Bank, number 6798, June. 2007.

Haeussermann, H., & Siebel, W. (1988). *Die schrumpfende Stadt und die Stadtsoziologie*. In J. Friedrichs (Ed.), *Soziologische Stadtforschung* (pp. 78–94).

He, Y. (2020). *Guimo hongli haishi gonggong fuwu? Zhongguo chengshi zhili jincheng ji yanjin jizhi* (Scale dividends or public service? China's urban governance process and its evolution mechanism). *Tsinghua Journal of Social Sciences, 2*(01), 203–226.

Li, H., & Shao, M. (2017). *Chenshi guimo, jineng chayi yu laodongzhe gongzi shouru* (City scale, gaps in skills, and income of laborers). *Management World, 2017*(08), 36–51.

Li, Q., Chen, Y., & Liu, J. (2012). *Zhongguo chengzhenhua "tuijin moshi" yanjiu* (Research on the "promotion mode" of urbanization in China). *Social Sciences in China, 2012*(07), 82–100.

Liu, Z., & Ma, Y. (2020). City distribution characteristics and economic relations of Guangdong-Hong Kong-Macao Greater Bay Area. *Journal of Urban Studies, 41*(04), 47–53.

Lu, M. (2016). *Great nation needs bigger city*. Shanghai People's Publishing House.

Lu, M. (2017). *Kongjian de liliang: Dili, zhengzhi yu chengshi fazhan (The power of space: Geography, politics, and urban development)*. London: Shanghai People's Publishing House.

Qin, C., Zhang, H., & Chao, M. (2011). *Quyu jingji xietiao fazhan: Gainian jiexi, panduan biaozhun yu pingjia fangfa* (Coordinated Development of Regional Economy: Conceptual Analysis, Judgment Standards, and Evaluation Methods). *Reform of Economic System, 2011*(04), 34–38.

Qiu, B., Yao, Y., Liu, Z., & Qin, Z. (2020). 2020), *Goujian mianxiang weilai de renxing chengshi* (Building resilient cities for the future. *Regional Economic Review, 06*, 1–11.

Shi, M., Jin, F., Li, N., Zhao, Z., & Jin, S. (2006). Inter-regional economic linkage and regional development driving forces based on an inter-regional input-output analysis of China. *Acta Geographica Sinica, 2006*(06), 593–603.

Sun, J., & Su, X. (2020). *Lun dongbei zhenxing zhong de chengshi jingming zengzhang zhanlue* (Discussion on the smart-growth strategy for cities involved in the revitalization campaign of the northeastern region). *Social Science Journal, 2020*(05), 50–62.

Sun, J., & Ye, Y. (2010). *Textbook of regional economics*. China Remin University Press.

Sun, J., & Ye, Y. (2020). *Textbook of regional economics*. London: China Remin University Press.

Sun, X., & Zhou, F. (2013). *Tudi caizheng yu fenshui zhi: Yige shizheng jieshi* (The land finance and tax sharing system: An empirical explanation). *Social Sciences in China, 2013*(04), 40–59.

Wang, B., Guo, B., & Li, H. (2020). Research on regional differences, scale distribution and urban efficiency of China—Empirical study based on 285 prefecture-level cities. *Inquiry into Economic Issues, 2020*(01), 114–127.

Wang, H., Miao, C., Ru, L., & Cui, C. (2012). The spatial framework and change of Chinese provincial region economic links. *Economic Geography, 32*(07), 18–23.

Wang, L. (2010). Urbanization path and city scale in China: An economic analysis. *Economic Research Journal, 45*(10), 20–32.

Wang, X., & Xia, X. (1999). Youhua chengshi guimo, tuidong jingji zengzhang (Optimize city scales and promote economic growth). *Economic Research Journal, 1999*(09), 22–29.

Wang, Y. (2012). Inverted "U" curve of urban scale efficiency and its policy implications——A comparative study on economic, social and environmental efficiency of prefecture-level cities and above in China. *Finance & Trade Economics, 2012*(11), 127–136.

Xu, B. (2019). Shrinking cities and smart development—intrinsic logical matching and reform and innovation space for government governance and market efficiency in shrinking small and medium-sized cities. *Economist, 2019*(12), 34–45.

Xiu, C., Wei, Z., & Wang, Q. (2018). Evaluation of urban resilience of Dalian city based on the perspective of "Size-Density-Morphology." *Acta Geographica Sinica, 73*(12), 2315–2328.

Xu, J. (2008). Poly-centric city-region: Reflections on POLYNET. *Urban Planning International, 2008*(01), 1–3.

Xu, Y., & Zhang, L. (2019). The Economic Resilience of Chinese Cities and Its Origin: From the Perspective of Diversification of Industrial Structure. *Finance & Trade Economics, 40*(07), 110–126.

Zhou, L. (2017). *Zhuanxing zhongde difang zhengfu: guanyuan jili yu zhili (Local governments in transformation: official incentives and governance)*. Shanghai People's Publishing House.

Zhang, P., & Liu, X. (2011). Urbanization, fiscal expansion and economic growth. *Economic Research Journal, 46*(11), 4–20.

Chapter 6
China's Integrated Urban–Rural Development and Rural Revitalization

The first section of this chapter summarizes the evolution of urban–rural relations in China, describing the change from the urban–rural dualism, the urban–rural coordination, and the urban–rural integration. So far, the country has achieved remarkable progress in this regard, upgrading its urban–rural coordinated development policy into an integrated development policy. Despite that, it is necessary to recognize that the urban–rural dualism is still a widely-existing problem and that China still has a long way to go in realizing the integrated development of its urban and rural areas. China's rural revitalization policy is essentially an effort to reshape its current urban–rural relations: In this process, the integrated development is a development goal as well as a development method. The second section of this chapter analyzes the content of China's rural revitalization strategy and the path to actualize the integrated urban–rural development for the goal of rural revitalization. Under the leadership of the CPC, China has already completed its poverty-alleviation targets, which has laid solid foundations for its rural revitalization endeavor.

1 The Evolution of Urban–Rural Relations in China and the Connotations of Its Integrated Urban–Rural Development

Urban–rural relations refer to the interactive, mutually-influential, and mutually-restrictive connections that exist widely between urban and rural areas under certain social and economic conditions. As a fundamental part of human society, urban–rural relations have rich connotations, including economic relations, social relations, class relations, ecological and environmental relations, cultural relations, and location relations.

J. Sun, *Regional Economic Development in China in 2020*,
https://doi.org/10.1007/978-981-95-3604-7_6

1.1 The Evolution of Urban–Rural Relations in China

At present, China's urban–rural relations are in a critical period of evolution which features the rapid transformation of its rural areas. A review of the urban–rural relations' evolution process facilitates a better understanding of the transformation of China's policy ideas and a better implementation of the rural revitalization. It is also of great significance for our endeavor to narrow our urban–rural gap and optimize our urban–rural relations.

1. **The Formation and Solidification of the Urban–Rural Dualism (1949–1977).** Urban–rural relations in this period can be summarized by the following two sentences: Agriculture nurtures industry; Rural areas support cities. In the early years of the PRC, China regarded the development of heavy industries as its top priority, striving to quickly narrow the gap between itself and developed countries and trying everything to rapidly turn itself into a major industrial country. To ensure the fast development of heavy industries in the urban areas, China adopted a series of policies and constantly enhanced them, which resulted in the formation of urban–rural dualism. A set of planning and management measures were popularized in the rural areas, taking the agricultural surplus value to support the development of heavy industries. Such measure actualized price scissors difference between industrial products and agricultural products, the centralized purchase and sales of major agricultural products, and the agricultural collective management, which were core economic foundations of the urban–rural dualism. Under such circumstances, the agricultural population whose economic benefits were harmed naturally wanted to flow to the urban areas. Nevertheless, the heavy industries could not provide so many jobs back then. To ensure the economic dualism, China built a corresponding social management system named the household-registration system to prevent too many rural people from flowing into the urban areas and ensure adequate labor supplies labor for agricultural production activities. Regulations of the People's Republic of China on Household Registration, promulgated in 1958, specified two types of registration: urban household registration and rural household registration. It also stipulated that rural residents could not move into urban areas unless permitted by the Chinese government. Later, a number of social security systems were built based on the household-registration system, enhancing the urban–rural dualism.
2. **The Urban–Rural Interactive Economic and Social Development (1978–2001).** This period saw initial breakthroughs in changing the urban–rural dualism. Nonetheless, the urban–rural gaps widened after shrinking for a period of time.

   First, the household contract responsibility system with income linked to output changed the way how people's communes organized agricultural production activities and weakened the government's control over rural economic activities. It allowed farming households to manage agricultural production on their own initiatives, effectively improving the internal incentive mechanism for agricultural production. In this period, the Chinese government liberalized the grain market and raised the purchase prices of agricultural products, providing external incentives for agricultural production. For the above-mentioned reasons, the output of agricultural products surged, and the rural people's income increased rapidly, which significantly decreased the urban–rural income gap. Based on the agricultural reform and development trends, the Chinese government promoted the reform of the agricultural product market system, abolished the system of unified purchase and sales, and step-by-step liberated agricultural product market.

Second, township enterprises thrived when the surplus labor force in rural areas gradually increased with the implementation of the household contract responsibility system and the reform of the agricultural product market and when the surplus labor force was not allowed to freely flow to the urban areas. The rural people were encouraged to engage in non-agriculture work in factories in their own hometowns instead of going to urban areas for opportunities. In 1988, the urban–rural income gap dropped to the lowest point after China adopted the policy of reform and opening-up.

Third, in the 1990s when the reform of urban state-owned enterprises was initiated and when the goal of establishing a socialist market economy was put forward, restrictions on the flow of labor between urban and rural areas were gradually relaxed. This period saw stronger urban–rural connections, but they usually featured one-way resource flows from rural to urban areas. It also witnessed a downward trend in the government's investment in rural areas, which led to the widening of urban–rural gaps in terms of education, employment, medical and health care, social security, culture, and entertainment. In this period, the urban–rural dualism still widely existed.

3. **The Coordinated Urban–Rural Development (2002–2016).** This period saw China's initial efforts to change the imbalanced urban–rural development. In 2002, the 16th National Congress of the CPC required to solve the urban–rural dualism through the coordinated urban–rural economic-and-social development. Among the five coordinated development goals proposed by the 3rd Plenary Session of the 16th CPC Central Committee, held in 2003, the coordinated urban–rural development was listed as the top priority. Its core was to solve the urban–rural development problems such as the widening urban–rural income gap, the imbalanced development between urban and rural areas, and the unequal access to public services for urban and rural residents. To implement this development plan, China began to build a policy system in 2004 regarding issues about agriculture, rural areas, and rural peoples. This system enabled the industrial development to boost the agricultural development and the urban development to promote the rural development.

   In 2005, the 5th Plenary Session of the 16th CPC Central Committee first proposed to build a new socialist countryside and promote the integration of urban and rural development. To coordinate urban and rural development and promote the building of a new socialist countryside, the CPC Central Committee called for the establishment of a new urban–rural integration paradigm and that of a long-term mechanism, which enabled the industrial development to promote the agricultural development and the urban development to boost the rural development, in its report to the 17th National Congress of the CPC in 2007. The

18th National Congress of the CPC in 2012 made it clear that the top priority of the Party's work was solving problems regarding agriculture, rural areas, and rural people, and that the fundamental solution to these problems was the integrated urban–rural development. The 3rd Plenary Session of the 18th Central Committee of the CPC further pointed out that the urban–rural dualism was the main obstacle restricting the integration of urban and rural development. It also urged the establishment of relevant systems and mechanisms: They should ensure the synergy between industrial and agricultural development and that between the urban the rural development; They should facilitate the formation of new-type relations between the urban and rural areas and those between workers and peasants, ensuring urban–rural equality in enjoying development opportunities and benefits brought by the modernization process.

4. **The Integrated Urban–Rural Development (2017–Present).** The CPC Central Committee's report to the 19th National Congress of the CPC includes the following statement regarding the pursuit of the rural vitalization strategy. "We must prioritize the development of agriculture and rural areas. To build rural areas with thriving businesses, pleasant living environments, social etiquette and civility, effective governance, and prosperity, we need to put in place sound systems, mechanisms, and policies for promoting integrated urban–rural development, and speed up the modernization of agriculture and rural areas." This is the first time that a Party document has clearly prioritize the development of agriculture and rural areas, which constitutes a turning point in China's urban–rural relations.

   From the coordinated urban–rural development strategy to the urban–rural integration or the integrated urban–rural development strategy, the central idea remains the same: Rural and urban areas have equal status, constitute an organic whole, and shall enjoy common prosperity. This period see historic changes take place in the urban–rural relations in China: The development of its urban and rural areas has entered a new stage.

1.2 Connotations of China's Integrated Urban–Rural Development

The urban–rural integration strategy regards the urban and rural areas as an organic whole and requires them to engage in barrier-free, in-depth interaction in terms of economy, social communication, and transportation, while retaining their characteristics. In the new era, this strategy has richer and more profound connotations.

First, the integrated urban–rural development entails the increasing factor mobility between urban and rural areas, and with the free flow of production factors, the market mechanism will play its role well. Since the beginning of the reform and opening-up, the cross-sectoral, cross-regional, and cross-industrial flows of the rural labor force have remained the most significant phenomena regarding the flow and allocation of urban and rural factors in China. Nonetheless, such rural labor flows are not effective enough because of the household-registration limitations and significant urban–rural differences in social securities. There is another reason for this one-way flow of people from rural to urban areas: There are still limitations in the flow of capital, technology, knowledge, information and other urban factors into rural sectors. Although China's current urban–rural factor mobility and redistribution efficiency have been significantly improved in comparison with the situation in the early years of the reform and opening up, the country still has a long way to go in this respect for its goal of building a modern socialist country. It is a must to eliminate institutional barriers that hinder the free flow and equal exchange of urban and rural factors and promote the flow of various factors to the rural areas, creating a virtuous circle of talents, lands, capitals, industries and information convergence in the rural area. It is essential to inject new kinetic energy into the rural revitalization campaign and create a new paradigm for the integrated development of urban and rural areas.

Second, the integrated urban–rural development involves the increase in urban and rural industrial diversity. From 1949 (the founding year of the PRC) to 1977, China's urban and rural industries had the following characteristics: The secondary and tertiary industries were concentrated in urban areas and the primary industry, in rural areas. This industrial distinction was changed by the rising township enterprises and the surging manufacturing industries in rural areas after China adopted the policy of reform and opening-up in 1978. With continuous increases in the scale and frequency of commodity transactions in rural areas, the tertiary industry began to thrive in rural areas during this period: The primary, secondary, and tertiary industries coexisted in rural areas. In the new era, when the principal contradiction is the contradiction between imbalanced and inadequate development and the people's ever-growing needs for a better life (including material, cultural, and ecological needs), there are growing demands for more diversified urban and rural industries. Therefore, it is necessary to diversify the functions of rural industries and promote the integration between urban and rural industries, which are important content of China's urban–rural integrated development.

Third, the urban–rural integrated development necessitates the increase in urban–rural spatial integration. Before 1978, due to the urban–rural industrial differences, i.e. the primary industry concentrated in rural areas and the secondary and tertiary industries concentrated in urban areas, the urban–rural boundaries were obvious. After 1978, with the advancement of urbanization and the rise of township enterprises, the boundaries between urban and rural areas began to become blurry: Some cities expanded to include their neighboring villages, turning them into special urban communities named urban villages (*chengzhongcun*); Some rural areas turned themselves into urbanized areas. In the new era, the integrated urban–rural development means that the degree of urban–rural spatial interlacing will further increase, and that their cooperation in various fields such as economy, culture, society, and ecological conservation will be enhanced. The urban–rural spatial boundaries will become more blurred. Some urban social and economic functions will be able to extend their influence to traditional rural areas. Rural complexes, featured towns, and modernized agricultural parks, which serve multiple purposes such as production, living, and ecological conservation, will become hubs connecting urban and rural areas. When urban public-transportation, communication, water-supply, gas-supply, and sewage-treatment networks expand to cover rural areas, a new pattern of urban–rural integrated development will take shape.

Fourth, the integrated urban–rural development encompasses urban–rural equality in social welfare. Before 1978, the urban–rural dualism created wide, long-existing income- and-consumption gaps between urban and rural residents. The different household registration types led to significant differences in the acquisition of public goods among the two groups of people. After 1978, when the factor allocation efficiency increased significantly under the system of the market economy, income and consumption increased year by year in urban and rural areas. In 2009, the urban–rural income-and-consumption gap began to narrow. After entering the twenty-first century, the New Rural Cooperative Medical System (NRCMS) and the New Rural Social Pension Insurance (NRSPI) program were implemented nationwide after pilot periods. This meant that China's rural residents began to receive formal social security benefits from the state, and the process of equalizing basic public services between urban and rural areas began. Despite the progress, it is necessary to see the still wide urban–rural social-welfare gap and grasp the ultimate goal of the integrated urban–rural development in the new era: deepening relevant reforms to allow urban and rural residents to equally enjoy China's development achievements and realizing the new development concept that highlights coordinated development and shared development.

1.3 The Current Status of and Problems in the Integrated Urban–Rural Development in China

With the enhancement of China's economic strength, it is time to let industrial development boost the agricultural development and the urban development promote the rural development. Given the state's increasing attention toward the urban–rural relations and the implementation of the coordinated and integrated urban–rural development strategies, constant improvements have been seen in the rural per-capita income, the rural basic public services, the urban–rural integration, and the urban–rural flow of production factors such as people and capitals. Nevertheless, it is essential to admit the wide existence of urban–rural dualism in China and the fact that China still has a long to go to realize its urban–rural integration goals.

1. **Narrowing Urban–Rural Income Gap but Still Large Absolute Value of the Gap.** The urban–rural income ratio is an important indicator to measure the urban–rural income gap and to some extent, it also indicates the degree of urban–rural integration. Figure 1 shows changes in the per-capita disposable income of urban and rural residents and those in the urban–rural income ratio after the beginning of the reform and opening-up. Given the efforts to restore rural vitality in the early years of the reform and opening-up, the period from 1978 to 1993 saw

rural income grow at a much faster rate in comparison with the urban income, bringing the urban–rural income ratio from 2.57 down to 1.82. This period boasts the smallest urban–rural income gap in the history of the PRC. After 1993, with the implementation of coastal development strategy, more preferential policies were given to the coastal areas, boosting the income increase of urban residents. Correspondingly, the urban–rural income ratio, which was 1.82 in 1993, jumped to 3.03 in 2002. After 2002, the state began attaching great significance to the optimization of urban–rural relations, successively formulating coordinated and integrated development strategies. After hitting a record high in 2007, which was 3.14, the urban–rural income ratio started to decline. The per-capita disposable income of rural residents began to grow faster than that of urban residents in 2013, taking the urban–rural income ratio down to 2.64 in 2019, which was only slightly higher than the ratio in the early years of the reform and opening-up. The decline of urban–rural income ratio is an encouraging fact, but the absolute value of the urban–rural income gap is still large.

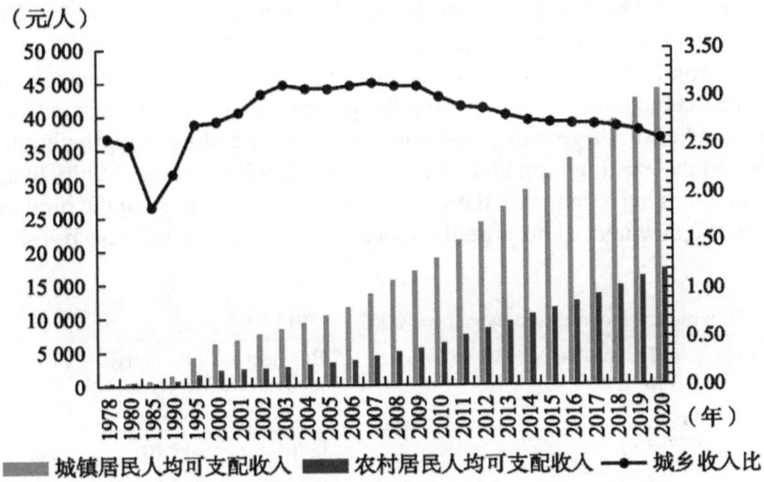

Fig. 1 Urban and rural per capita disposable incomes and urban–rural income ratios in China since adopting the policy of reform and opening-up. *Data source* The data in this Figure are extracted from "Statistical Bulletin on National Economic and Social Development of the People's Republic of China 2020" (*Note* The disposable income data from 2013 to 2019 come from surveys about household incomes and expenses and living conditions, and the those from 1978 to 2012 are calculated based on historical data offered by similar surveys. The disposable income data in this Figure are market prices at specific points of time, but the index is calculated at comparable prices)

城镇居民人均可支配收入	Per capita disposable income of urban residents
农村居民人均可支配收入	Per capita disposable income of rural residents
城乡收入比	Urban–rural income ratio
(年)	(year)

(continued)

(continued)

(元/人)	(yuan per person)

As shown in Table 1, which is about the income compositions for urban and rural residents, urban residents outshone rural residents in terms of wage income, property income, and transfer income. The urban–rural wage income ratios remained above 4 except for two years: 3.97 for 2018; 3.88 for 2019. The urban–rural property income ratios remained above 10, and the ratio for 2019 even hit 11.64. This shows that in comparison with urban residents, rural residents have less property investment and weaker ability to deal with risks. The urban–rural transfer income ratios remained between 2 and 3, and the ratio for 2019 was 2.29. As to the business income, this is the only category where rural residents did better than urban residents.

Figure 2 is a breakdown of rural residents' disposable income in 2019. The top two income sources for rural residents were the wage income and the business income. The former faces challenges brought by the new normal of Chinese economy (The economy may grow at the mid to high speed for some years, and the economic structure will undergo comprehensive and fundamental changes), the macroeconomic slowdown, and the partial replacement of manual labor by robots. The latter's growth is restricted by the price ceiling for agricultural products and the existing farm land system featuring collective ownership. In a word, it is challenging to maintain the years-long growth advantage of the rural income under the new normal because the space for rural income increase has shrunk.

Table 1 Urban–rural income ratio by category, 2013–2019

Urban–rural income ratio Year	Disposable income	Wage income	Business income	Property income	Transfer income
2013	2.81	4.55	0.76	13.10	2.62
2014	2.75	4.32	0.77	12.66	2.57
2015	2.73	4.20	0.77	12.10	2.58
2016	2.72	4.12	0.80	12.02	2.54
2017	2.71	4.04	0.81	11.90	2.51
2018	2.69	3.97	0.83	11.77	2.39
2019	2.64	3.88	0.84	11.64	2.29

Data source The data in this Table are extracted from China Statistical Yearbooks 2020.

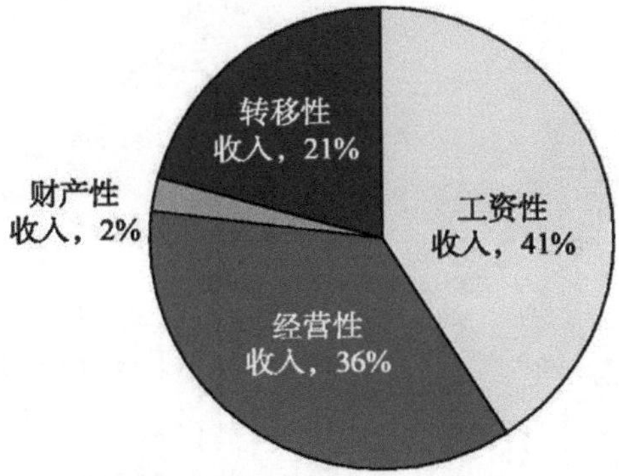

Fig. 2 A breakdown of rural residents' disposable income in 2019 (*Data source* The data in this Figure are extracted from China Statistical Yearbook 2020)

Transfer income (21%)	Wage income (41%)	Business income (36%)	Property income (2%)

2. **Increasing Urban–Rural Population Mobility and Lagging Urban Household Registration Growth.** With the reform of the household-registration system, transferred rural population can get their urban residence permits quicker than before. Given this, a increasing amount of rural population move to urban areas, significantly boosting the urbanization rate. As shown in Fig. 3, China's urban population was only 170 million in 1978, and from 1978 to 2011 its urbanization rate of permanent residence only increased from 17.92% to about 50%. Nevertheless, within the several years from 2011 to 2019, its urban population quickly expanded to 860 million and its permanent residence urbanization rate jumped to 60.6%.

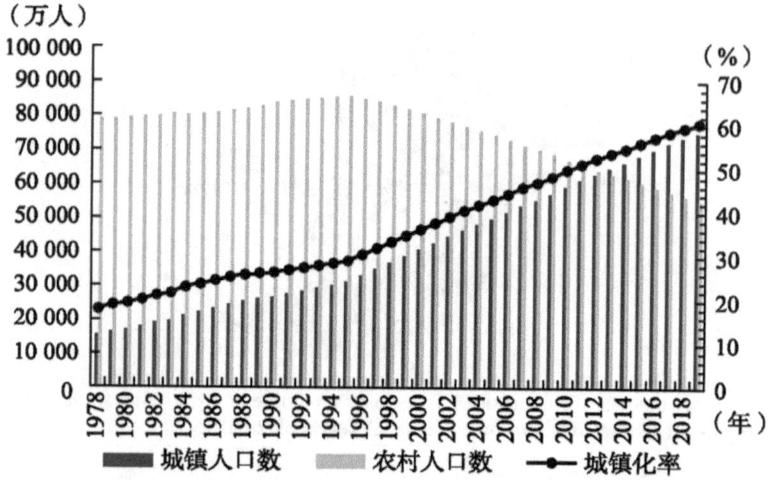

Fig. 3 Changes in the Chinese population since adopting the policy of reform and opening-up. *Data source* The data in this Figure are extracted from China Statistical Yearbooks 2020 (*Note* Data before 1981 are extracted from household registration statistics; the data of 1982, 1990, 2000 and 2010 are estimated based on corresponding national census data; the rest data are estimated according to corresponding sample surveys)

城镇人口数	Urban population
农村人口数	Rural population
城镇化率	Urbanization rate
(年)	(Year)
(万人)	(10,000 people)

It is worth noting that the urbanization rate of household-registered population in China did not increase as rapidly as its permanent residence urbanization rate did. In 2012, China's household-registered population urbanization rate was only 35.3% while the corresponding permanent residence urbanization rate hitting 52.6%: There was a difference of 17.3 percentage points. In 2019, the difference was narrowed to 16.22 percentage points, with the household-registered population urbanization rate and the corresponding permanent residence urbanization rate reaching 44.38% and 60.6%, respectively. Because basic public welfare systems such as those about education, medical care, and old-age care are based on the household-registration system, local governments are reluctant to push forward reform in the household-registration system. In addition, the absence of protection mechanisms for rural rights and interests also stops rural people from getting urban household registration since they are worried about losing their rights and interests after giving up their rural household registration. In the drive to promote the new-type urbanization, the top priority is to increase the urbanization rate of household-registered population, which underpins the establishment of the new paradigm for urban–rural integrated development.

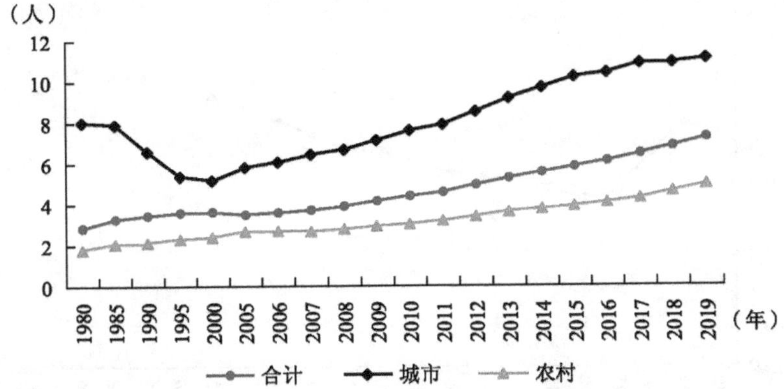

Fig. 4 Health-care technicians per 1,000 people (*Data source* The data in this Figure are extracted from China Statistical Yearbook 2020)

3. **Improved Basic Public Services in Rural Areas but a Still Big Urban–Rural Gap in Public Services.** Since the 18th National Congress of the CPC, held in 2012, China has pursued the equalization of basic public services between urban and rural areas, and the national public finance has increased its support for rural areas and the agricultural sector. Basic public-service networks regarding medical insurance, compulsory education, and basic old-age insurance have extended to cover all rural areas. In the medical sector, as shown in Fig. 4, 5, 6, and 7, rural areas have seen continuous improvement in medical and public health services but there is still a big urban–rural gap in this regard. Take the data of 2019 as an example: The urban number of health-care technicians per 1000 people was 2.24 times the rural number; The urban number of professional (assistant) physicians per 1,000 people was 2.09 times the rural number; The urban number of registered nurses per 1,000 people was 2.62 times times the rural number; The urban number of beds in medical and health institutions was 1.83 times times the rural number.

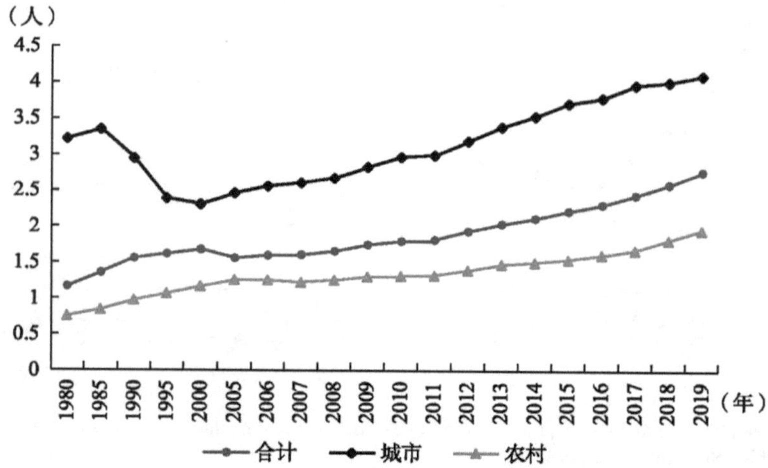

Fig. 5 Professional (assistant) physicians per 1,000 people (*Data source* The data in this Figure are extracted from China Statistical Yearbook 2020)

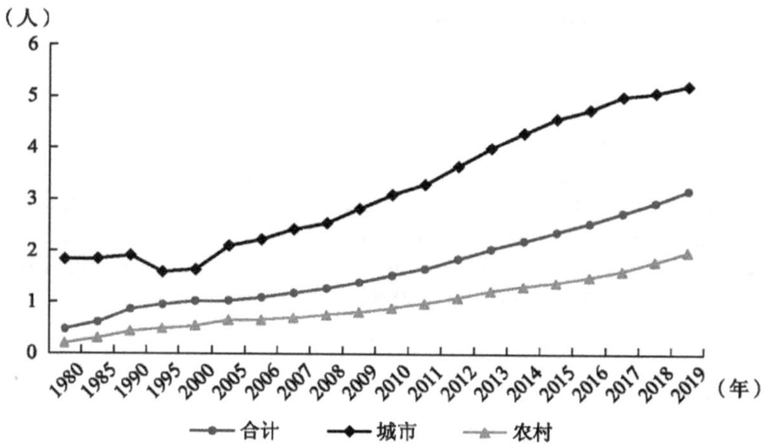

Fig. 6 Registered nurses per 1,000 people (*Data source* The data in this Figure are extracted from China Statistical Yearbook 2020)

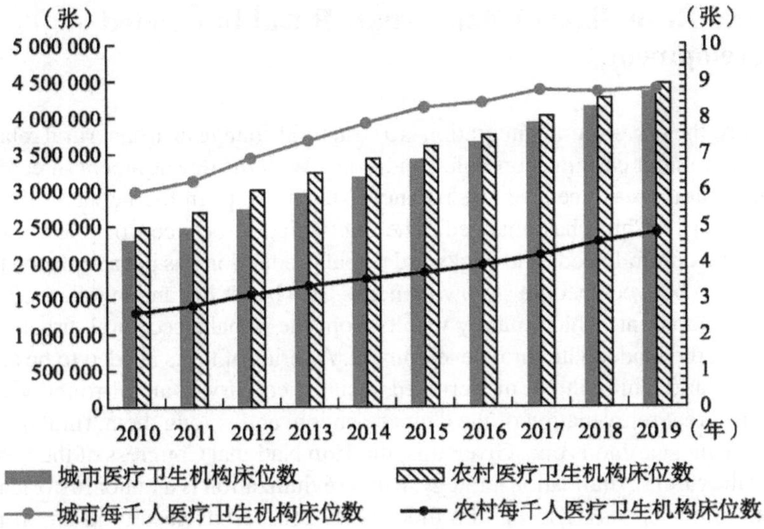

Fig. 7 Number of beds in medical and health institutions (*Data source* The data in this Figure are extracted from China Statistical Yearbook 2020)

城市	Urban numbers
农村	Rural numbers
合计	Total numbers
(人)	(person)
(年)	(year)
城市医疗卫生机构床位数	The number of beds in urban medical and health institutions
农村医疗卫生机构床位数	The number of beds in rural medical and health institutions
城市每千人医疗卫生机构床位数	The number of beds per 1,000 people in urban medical and health institutions
农村每千人医疗卫生机构床位数	The number of beds per 1,000 people in rural medical and health institutions
(张)	(beds)

Comparisons based on time sequences prove the trends of eliminating urban–rural dualism and realizing urban–rural integration, which have existed since the beginning of the reform and opening up. Nevertheless, spatial comparisons reveal the fact that the urban–rural dualism is still widely embodied in China 's modernization process, reminding us that China still has a long way to go to solve this dualism problem.

2 Rural Revitalization and Urban–Rural Integrated Development

Essentially, the urban–rural integration is a structural change in urban–rural relations, which occur under certain economic conditions. With the development of economy, urban and rural areas need a more balanced relationship. In the new era, the main contradiction in China has changed. The contradiction between people's growing material and cultural needs and backward social production has been replaced by the contradiction between people's growing needs for a better life and imbalanced, inadequate development, which mainly results from the imbalanced rural–urban development and the inadequate rural development. A series of tasks needed to be accomplished to build China into a modernized socialist country in an all-round way and realize the great rejuvenation of the Chinese nation, and among them, rural development is the most arduous one. Given this, the 19th National Congress of the CPC has adopted the rural revitalization strategy. Rural revitalization is a major historical task included in our endeavor to build a moderately prosperous society in an all-round way and a socialist modern country in all respects. It is also a central task for our work regarding agriculture, rural areas, and rural people in the new era.

2.1 The Content Framework of Rural Revitalization Strategy

The strategic thinking on rural revitalization entails a coherent body of ideas, which expound upon the development goals, general requirements, and key tasks of rural revitalization, and profoundly explain the purposes, leaders, and methods of the rural revitalization and other theoretical and practical issues.

1. **Development Goals.** Similar to the two-step strategic plans made for the building of a moderately prosperous society in all respects and the realization of China's second centenary goal, i.e. building a modern socialist country in all respects, the implementation plan formulated for the rural revitalization strategy also includes two stages. In the first stage, which last till 2020, remarkable progress should be made in rural revitalization: The institutional framework and policy system should be basically formed; The comprehensive agricultural production capacity should be steadily increased., including the improved quality of agricultural supply system and the integrated development of the primary, secondary, and tertiary industries in rural areas; The income sources for rural residents should be further increased to continuously narrow the urban–rural gap in living standards; All rural poor people and poverty-stricken counties, under the current standard, should be lifted out of poverty, eliminating regional poverty problems; Rural infrastructure construction should be promoted for a more livable and beautiful environment; Equalization of basic public services between urban and rural areas should be promoted, with the establishment of relevant systems and mechanisms of urban–rural integrated development; The rural areas' appeal to talents needs

to be gradually enhanced; The rural ecological environment should be obviously improved, increasing their agricultural and ecological functions; The rural grass-roots organizations with CPC organizations as the core should be strengthened, optimizing the rural governance system; The CPC's leadership system and mechanism for rural work should be improved; The ideas and measures for the implementation of the rural revitalization strategy should be initially established in all regions and departments across the country. The second stage will last till 2035. It will witness decisive progress in rural revitalization and the actualization of the agricultural and rural modernization. The agricultural structure shall be fundamentally optimized with improvements made in the rural employment quality and the alleviation of relative poverty, which means a solid step towards the goal of common prosperity; Equalization of basic public services between urban and rural areas shall be basically realized, with progress in the urban–rural integrated development systems and mechanisms; Etiquette and civility will be highlighted in the rural areas, and the rural governance system will be more optimized; The rural ecological environment shall be fundamentally improved, and beautiful and livable villages shall be basically actualized. By 2050, rural vitalization will be achieved. Rural areas should have robust agriculture, beautiful landscapes, and prosperous farmers.

2. **General Requirements.** When delivering its report to the 19th National Congress of the CPC, the CPC Central Committee put forward its general requirements for rural revitalization: "To build rural areas with thriving businesses, pleasant living environments, social etiquette and civility, effective governance, and prosperity."
3. **Key Tasks.** The rural revitalization strategy encompasses ideas about how to revitalize the rural areas. As early as 2017, the annual central rural work conference expounded upon seven methods for the socialist rural revitalization with Chinese characteristics. First, it is a must to reshape our urban–rural relations and pursue integrated urban–rural development. Second, it is essential to consolidate and improve our basic rural management system and strive for common prosperity. Third, it is necessary to deepen the supply-side structural reform in the agricultural sector, promoting high-quality agricultural development. Fourth, it is important to adhere to principles such as the harmonious coexistence between man and nature, and the green development. Fifth, rural civilization needs to be inherited and revamped, and rural culture needs to be promoted. Sixth, the rural governance system shall be optimized for good rural governance. Seventh, targeted-poverty-alleviation tasks must be completed, and poverty-reduction path with Chinese characteristics shall be upheld. On March 8, 2018, at the deliberation session of the Shandong delegation to the 1st Session of the 13th National People's Congress, Xi Jinping put forward five requirements for rural revitalization, requiring comprehensive arrangements for and scientific implementation of the rural revitalization strategy and the all-round revitalization in terms of industries, talents, culture, ecological conservation, and grass-roots organizations. The above facts prove that the strategic thinking of rural revitalization is constantly developed and enriched as the time goes by.

2.2 Rural Revitalization: The Path of Integrated Urban–Rural Development

The integrated urban–rural development and the rural revitalization are two endeavors that complement each other: The former is a development goal and method for the latter; The latter is fundamentally an attempt to reshape the urban–rural relations.

At the Central Rural Work Conference held from December 28 to 29, 2020, General Secretary Xi Jinping emphasized the necessity of fully implementing the rural revitalization strategy after completing the targeted-poverty-alleviation goals. This marked a historic shift in the focus of work on agriculture, rural areas, and rural residents. In terms of depth, extensiveness, and difficulties, the rural revitalization endeavor is no simpler than the poverty alleviation campaign: Strong top-level design is needed for the adoption of effective measures and the pooling of resources. First, the development of rural industries need to be sped up based on the industrial development laws and the local resource advantages, and the industrial layout and interest linkage mechanism should be improved to let rural residents benefit more from their local industrial development. Second, the socialist spiritual civilization shall be promoted in the rural areas for the improvement of rural netiquettes and civility, the implementation of socialist core values, and the popularization of scientific knowledge. Third, the ecological civilization principle shall be highlighted and committed efforts shall be exerted to restore the ecology and solve relevant environmental issues such as the agricultural non-point source pollution, soil pollution, groundwater over-exploitation, soil erosion, etc. Fourth, rural reforms, especially those in key areas and key links, need to be deepened for the vitality of rural resource factors and the complete agricultural support and protection system, and the creativity of the grassroots level and the masses shall be respected for new breakthroughs in the reforms. Fifth, constant efforts shall be made and proper guidance shall be given for the equalization of basic public services between urban and rural areas, realizing goals such as accelerated rural public infrastructure construction, stronger guarantees for basic livelihoods in the rural areas, better living environment, cleaner toilets, improved sewage and garbage disposal systems, and optimized village layouts with traditionally rural characteristics. Sixth, solid work shall be done to promote the integrated urban–rural development, the enhancement of relevant development systems and mechanisms, and the urbanization of transferred rural population: The work focus shall be placed on the strengthening of the comprehensive service capacities of counties, giving counties more autonomy in pooling and utilizing resources. Seventh, efforts shall be exerted to improve rural governance, including establishing the CPC-led rural governance systems, promoting the campaign of building safe villages, and innovating rural governance methods.

It can be seen from the above that one of the practical methods to realize rural revitalization is to establish and improve relevant institutional mechanisms and policy systems. Firstly, these mechanisms and systems remove barriers hindering the free flow and equal exchange of urban and rural factors and encourage various factors to flow to rural areas and form a virtuous circle of talents, lands, capital, industries, and information aggregation collection there, injecting new kinetic energy into the rural revitalization campaign. Some development mechanisms need special attention: the mechanism that urbanizes transferred rural population; the mechanism that encourages urban talents to develop in rural areas; the rural land contract system; the rural homestead system; the system that allows collective construction land for commercial use to enter the market; the financial input guarantee mechanism; the rural financial service system; the mechanism that introduces industrial and commercial capitals into rural areas for development purposes; the mechanism for the transformation of scientific and technological achievements in rural areas. Secondly, these institutions and mechanisms are conducive to the equalization of basic public services between urban and rural areas and the extension of public-service systems and social undertakings to rural areas and the coverage of social undertakings in rural areas. This helps to establish a basic public-service system that covers all the people and implements unified standards in urban and rural areas. The public services and benefits included in this unified system are as follows: educational services, medical and health services, cultural services, social insurance benefits, social assistance, and social governance. Thirdly, these mechanisms and systems facilitate the integrated development of urban and rural infrastructure. The focus of public infrastructure construction work shall be put in the rural areas, and the principle of mechanisms established before the construction projects shall be adhered to. The upgrading of the rural infrastructure shall be accelerated: This underpins the actualization of the unified planning, construction, management, and maintenance of urban and rural infrastructure and the establishment of relevant planning, construction, management, and maintenance mechanisms. Fourthly, these mechanisms and systems promote the diversified development of the rural economy and the whole-industrial-chain-development of agriculture. This calls for the enhancement of specific mechanisms regarding the following aspects: agricultural support and protection, new-industry and new-business development, ecological-product value realization, rural-culture protection and utilization, coordinated development of urban and rural industries of rural culture, and the overall planning of urban and rural areas. Fifthly, these mechanisms and systems ensure the sustained income growth of rural residents and the increase in their income sources, continuously narrowing the urban–rural gap in living standards To be more specific, they create an environment that ensures the continuous wage income increase and business income increase for rural people; they give rural people opportunities to enjoy property income increases and transfer income.

2.3 Breakthrough Points in Integrated Urban–Rural Development: Counties

Counties become the main implementers of the rural revitalization strategy, which is determined by their positions and functions in the state administration system.

From the perspective of urban–rural relations, counties are administrative units connecting urban and rural areas in China: They are often an urban–rural combination, boasting spatial characteristics of urban–rural integration. Essentially, rural revitalization is the reshaping of urban–rural relations in the new era: The "reshaping" here refers to the integrated development. From the perspective of China's urbanization practice after the beginning of reform and opening-up, China's big cities have developed far faster than its rural areas have. Big cities in China are not much different from their counterparts in developed countries, and some of them are even better than some famous international cities in multiple respects. Nevertheless, its rural development has relatively lagged behind, which calls for the establishment of a new-type urban–rural integrated development mechanism and the common progress of urban and rural areas.

From the perspective of spatial distribution and administrative roles, only counties can combine urban and rural functions in China. As to the spatial distribution, counties in China are often connected to their neighboring villages through towns (townships) that have once served as market places within walking distances for villagers. With the changes of the times, the size of towns (townships) changes, but such changes are only within a certain spatial range. That is, towns (townships) are not real urban communities. Regarding administrative roles, towns (townships) can provide only basic administrative and living services for surrounding villages, without the administrative, economic, political, cultural, and social functions needed for a city. Therefore, the implementation of the rural revitalization strategy has to rely on counties' functions of integrating urban–rural development space and promoting the joint development of urban and rural areas.

Compared with towns (townships), counties have information and efficiency advantages when making overall development arrangements for the all-round rural revitalization in terms of industries, talents, culture, ecology, and grassroots-level organizations.

References

Gao, F. (2019). The inherent meaning and poucy implication of China's new stage of urban-rural integration development. *Journal of Guangxi University of Finance and Economics*, 32(01), 1–12 & 35.

Jin, S., Cao, D., & Lin, X. (2019). Cong chengxiang eryuan dao chengxiang ronghe: Xin zhongguo chengli 70 nian lai chengxiang guangxi de yanjin ji qishi (From urban-rural dualism to urban-rural integration: The evolution of urban-rural relations in the 70 years following the founding of the PRC and its relevant enlightenment). *Economic Review Journal, 2019*(08), 13–19.

Li, A. (2019). The process, problems and paths of urban-rural integration development in China. *Macroeconomic Management, 2019*(02), 35–42.

Nian, M. (2020). The evolution of China's urban-rural relations, fusion obstacles and supporting policies. *Economist, 2020*(08), 70–79.

Wang, L. (2020). Taking county as unit to make overall advancement: The methodology of the rural vitalization strategy. *Journal of China Executive Leadership Academy Pudong, 14*(04), 119–126.

Yao, Y., & Liang, M. (2020). *Xin zhongguo chengli yilai de chengxiang guanxi: Licheng, luoji yu zhanwang* (Urban-rural relations since the founding of the PRC: History, logic and prospect). *Jilin University Journal (Social Sciences Edition), 60*(01), 120–129 & 222.

Zhang, H., Gao, L., & Yan, K. (2018). Strategic thinking on rural revitalization strategy: Theoretical origin, main innovation and realization path. *Chinese Rural Economy, 2018*(11), 2–16.

Zhang, H. (2019). The evolution of China's urban-rural relations in the past seven decades: From separation to integration. *Chinese Rural Economy, 2019*(03), 2–18.

Zhang, Y., Long, H., Ma, L., Tu, S., & Chen, K. (2019). Research progress of urban-rural relations and its implications for rural revitalization. *Geographical Research, 38*(03), 578–594.

Part III
New Paradigm for China's Regional Economic Development

Chapter 7
China's "Dual-Circulation" Paradigm: Its Background and Philosophy of Development

In May 2020, the Standing Committee of the Politburo of the CPC Central Committee first put forward the following requirement: "It is essential to deepen the supply-side structural reform and fully leverage the strengths of China's enormous market and domestic demand potential, establishing a new "dual-circulation" development paradigm with domestic and international circulations complementing each other." The 5th Plenary Session of the 19th Central Committee of the CPC incorporated the idea "accelerating the establishment of a new 'dual-circulation' development paradigm with domestic circulation as the mainstay and domestic and international circulations promoting each other" into the "Proposals of the CPC Central Committee on the Formulation of the 14th Five-Year Plan (2021–2025) for National Economic and Social Development and the Long-Range Objectives Through the Year 2035 (Vision 2035)." These facts suggest that the "dual-circulation" paradigm is a major strategic arrangement that the CPC Central Committee has made to help the nation adapt to significant changes in internal and domestic environments and upgrade its opening-up endeavor. This chapter focuses on the development logic of the "dual-circulation" paradigm.

1 Background for the Changing Development Paradigm

1.1 The International Background

1. **The COVID-19 Pandemic's Impact on Important Economies of the World.** Given the COVID-19 pandemic's impact on the global economy, 2020 was a very special year. That year, public health measures led to economic shutdown in various countries, bringing their quarterly GDP performance to all-time low points. According to the "Statistical Communique of the People's Republic of China on the National Economic and Social Development 2020," which was

J. Sun, *Regional Economic Development in China in 2020*,
https://doi.org/10.1007/978-981-95-3604-7_7

released by the official website of the NBS of the PRC on February 28, 2021, China was the only major economy in the world to achieve positive economic growth in 2020. At the end of 2020, World Economic Outlook update from the International Monetary Fund (IMF) projected a 4.4% decrease in world economy, which was nearly 7 times the economic decline caused by the 2008 Global Financial Crisis and the worst recession after the Great Depression of the 1930s. Major economies of the world were all deeply troubled by the pandemic, but their performances in economic recovery varied. This had a far-reaching impact on the evolution of the world economic structure.

(1) ***The Overall-shrinking European and U.S. Economies.*** Despite their advantages in medical technology and medical systems, the United States and European countries took inappropriate prevention and control measures during the pandemic and thus experienced sharp drops in their economies in each quarter of 2020, as shown in Table 1. The United States achieves positive growth in the first quarter but the forecast values for the following three quarters are all negative. It is worth noting that although its unprecedented economic stimulus package consisting of monetary policy and fiscal rescue measures fails to control the pandemic, they do make the country's economic data look slightly better than those of major European countries. Major European countries have negative growth in the first three quarters of 2020: The cliff-like declines, i.e. double-digit decline rates, occurred in the second quarter, with the highest one reaching −21.6%. The forecast values for these countries' economic growth in the fourth quarter are negative. These data show that the European and the U.S. economies experience overall contraction in 2020.

(2) **The Gradually Recovering Major ASEAN Countries.** As shown in Table 2, Vietnam outshines the other ASEAN countries: It is the only major ASEAN country that maintains positive growth in the four quarters of 2020.

   Malaysia, Thailand, and the Philippines see double-digit decline rates in the second quarter: The largest decline, which is −17.1% occurs in Malaysia. Nevertheless, in the third quarter, they realize economic recoveries and this trend continues into the fourth quarter. Although they have not yet turned their negative growth rates into positive ones in the fourth quarter, their overall economic recovery and growth are significant.

   In comparison with the United States and the European countries, the major ASEAN countries are export-oriented economies that are prone to international vicissitudes. Their economic fluctuations are usually synchronized with those of their major trading partners. They have also been adversely-affected by the pandemic, but with strong endogenous economic growth momentum, they recover faster than the United States and the European countries do.

(3) **The Dividing BRICS Countries.** As shown in Table 3, there are great economic-performance differences among the BRICS countries in 2020. China adopts strict measures for epidemic prevention and control in the

Table 1 Year-over-year GDP growth rates of the United States and Europe 2019–2021

Country (Region) Period	The United States	The United Kingdoms	The Eurozone	France	Germany	Italy	Spain
2019	2.2%	1.3%	1.3%	1.5%	0.6%	0.3%	2.0%
2020 (forecast data)	(3.7%)	(9.8%)	(7.4%)	(10.7%)	(5.3%)	(8.9%)	(10.8%)
Q1 of 2020	0.3%	(2.3%)	(3.0%)	(5.6%)	(2.2%)	(5.8%)	(4.2%)
Q2 of 2020	(9.0%)	(21.0%)	(14.9%)	(18.6%)	(11.3%)	(18.2%)	(21.6%)
Q3 of 2020	(2.9%)	(8.7%)	(4.4%)	(3.7%)	(4.0%)	(5.2%)	(9.0%)
Q4 of 2020	(2.4%)	(7.8%)	(7.3%)	(4.9%)	(3.6%)	(6.6%)	(9.1%)
2021	3.2%	5.9%	3.6%	6.0%	2.4%	4.0%	4.5%

Data source The data included in this Table are from official statistics of the involved countries and region and forecasts made by authoritative institutions such as the IMF; data in parentheses are negative numbers.

Table 2 Year-over-year GDP growth rates of major ASEAN countries 2019–2021

Country period	Malaysia	Indonesia	Thailand	Philippines	Vietnam
2019	4.3%	5.0%	2.4%	6.0%	7.0%
2020 (forecast data)	(4.4%)	(0.6%)	(4.9%)	(6.4%)	2.8%
Q1 of 2020	0.7%	3.0%	(2.1%)	(0.7%)	3.8%
Q2 of 2020	(17.1%)	(5.3%)	(12.1%)	(16.9%)	1.8%
Q3 of 2020	(2.6%)	(3.5%)	(6.4%)	(11.4%)	(2.1%)
Q4 of 2020	(3.5%)	(2.2%)	(4.2%)	(8.3%)	(2.9%)
2021 (forecast data)	6.5%	8.0%	5.5%	7.5%	6.5%

Data source The data included in this Table are from official statistics of the involved countries and forecasts made by authoritative institutions such as the IMF; data in parentheses are negative numbers.

first quarter that sees the outbreak of the epidemic, resulting in economic shutdown in some areas and and an economic downturn. Nevertheless, its success in controlling the epidemic underpins its strong economic recovery that begins in the second quarter. Its economy is expected to grow by more than 6% in the fourth quarter. Among the world's major economies, China is the only one that is projected by the IMF to achieve positive growth in 2020. India's insufficient measures for epidemic prevention and control failed to stop the widespread spread of the virus: Its cumulative confirmed COVID-19 cases exceed 10 million, and economic activities are widely suspended.

As a major energy exporter, Russia experience an economic downturn when the COVID-19 pandemic brings the cause energy prices down and hinders its energy export trade.

In Brazil, the service sector contributes around 65% to its GDP, which is the highest among the BRICS countries. Its business activities are widely impacted by the quarantine and isolation measures taken due to the pandemic, and thus its GDP drops by 10.9% year-on-year in the second quarter. In this period, it is ranked fourth by the Organisation for Economic Co-operation and Development (OECD) in the investment decline list for the 31 developed and emerging economies of the world. This ranking can be proved by the data released by the Brazilian Institute of Geography and Statistics (IBGE), which show that Brazil's investment rate decreased by 15.4% year-on-year in the second quarter. The IMF predicts that Brazil will have the biggest economic recession in nearly 120 years in 2020 in its World Economic Outlook. Deeply troubled by high indebtedness and high unemployment rates, South Africa's economy has been weak for years. The impact of the pandemic further worsens its economy. According to the data released by Statistics South Africa (Stats SA), the country's GDP has shrunk continuously in the third and fourth quarters of 2019, with an annual growth rate of only 0.2%; it drop by 0.5% month-on-month in the first quarter of 2020 and 16.4% in the second quarter. Its decline in the second quarter is second only to that of India in the world. Although its GDP increases by 13.5% in the third quarter, its unemployment rate reaches all-time high at the end of the quarter: 30.8%. Its economy is estimated to shrink by 6.7% in 2020 and grow by only 3.2% in 2021: The projected growth rate in 2021 is not even half of the estimated decline rate in 2020.

2. **The Global Economic Growth Sliding Into Long-Term Stagnation.** Figure 1 shows the overall situation of global economic growth in recent 15 years. The economic-growth rate in 2019 is around 2.5%, lower than those of 2017 and

Table 3 Year-over-year GDP growth rates of the BRICS countries 2019–2021

Country Period	India	Russia	Brazil	South Africa	China
2019	4.7%	1.3%	1.4%	0.2%	6.1%
2020 (forecast data)	(8.1%)	(3.7%)	(5.1%)	(6.7%)	1.9%
Q1 of 2020	3.1%	1.6%	(0.3%)	0.1%	(6.8%)
Q2 of 2020	(24.4%)	(8.0%)	(10.9%)	(17.5%)	3.2%
Q3 of 2020	(7.4%)	(3.6%)	(3.9%)	(6.0%)	4.9%
Q4 of 2020	0.4%	(4.8%)	(1.1)%	(3.2%)	6.5%
2021 (forecast data)	5.5%	3.0%	(3.2%)	(3.2%)	8.5%

Data source The data included in this Table are from official statistics of the involved countries and forecasts made by authoritative institutions such as the IMF; data in parentheses are negative numbers.

2018, and the global economic-growth rate has been declining since 2011. The an average growth rate of recent five years is only 2.88%, which is far lower than that of the pre-crisis (the 2008 Financial Crisis) period from 2002 to 2007: 4.8%. It seems that the 2008 Global Financial Crisis has significantly influenced the global development pattern, and the COVID-19 pandemic is projected by the IMF to turn the global economic-growth rate into a negative number in 2020. All this suggests that the global economy will slide into long-term stagnation.

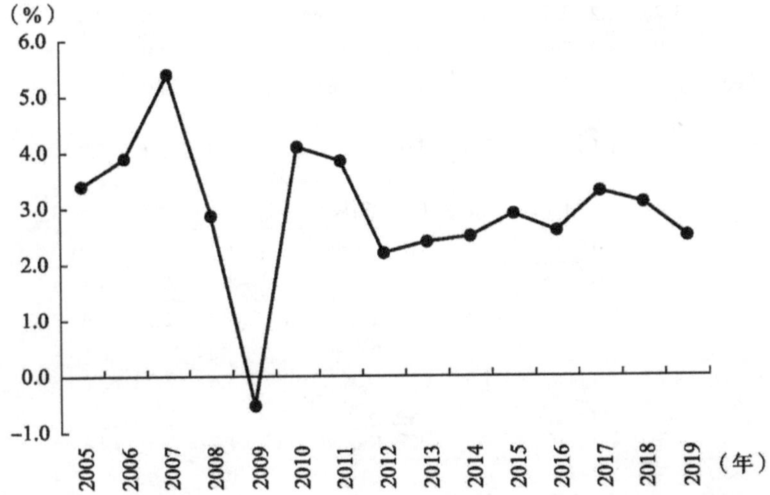

Fig. 1 Global economic growth rate 2005–2019 (*Data source* the World Development Indicators (WDI) of the World Bank)

(年)	(year)
(%)	(%)

Major economies of the world show signs of slowing down. From Table 4, it can be seen that the growth rates of these major economies have been declining over the past five years. Except for a short-term rise (3.2%) in 2018, the growth rate of the U.S. economy has dropped from 2.9% in 2015 to 2.3% in 2019. The growth rate of the Chinese economy, which is one of the fastest-growing economies of the world, has been decreasing over the last five years, from 8% in 2015 to 5% in 2019. Although the economy is still in an transition period, it remains a main driving force of the global economic growth. The growth rate of the Indian economy, which is also an Asian economy, is in the top echelon. The growth rates of the traditional capitalist powers, such as Japan, Germany, Britain, France, and Italy are below 2%, and some of them, even much lower than 1%. That is,they are basically in a state of economic stagnation. From 2015 to 2019, a total of 7 countries experienced negative economic growth, but in 2020, all countries are projected to experience it except China.

Table 4 GDP growth rate of major countries in the world 2015–2019

Period country	GDP Growth Rate (%)					Period country	GDP Growth Rate (%)				
	2015	2016	2017	2018	2019		2015	2016	2017	2018	2019
The United States	2.9	1.6	2.2	3.2	2.3	Russia	−2.0	0.2	1.8	2.5	1.3
China	7.0	6.8	6.9	6.8	6.1	South Korea	2.8	2.9	3.2	2.7	2.0
Japan	1.2	0.5	2.2	0.3	0.7	Spain	3.8	3.0	2.9	2.4	2.0
Germany	1.7	2.2	2.5	1.5	0.6	Australia	2.2	2.8	2.4	2.9	1.9
India	8.0	8.3	7.0	6.1	5.0	Mexico	3.3	2.9	2.1	2.1	−0.1
The United Kingdoms	2.4	1.9	1.9	1.3	1.4	Indonesia	4.9	5.0	5.1	5.2	5.0
France	1.1	1.1	2.3	1.8	1.5	The Netherlands	2.0	2.2	2.9	2.6	1.8
Italy	0.8	1.3	1.7	0.8	0.3	Turkey	6.1	3.2	7.5	2.8	0.9
Brazil	−3.5	−3.3	1.3	1.3	1.1	Poland	3.8	3.1	4.9	5.3	4.1
Canada	0.7	1.0	3.2	2.0	1.7	Thailand	3.1	3.4	4.1	4.2	2.4
Argentina	2.7	−2.1	2.7	−2.5	−2.2	The Czech Republic	5.3	−2.5	4.4	2.8	2.6
Nigeria	2.7	−1.6	0.8	1.9	2.2	New Zealand	3.6	3.7	3.2	3.8	2.2
Israel	2.3	4.0	3.5	3.5	3.5	Kazakhstan	1.2	1.1	4.1	4.1	4.5
Philippines	6.3	7.1	6.9	6.3	6.0	Ukraine	−9.8	2.2	−2.5	3.4	3.2
Singapore	3.0	3.2	4.3	3.4	0.7	Sri Lanka	5.0	4.5	3.6	3.3	2.3
Malaysia	5.1	4.5	5.7	4.7	4.3	Myanmar	7.0	5.8	6.4	6.8	2.9
South Africa	1.2	0.4	1.4	0.8	0.2	Cambodia	7.0	7.0	6.8	7.5	7.1
Egypt	4.4	4.3	4.2	5.3	5.6	Laos	7.3	7.0	6.9	6.2	4.7
Bangladesh	6.6	7.1	7.3	7.9	8.2	Mongolia	2.4	1.2	5.3	7.2	5.1
Pakistan	4.7	5.5	5.6	5.8	1.0	Brunei	−0.4	−2.5	1.3	0.1	3.9
Vietnam	6.7	6.2	6.8	7.1	7.0						

Data source the WDI of the World Bank.

We collate data about the top 20 countries by economic aggregate in 2019 to find how many of them have grown faster than the global average in recent 15 years. Please see Fig. 2 for the statistical results. Statistics show that 10 countries' economic-growth rates exceed the global average in 2016 and that the numbers for the other years are below 10. In recent years, the downward trend in this number is obvious and the lowest record is 4 for the year 2019. This suggests that these top economies are troubled by weak economic performance and inadequate momentum, which reduces their capacities for driving the global economic growth.

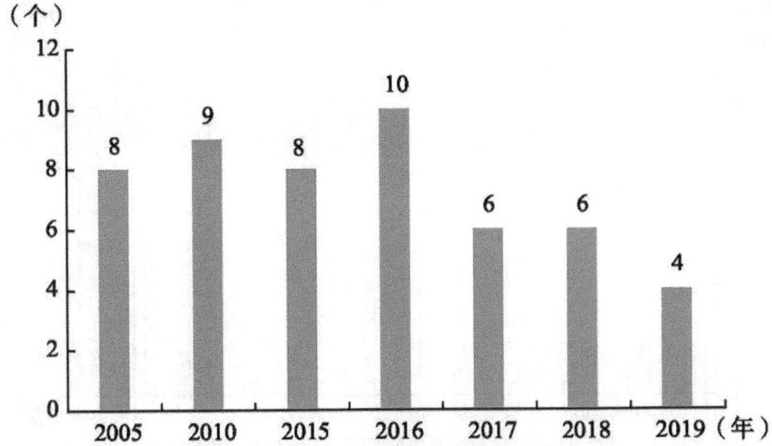

Fig. 2 The number of countries with growth rates exceeding the global economic growth rate among the top 20 countries by economic aggregate (*Data source* the WDI of the World Bank)

(年)	(year)
(个)	(The number of countries)

2 The Prevailing Anti-Globalization Movement

(1) ***The Rise in Trade Protectionism.*** In recent years, trade protectionism has prevailed all over the world, and the most representative event is the Sino-US trade dispute. On March 8, 2018, the U.S. President Donald Trump announced 25% and 10% tariff increases respectively for imported Chinese steel and aluminum products, initiating the Sino-US trade war. After that, the United States imposed tariffs on imported Chinese goods worth $200 billion and then on imported Chinese goods worth $300 billion. In total, the United States imposed tariffs on Chinese goods worth a total of $550 billion. (The total value of US imports from China was $506 billion in 2017), covering all US imports from China. This was a trade war of an unprecedented scale. Figure 3 shows the number of trade-protection measures introduced globally from 2009 to 2020: The number of adopted trade-protection measures has kept rising since the 2008 Financial Crisis, and it has increased sharply after 2018. As of December 2020, there are 1,862 newly-adopted trade-protection measures, which are 50% more than the annual number of 2019.

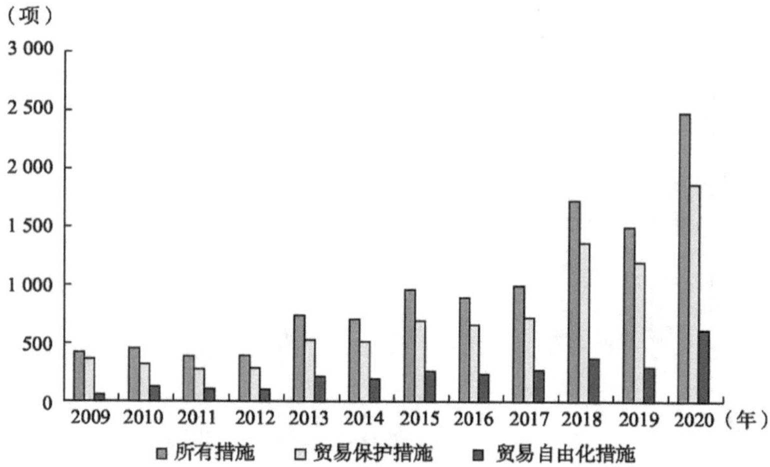

Fig. 3 The number of trade-protection measures introduced globally 2009–2020 (*Data source* Global Trade Alert)

所有措施	All types of measures
贸易保护措施	Trade protection measures
贸易自由化措施	Trade liberalization measures
(项)	(The number of measures)
(年)	(year)

(2) ***Anti-Globalization Sentiments Enhanced by the COVID-19 Pandemic.*** The COVID-19 pandemic has brought serious difficulties to the entire world. Countries have generally adopt self-blockade measures to prevent and control the virus, which causes a sharp drop in global trade volume and interruptions in the global industrial chains. Under such circumstances, they begin to reflect on the advantages and disadvantages of the globalized industrial chains. The globalized industrial-chain system maximizes global production efficiency and productivity, but under this system, most countries have incomplete industrial chains. So far, only China has established a complete industrial system, and other countries just have special strengths in some aspects of the system. This structure can maximize efficiency and profits when the world is peaceful and prosperous, but its disadvantages will be magnified in the case of a global pandemic, war, or economic recession. The COVID-19 pandemic remind some countries of this point, increasing their worries for the disadvantages. Therefore, the anti-globalization movement may enjoy a strong short-term support across the globe.

(3) ***The Shrinking Global Market.*** Given the global economic downturn, the COVID-19 pandemic, and the surging trend of anti-globalization, the global trade market has shrunk dramatically. At the beginning of the fourth quarter of

2020, the "Global Trade Outlook and Statistics" released by the World Trade Organization (WTO) has estimated that the global trade volume of goods will shrink by 9.2% in 2020. The WTO has made it clear that it is highly uncertain of this estimated data since the performance of global trade depends on the development of the pandemic and the measures adopted by the governments for the prevention and control of the virus: When the previously-suppressed market demands are released and the enterprises' inventories are replenished, the recovery and growth momentum of trade will be weakened; Provided that the COVID-19 infections rebound in the fourth quarter, and trade will suffer negative impacts. This WTO document also predicts a 7.2% increase in the global trade in goods in 2021, which is much lower than the previous growth rate of more than 20%. It points out that in a long time, the trade scale will be far lower than the normal scale for periods before the COVID-19 pandemic.

Figure 4 shows the total global imports and exports in each quarter from 2013 to 2020. Except for the period lasting from 2016 to 2018, all the other periods see downward trends. The total imports and exports nearly hit all-time low in the second quarter of 2020. After that, the global trade picks up slowly until suffering a new blow.

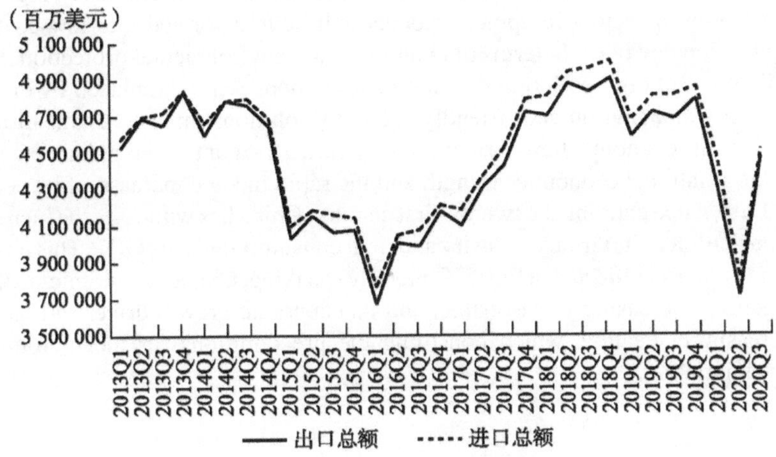

Fig. 4 Quarterly global imports and exports 2013–2020 (*Data source* the WTO Trade Database)

(百万美元)	(million USD)
出口总额	Total exports
进口总额	Total imports

2.1 The Domestic Background

1. **Changed Domestic Development Paradigm: From High-Speed Growth to High-Quality Development.** The 19th National Congress of the CPC, held in 2017, makes the following ideas clear: China will stop emphasizing high-speed economic growths and start pursuing high-quality development; Its economy is in a critical transitional period that calls for an upgraded development model, an optimized economic structure, and new growth drivers; The establishment of a modernized economic system is an urgent task to be completed for the successful economic transformation and a strategic development goal of China. The pursuit of the high-quality development discussed here means that China will replace its extensive economic-growth model that is driven by increases in factor input with an intensive economic-growth modell that is underpinned by technological progress, management improvement, and highly-skilled workers. In terms of industrial structure, it means that technology-intensive and knowledge-intensive industries will replace resource-intensive and labor-intensive industries as China's main industries. In terms of product structure, it means the replacement of products with low technical content and low added value with those with high technical content and high added value. In terms of economic efficiency, high-cost and low-efficiency development models will be transformed into low-cost and high-efficiency ones. In terms of ecological and environmental protection, high-emission and high-pollution development methods will be replaced by circular economy and environment-friendly economy solutions. In a word, this ultimate goal of this economic transformation is to realize the continuous enhancement of China's national economic strength and the rapid income increase of its people.

   First, since entering the twenty-first century, China has witnesses the increased contribution of its tertiary industry and final consumption to its GDP. This upward trend is clearly reflected in Fig. 5. Generally speaking, China's economic structure is getting increasingly reasonable, and its economic growth drivers are getting increasingly healthy, which constitute the preconditions for the high-quality development.

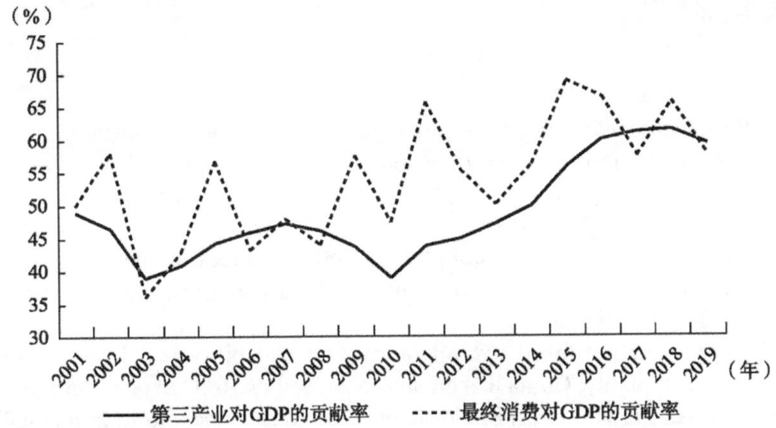

Fig. 5 Contribution rates of the tertiary industry and the final consumption to China's GDP (*Data source* China Statistical Yearbooks)

第三产业对GDP的贡献率	Contribution Rate of the tertiary industry
(年)	(year)
最终消费对GDP的贡献率	Contribution Rate of the final consumption

Second, China's economic-growth model has become more intensive. As shown in Fig. 6, China's energy consumption per unit GDP decreased from 1.61 metric tons of standard coal per 10,000 yuan of GDP in 2001 to 0.55 metric tons of standard coal per 10,000 yuan of GDP in 2019, a decrease of 65.8%.

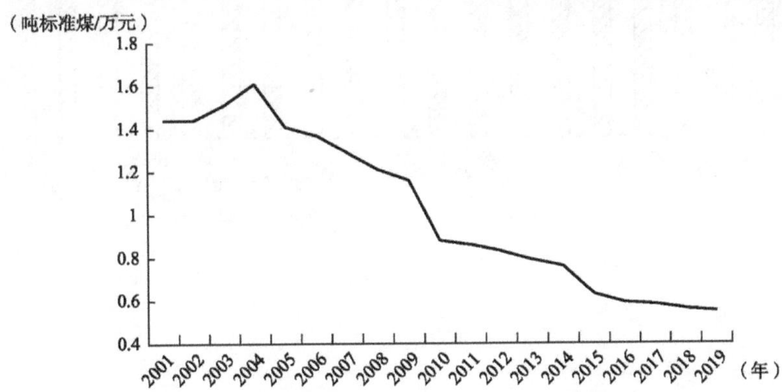

Fig. 6 Total energy consumption per unit of GDP in China 2001–2019 (*Data source* China Statistical Yearbooks)

(吨标准煤/万元)	(metric ton of standard coal/10,000 yuan)

(continued)

(continued)

(年)	(year)

Third, there have been positive changes in China's energy consumption structure. As shown in Fig. 7, the proportion of coal has decreased year by year, from 68% in 2001 to 57.7% in 2019; the proportion of primary energy and that of natural gas have increased from 8.4% to 15.3%, and from 2.4% to 8.1%. The upgrading of the energy consumption structure reduces the emission of pollutants and the cost of air pollution control, which underpins the development of an ecological civilization.

Fourth, with improved overall innovation capability and better economic development quality, China has greatly increased its R&D expenditure in recent 20 years. As shown in Fig. 8, the share of its R&D expenditure in its GDP has increased from less than 1% in 2001 to 2.23% in 2019, an increase of 220% within 19 years. Nonetheless, compared with developed countries, China still has a long way to go in this regard.

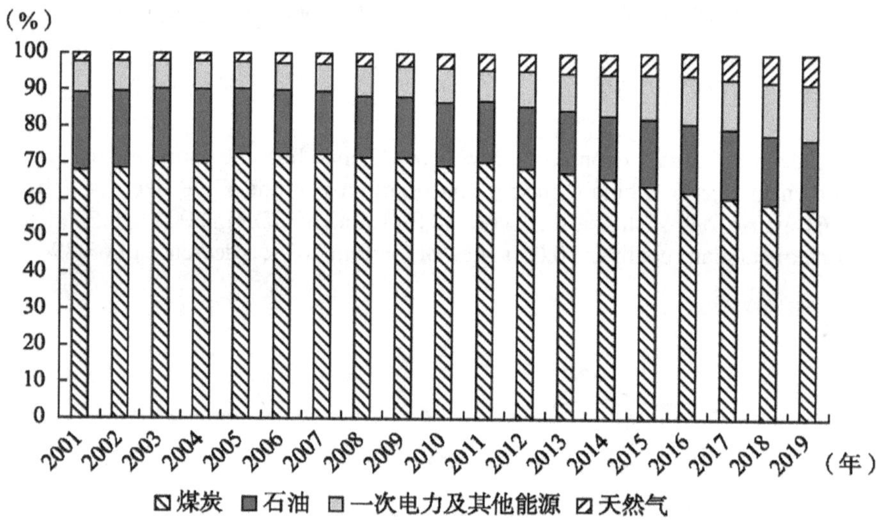

Fig. 7 Energy consumption structure 2001–2019 (*Data source* China Statistical Yearbooks)

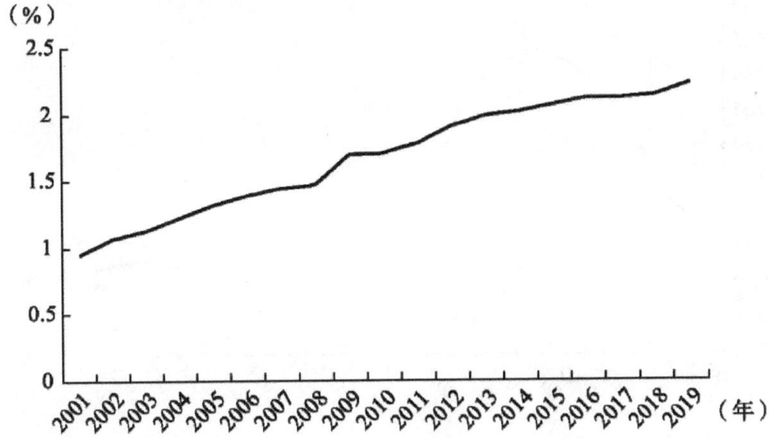

Fig. 8 R&D expenditure as a proportion of China's GDP 2011–2019 (*Data source* China Statistical Yearbooks)

煤炭	Coal
石油	Oil
一次电力及其他能源	Primary energy
天然气	Natural gas
(年)	(year)

2. **Decreased Reliance on Foreign Trade.** The Chinese economy's reliance on foreign trade, one of the country's three major economic growth drivers, increased significantly after 2001 when China joined the WTO. As shown in Fig. 9, the foreign trade's share in its economy rose from 38% in 2001 to 64% in 2006. After 2006, the country's dependence on foreign trade declined. As of 2019, the proportion of foreign trade has decreased to 32%. As to China's economic performance after joining the WTO, the country has developed itself from the sixth largest trader in the world to the largest one and increased its total imports and exports by about eight times. Its GDP ranked the sixth in the word in 2001, but in 2019, it ranked the second.

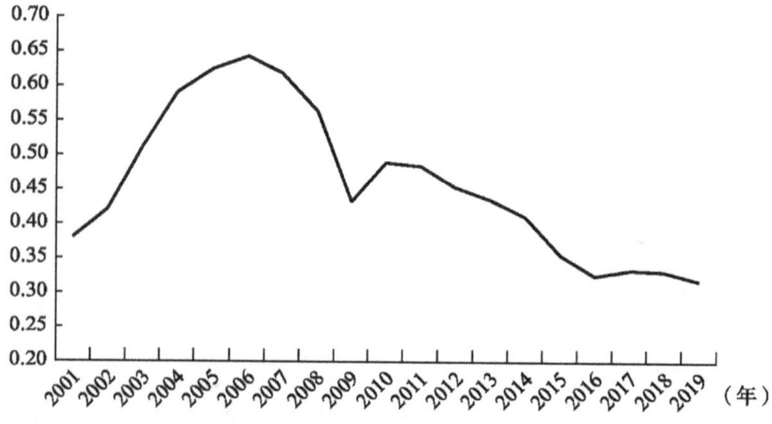

Fig. 9 China's reliance on foreign trade (*Data source* China Statistical Yearbooks)

(年)	(year)

Dependence on foreign trade is an index to measure the economic contribution of imports and exports, the degree of economic dependence on the international market, and the status of opening-up. The continuous decline in China's dependence on foreign trade is the result of domestic and international influences. Domestically, the Chinese economy faces changes in objective conditions and policies such as the shrinking demographic dividend. Internationally, it is influenced by the weak global economic performance caused by economic fluctuations. The 2008 Financial Crisis of the United States and the European Sovereign Debt Crisis plunged the global economy into a quagmire. This difficult situation lasted until 2015, when the United States took the lead in overcoming the impact of the 2008 Financial Crisis and finally replaced its 10-year-long zero-interest-rate policy with an interest-rate hiking cycle. The global economic recession has a certain impact on the imports and exports of all countries in the world, including China. To summary, it is the domestic and international influences that have reduced the country's reliance on foreign trade and decreased the proportion of imports an exports in its GDP.

The declining reliance on foreign trade suggests that China's endogenous drivers for economic development are getting increasingly strong and that the contribution of its domestic demand, especially its domestic consumption demand, is getting increasingly great. This has provided a favorable domestic environment for China's endeavor to create a "dual-circulation" development paradigm.

3. **Low Per-Capita GDP.** Since setting the goal of building a socialist market economic system in 1992, China has seen its per-capita GDP growing at a rate far exceeding the world average. According to the statistics of the World Bank, the growth rates of per-capita GDP in high-income, upper-middle-income, and

middle-income countries have been decreasing since 2015, and some of these countries have even experienced negative growth in per-capita GDP. China's per-capita GDP maintained positive growth in 2015 and 2016, growing by 5% and 1%, respectively. Nevertheless, we must acknowledge an unsatisfactory fact: our per-capita GDP is still very low. Table 5 shows the comparison of China's per-capita GDP with the global average and with the per-capita GDPs of high-income countries, upper-middle-income countries, middle-income countries, lower-middle-income countries, and low-income countries from 1960 to 2019. China's per-capita GDP exceeded $10,000 in 2019, which was only equivalent to the average level of high-income countries in the 1980s. In addition, due to China's large population, unbalanced regional economic development, and wide income gaps, its per-capita purchasing power is still far lower than that of the developed countries. This constitutes a problem that needs to be solved urgently for the success of the "dual-circulation" development paradigm, since it hinders China from quickly increasing its effective domestic demand.

4. **Industrial Structure Dominated by Primary Processing and Primary Industries.** Usually, a county will follow such a path to develop itself into a major industrial country: It begins with industries featuring high consumption, high pollution, and low technological content, and then replace them with new industries boasting low consumption, low pollution, high technological content, and high profit margin. Known as the world's factory, China has the most complete industrial chain in the world and a strong ability to mobilize and convert production capacities. Nevertheless, due to its short development time and insufficient technology accumulation, China mainly produces low-end commercial products such as clothes, toys, timber, etc. Although the output is high, the profit is usually very low. Besides, such production activities can cause serious pollutions. With regard to the production of high-end products, China still mainly focuses on processing businesses. Take its automobile industry as an example, many core components of the automobiles manufactured in China depend on imported technologies, and thus most of the profits are taken away by the European or U.S. enterprises with technology monopoly. Another example is China's mobile phone industry. Chinese manufacturing companies are usually only responsible for the low-technical assembly work, and the high-end chip design work is monopolized by a few foreign countries. The key to solving these problems lies in scientific and technological innovation: It is the essential part of China's industrial-transformation-and-upgrading endeavor, and the only path for the country to achieve high-quality economic development. In a word, to realize the "dual-circulation" paradigm, China must change the existing industrial structure dominated by primary processing and primary industries, enhance the competitiveness of its high-end industries, and promote the common progress of its primary and high-end industries for a better "domestic circulation."

Table 5 Per-capita GDP of China and other countries 1960–2019

Region Year	China	High-income countries	Upper-middle -income countries	Middle-income countries Country	Lower-middle -income countries	Low-income countries	Global average
1960	89.52	1,363.92	181.61	148.04	101.10	144.35	451.75
1965	98.49	1,867.03	208.54	178.02	133.73	173.71	591.72
1970	113.16	2,726.39	266.34	218.21	150.73	212.42	804.00
1975	178.34	5,106.65	551.84	420.35	241.97	408.68	1,457.19
1980	194.80	9,349.43	923.92	706.37	419.79	681.92	2,532.82
1985	294.46	10,256.70	969.13	736.60	440.90	708.00	2,643.80
1990	317.88	18,440.70	1,225.70	898.98	493.64	859.19	4,285.30
1995	609.66	24,273.19	1,680.92	1,136.94	490.79	1,071.13	5,412.47
2000	959.37	25,103.29	1,878.61	1,264.83	565.86	1,189.52	5,499.15
2005	1,753.42	33,172.96	2,955.15	1,952.91	860.30	1,821.38	7,298.62
2010	4,550.45	38,653.90	6,091.87	3,867.49	1,542.73	3,577.51	9,553.18
2011	5,618.13	41,538.63	7,227.93	4,518.54	1,708.70	4,169.29	10,490.05
2012	6,316.92	41,344.81	7,680.99	4,771.64	1,775.97	4,396.13	10,607.44
2013	7,050.65	41,629.15	8,073.30	4,992.77	1,842.81	4,596.19	10,783.85
2014	7,678.60	42,298.55	8,261.03	5,119.58	1,929.08	4,708.91	10,952.68
2015	8,066.94	39,735.36	7,698.21	4,804.80	1,884.58	4,417.79	10, 249.09
2016	8,147.94	40,343.82	7,632.39	4,780.83	1,921.07	4,383.46	10,286.38
2017	8,879.44	42,013.89	8,368.53	5,196.36	2,037.37	4,753.28	10,827.70
2018	9,976.68	44,417.87	8,906.24	5,478.72	2,091.39	5,001.70	11,385.68
2019	10,261.68	44,617.48	9,036.77	5,573.24	2,174.44	5,080.88	11,441.73

Data source: the World Bank.

3 The Development Philosophy for the "Dual-Circulation"

Internationally, today's world is undergoing great changes that have never been seen in a hundred years, and the international political and economic structures have undergone profound adjustments. Protectionism and populism are on the rise across the globe, impacting the international trade and investment and the global industrial chain. That is, the international environment for China's development becomes increasingly challenging. This requires China to maintain its independence in key areas, which guarantees its ability to deal with external uncertainties. Domestically, China is shifting its focus from high-speed growth to high-quality development. As its enters the new era of socialism with Chinese characteristics, China sees an impressive increase in its overall productivity and an increasingly prominent problem: its unbalanced and inadequate development can hardly meet the people's growing needs for a better life. To solve this problem, it is a must to achieve high-quality development through the "dual-circulation" paradigm. The paradigm allows the domestic and overseas circulations to reinforce each other, with the domestic circulation as the mainstay. It should be noted that this paradigm is not the result of a passive strategic contraction but an actively-made strategic choice: China plans to realize the benign interaction between its internal and external circulations through more in-depth reforms.

3.1 "Domestic Circulation" As the Mainstay

The main reasons for the promotion of the "domestic circulation" include the following: China's economic transformation and upgrading, and its focus shifted from high-speed growth to high-quality development; the weak performance of the global economy and the rise of trade protection. Under such circumstances, it is essential to enhance the Chinese economy in terms of quality and aggregate. Unfortunately, in the current stage of development, it is impossible to rely on external demand and investment to boost economic development. What China needs now is the establishment of a sound domestic demand system and that of a long-term mechanism for the promotion of consumption and investment. It must exert efforts to expand its domestic demand and boost innovation.

1. **Efforts to Expand Domestic Demand.** Efforts to expand domestic demand include two aspects: boosting investment and increasing consumption. The plan to promote the "domestic circulation" is different from short-term policies adopted by the country before to stimulate consumption and investment. It aims to establish long-term mechanisms that encourages all kinds of market players to conduct consumption or investment.

   First, adjustments in income distribution and the reform of the consumption structure must be carried out to increase residents' spending power and the quality of their consumption. As to income distribution, labor income shall be increased, especially for the low-income groups. This is because the marginal propensity to consume has long been limited by undervalued labor prices, soaring house prices, and a widening gap between rich and poor. Without income increases for low-income urban residents and migrant workers, it is impossible to expand domestic demand. This can be proved by a simple calculation. In 2019, China's urbanization rate of household-registered population was 44.38%. and its total population was about 14 billion. That is, the country still has around 780 million rural-registered population. If their monthly income increases by 1,000 yuan, the total annual income of the rural-registered population will increase by 9.4 trillion yuan, which will be a huge increase in the domestic consumption power. Regarding the consumption structure, relevant consumption-supporting policies shall be renewed or formulated, such as those aimed to reduce the residents' spending on education, medical care, and old-age care by enhancing public services and raising fiscal expenditure on social welfare. Through adjustments to the tax collection system, the collection of consumption taxes can be moved from the production links to the consumption links, which will encourage local governments to improve their consumption infrastructure. The "housing without speculation" policy is also an effective measure to optimize the consumption structure. At present, the negative influence of high housing prices are very prominent, especially their impact on daily consumption. Let's illustrate this issue by the following calculation. In July 2020, the average price per square meter of the sample residential real states in 100 Chinese cities is 15,500 yuan. Provided that the minimum area needed by a family of three (two parents plus one child) is 90 square meters, the family will have to spend at least 1.395 million yuan to buy a home. Nevertheless, the per-capita disposable income of the Chinese urban residents is only 42,300 yuan (in accordance with data from 2019). If the house prices rise by 3% annually, the price increase will be higher than the annual income of the family. Under such circumstances, the consumption power of the population who have bought or plan to buy a home will be suppressed, and the risk exposure of the financial system will be increased.

Second, funds shall be encouraged to flow into the manufacturing sector. The most striking feature of the manufacturing investment in China is that it is dominated by private enterprises. Private enterprises are the most market-oriented market players in China, and entrepreneurs' confidence in the future determines China's manufacturing investment. To enhance entrepreneurs' confidence in the future, it is essential to do well in the following aspects: improving the business environment and giving full play to the role of the market; implementing policies about tax and fee reductions and reducing production and operation costs for enterprises; helping exporting companies to increase their domestic sales and enter into new international markets.

Third, use new-type urbanization construction projects to drive the growth of investment in fixed assets. Around the beginning of the twenty-first century, China successively put forward the development campaign of the western regions and the plans to revitalize the northeast and the central regions. In essence, these regional development strategies represent the intervention of administrative power in economic activities and a pursuit of egalitarianism. Facts have proved that these policies have not achieved their expected results: The central and western provinces' proportions in the national GDP did not increase significantly, and the gaps of per-capita GDP among different regions continued to expand. The two major methods to promote regional egalitarianism, namely the financial transfer payment system and the preferential policy for backward areas, even led to problems such as distortions in incentive mechanisms and mismatch of resources. Faced with these problems, China has gradually abandoned its pursuit of egalitarianism and actively adjusted its regional development strategies since 2013. At present, it plans to enhance its economic growth poles that are based on it major city clusters, and adjust its economic structure, injecting new vitality into China's overall economic development. During this period, a new trend of population mobility emerged: Since 2010, its population has tended to concentrate in developed areas such as the first- and second-tier cities, the YRD, and the PRD. Given this, China has adjusted its urbanization strategy and put forward the new-type people-oriented urbanization policy. In resent years, it has kept renewing its land system and household registration system to adapt to the new trend of population mobility, ensuring that public resources are allocated based on the population sizes of different localities. The "Opinions of the CPC Central Committee and the PRC State Council on Perfecting the System of the Socialist Market Economy", issued in May, 2020, even encourages city clusters to explore new systems that allow quick changes of registration household registration and realize mutual recognition of residence permits within themselves. This proves that the focus of the new-type urbanization is on the development of city clusters.

2. **Efforts to Promote Innovation.** The "domestic circulation" shall be enhanced in terms of economic aggregate and quality. The quality improvement mainly relies on the progress in scientific and technological innovation, which can be achieved through development methods such as those used to increase financial support for innovation; strengthen basic research; and enhance collaboration among industries, universities, and research institutes. These development methods are also conducive to the growth in the overall strength of the country.

First, it is necessary to increase financial support for innovation. Given the uncertainty of innovation activities, the financial support needed by these activities are different from traditional financial services based on indirect-funding practices. That is, a new financial ecology that is more conducive to scientific and technological innovation needs to be established to facilitate such activities. This new financial ecology calls for a multi-level capital market that finances innovative activities and provides innovative enterprises with the most direct stimulus. It also requires the expansion of the venture capital market by enhancing local venture capital institutions and private equity firms and introducing more international venture capital investors, which is conducive to the development of scientific and technological unicorn enterprises. A renewed banking business model can also contribute this new financial ecology by improve the support of indirect financing for scientific and technological innovation.

For instance, policies can encourage credit institutions and venture capital investors to jointly provide financing support for science- or technology-based start-ups work. For the time being, we call this model investment-loan-linkage model. It allows venture capital investors and banks to make up for each other's shortcomings: The former helps the latter in evaluating investments and screening risks; the latter help the former provide a more sustained and stable financial support.

Second, it is essential to strengthen basic research. It is difficult to make a breakthrough in basic research, but once such a breakthrough is realized, huge industrial technological progress will be achieved. Without such breakthroughs, it is hard to grasp the development initiative. China has decided on taking the path of independent innovation. To this end, China must acquire a strong original innovation ability, which requires it to enhance its basic research, the original of scientific and technological innovation. To improve the basic research, China must to do well in the following aspects: strengthening the governments' role in promoting scientific research activities, especially increasing their fiscal expenditure on relevant sectors; giving full play to the innovation advantages of enterprises and guiding them to appropriately increase their R&D investment in basic research fields with relevant fiscal arrangements and taxation measures.

Third, it is a must to upgrade the collaboration among industries, universities, and research institutes. According to the experience of developed countries, such a collaboration is often an efficient innovation model. Nevertheless, such collaborations in China usually feature low efficiency: They are usually initiated by universities that want to have their research results commercialized and seek cooperation with enterprises, but many research results of the universities may not be suitable for commercial application by enterprises. Although this kind of collaboration is often carried out for market purposes, the government can still play an important role in facilitating such collaborations by improving the institutional environment, optimizing laws and regulations, and giving encouragement to relevant parties.

3.2 "International Circulation" and "Domestic Circulation" Reinforcing Each Other

In the "dual-circulation' paradigm, the "domestic circulation" is closely linked to the "international circulation" while being relatively independent from it. For instance, the opening-up drive can enhance our scientific and technological exchanges with foreign countries while boosting our scientific and technological innovation abilities, promoting the progress of our enterprises, and expanding the domestic development space for our technologies, industries, and capitals. Relevant international cooperation often increases the efficiency and flexibility of our domestic technological innovation and application activities. In this sense, the "international circulation" involves and serves the "domestic circulation." In a word, further opening-up efforts and adaptation to the global market can better meet the needs of our domestic production and facilitate our industrial-structure improvement and technological progress. In addition, the enhancement of our international trade and investment and the participation in globalization are also preconditions for the RMB to become a global currency and achieve a higher level of internationalization.

1. **Enhanced "Domestic Circulation" for Economic Stability.** In 2020, the rising trade protectionism and the COVID-19 pandemic have caused the global economy to suffer heavy losses. Because of this once-in-a-century drastic change, China's industry chains for foreign demands have changed from an economic driving force into an economic weak point. Under such circumstances, China hopes to replace its reliance on the foreign demand with that on the domestic demand within a certain period. That is, China needs to promote its "domestic circulation." Firstly, the "domestic circulation" can digest the overcapacity caused by trade barriers and the weak performance of the international market. With measure to turn export sales into domestic sales, adjust our industrial restructure, and develop our domestic market, the virtuous circle of domestic supply and demand will be established, and the equilibrium of the commodity market will be achieved. Secondly, the "domestic circulation" calls for the synergistic growth of income and living standards, and the balance in the labor market, which can be realized through the efforts to stabilize the employment market, improve the social security systems, optimize the income distribution mechanism, and effectively increase the income of low-income groups. Thirdly, the "domestic circulation" requires a proper relationship between real and virtual economies, which is actualized through measures to reform our financial system, boost our real economy, and ensure healthy interactions between our economic and financial sectors and the balance in our capital market.
2. **Upgraded Domestic Supply and Demand for Better Economic Balance.** In the long run, the "domestic circulation" of China's will have larger volume and higher quality: With the help of the advanced opening-up endeavor, it will be closely linked to the country's "international circulation," and these two circulations will reinforce each other. This process is roughly divided into three steps. In the first step, with the improvement of productivity, China's economy will grow at a relatively high speed, and its hinterland areas' economic potential will be released. In the meantime, the economic progress will increase people's income and consuming power. In this way, domestic supply and demand will be expanded. In the second step, driven by innovation activities, the quality of the "domestic circulation" and that of the economy will be improved. In the third step, the China's openness will be enhanced. The "domestic circulation" does not mean that China plans to cut itself from the outside world. Instead, it aims to realize higher-level opening-up goals based on China's economic progress in aggregate and quality and the country's strengthened economic resilience that enables it to better handle external impacts. Only with these improvement can China better fit in with the global economic system, and the expanded Chinese domestic can help to form a larger international market. In the foreseeable future, China will push forward reforms in many aspects through various policies. Different

from its previous measures used to expand domestic demand, these reforms are intended to realize the synergistic growth of supplies and demands and remove barriers in economic operation, especially those caused by problems in current institutions and benefit-distribution systems. This is where the digital economy can help since it can empower the upgrading of supplies and demands in an all-around way. It can be predicted that in the future, more supportive policies will be formulated to facilitate breakthroughs regarding core technologies, and that the industrialization of digital economy and the digitization of traditional industries will be vigorously promoted. Correspondingly, China's Digital Currency Electronic Payment (DCEP) and digital finance will be implemented, realizing the digital upgrading of the national governance.

3. **Greater Efforts to Promote Industrial Upgrading and Move Up the Global Value Chain.** At present, from a global perspective, the Chinese economy is still in the middle-lower reaches of the global value chain: It relies on the demands and technologies from the upper reaches and provides products with medium and low added value. When promoting the "dual-circulation" paradigm, China must focus on the overall upgrading of its "domestic circulation," seeking opportunities in the crisis unseen in a hundred years. In the meantime, it has to adjust its position in the "international circulation" from the following three aspects. First, as the country with the most complete industrial system in the world, China is currently the supply center in the global value chain. In the future, it shall become a supply and demand center in the global value chain and actively participate in and promote the reshaping of the global value chain. Second, "made in China" shall be upgraded into "intelligent manufacturing in China": This requires improvements in R&D, scientific and technological innovation, intelligent manufacturing and brand marketing. Third, China has to move closer to the world's center stage, changing itself from a passive recipient of the global economy and governance systems into a leader of these systems. This will enable China to fundamentally improve its external environment and prevent risks caused by cut supply chains and adverse spillover effects.
4. **Further the Opening-Up Drive for Diversified Involvement with the Global Value Chain.** China must actively promote the BRI-related projects for a new westward opening-up paradigm, which is conducive to the development and upgrading of its hinterland economy and the internationalization of RMB. With this progress, China will be able to further its participation in globalization and see its"dual circulations" interact with each other in a diversified, flexible, and steady manner. It also needs to deepen its exchanges and cooperation with more foreign countries through the development of the free trade zones. By establishing trade ties with more partners, it will become better at resisting the impact of unilateralism.

In a word, the "dual-circulation" development paradigm is aimed to realize high-quality development and establish a new economic system for China when it actively adapts to the great changes that have never happened in a hundred years. Domestically, various reforms are pressed ahead to promote the industrial upgrading and the consumption expansion, creating a large-scale, well-structured, and high-quality domestic circulation. Internationally, China steps up its opening-up efforts and moves up the global value chain by actively participating in global governance and optimizing its external environment, trying to realize the common progress of the Chinese economy and the global economy. Similar to the policy of reform and opening-up, the "dual-circulation" development paradigm propels China to move closer to the world's center stage.

Chapter 8
Synergies Between the "Dual-Circulation" Paradigm and Metropolitan Areas

At present, the Chinese economy faces profound changes in its domestic and international environments. Domestically, the focus of the economic development has been shifted from high-speed growth to high-quality development, and obvious changes regarding the development drivers, development model, and the country's comparative advantage have occurred. As to the development drivers, the overseas demands have shrunk since the 2008 Global Financial Crisis, and thus the domestic demands have gradually replaced the overseas ones in providing growth momentum for China. This point is verified by China's decreasing dependence on foreign trade: The share of its foreign trade in its GDP falls year by year after reaching its peak of the reform-and-opening-up period in 2007 (See Fig. 1).

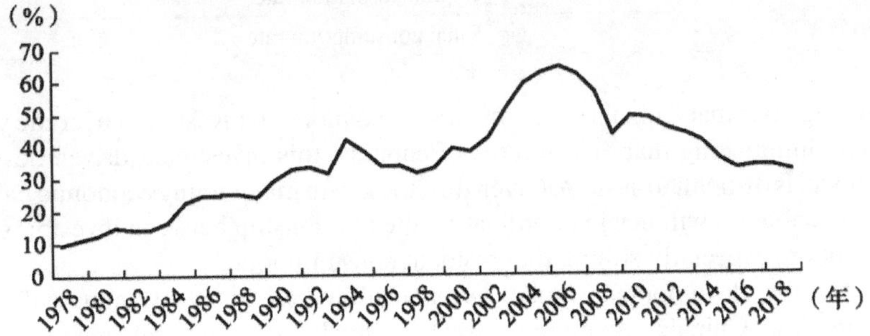

Fig. 1 Changes in China's reliance on foreign trade since the beginning of the reform and opening-up (*Data source* The data of this Figure comes from the EPS database and estimates based on data from this database)

(年)	(year)

J. Sun, *Regional Economic Development in China in 2020*,
https://doi.org/10.1007/978-981-95-3604-7_8

It is worth noting that although domestic demands have become the main growth engine for China, there is a structural imbalance between investment and consumption regarding these demands. As shown in Fig. 2, the proportion of capital formation or investment remains to be higher than that of consumption or final consumption. The former reached its peak in 2011: 47%. The latter dropped to its lowest point in 2010, which is 49.3%.

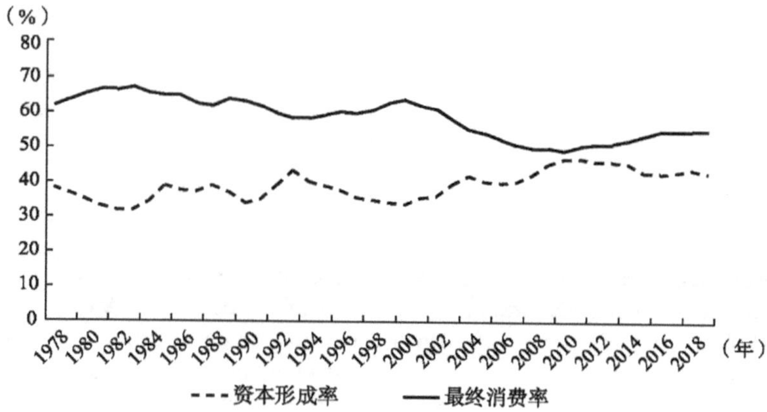

Fig. 2 Changes in China's capital-formation and consumption rates since the beginning of the reform and opening-up (*Data source* the EPS database)

(年)	(year)
资本形成率	Capital-formation rate
最终消费率	Final consumption rate

This suggests that China's economy has relied more on investment over the years. With the diminishing marginal returns of capitals, this investment-driven development model is difficult to sustain. Given this, the key to maintaining economic balance and sustainable growth lies in coordinating the relationship between investment and consumption, especially stimulating resident consumption.

As for the development model, urbanization has replaced industrialization in promoting China's economic growth, which is a natural result of the country's economic development. Generally speaking, China has entered the post-industrialization stage wherein the GDP share of its tertiary industry in GDP has surpassed that of the secondary industry. This surpassing first appeared in 2012, and in 2019, the tertiary industry accounted for 53.9% of the national GDP, far exceeding the 38.9% proportion of the secondary industry. As shown in Fig. 3, China's urbanization rate has increased rapidly since the beginning of the reform-and-opening-up period and reached 60.6% in 2019, exceeding the global average of 55%. Over the years, the rural population has kept flowing into urban areas, and the urban space has kept expanding. Despite that, there is still much room for growth regarding China's urbanization because its current urbanization is still hindered by a number of institutional constraints such as those regarding the household registration system, the land system, and the social security system.

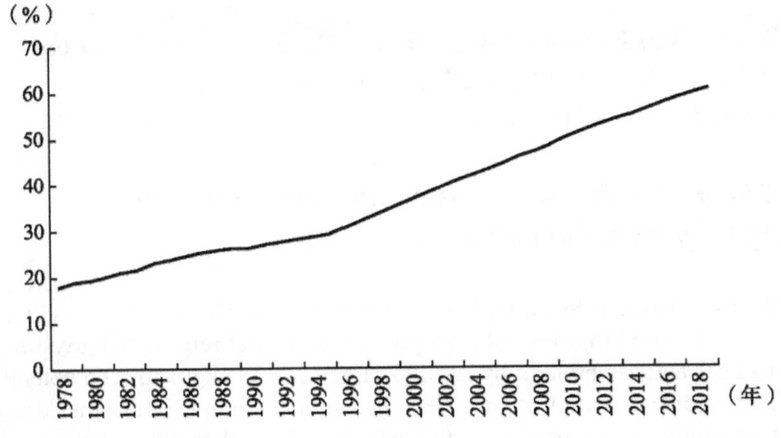

Fig. 3 Changes in the urbanization rate of china since the beginning of the reform and opening-up (*Data source* the EPS database)

(年)	(year)

With regard to China's comparative advantage, China's super-large-scale economy has gradually become its greatest comparative advantage, even absolute advantage. This changes has taken place gradually with the increase in China's per-capita GDP. It has exceeded $10,000 mark 2019, which means that the country's advantage in labor cost has been weakened.

Internationally, China faces challenges brought by the rising populism and counter-globalization trends as well as the trade frictions and technology blockades imposed by Western countries led by the United States who want to suppress China's development. The COVID-19 pandemic in 2020 has aggravated the anti-globalization trend and exerted adverse effects on the global industrial chain and supply chain. Under such circumstances, China has adjusted and improved its economic development strategy, proposing to establish a new development paradigm with the "domestic circulation" as the mainstay and the "domestic circulation" and 'international circulation" promoting each other. This "dual-circulation" paradigm is underpinned by high-quality regional development, and the development of metropolitan areas is a new trend and feature of the regional economic development. This chapter discusses China's metropolitan areas. It discovers the importance of the metropolitan areas for the "dual-circulation" paradigm by analyzing the theoretical connotation and practical characteristics of the new paradigm and the theoretical and practical characteristics of metropolitan areas, points out the specific paths that metropolitan areas can take to promote the "dual circulations," and give suggestions on fostering and developing modernized metropolitan areas under the new paradigm.

1 Theoretical Explanations on and Regional-Development Problems Regarding the New "Dual-Circulation" Development Paradigm

1.1 Theoretical Explanations on the "Dual-Circulation" Development Paradigm

On May 14, 2020, the Standing Committee of the Politburo of the CPC Central Committee proposed to deepen the supply-side structural reform and give full play to China's advantages brought by its ultra-large-scale market and high potential of its domestic demand, building a new development paradigm in which the "domestic and international circulations" promote each other. Since then, the new development paradigm has been repeated discussed in multiple important meetings. Table 1 include the Chinese authorities' statements on the paradigm made in 2020.

So far, Chinese academic circles have interpreted the new development paradigm mainly from the following seven perspectives.

Firstly, some scholars focus on analyzing the primary-and-secondary relation of the two circulations. In their opinion, China used to rely mainly on its "international circulation" and regard its "domestic circulation" as the secondary circulation, especially before the 2008 Financial Crisis; but it has to change this development paradigm and take its "domestic circulation" as the mainstay given the changes regarding its development stage, market conditions, external environment, and the impact of the COVID-19 pandemic, which makes it hard to rely mainly on the "international circulation" for economic growth. Huang (2020a, 2020b) emphasized that taking the "domestic circulation" as the mainstay did not mean a denial of participating in the "international circulation," and that it was a must to have the "international circulation" and "domestic circulation" to promote each other, especially boosting the "international circulation" with a strong "domestic circulation." Quan (2020) and Dong (2020) agreed that the "international circulation" and "domestic circulation" are dialectically unified, complementary, and inseparable, and that they promote and restrict each other.

Secondly, some scholars highlights the concept of circulation. They point out that in addition to the primary-and-secondary relationship, the most important issue regarding the paradigm is how to smooth out difficulties in the five circulation links, namely, investment, production, distribution, circulation, and consumption. They stress that only when these links are connected smoothly can the new paradigm be truly formed, and their research focus is usually placed on the function of and the connections among these links.

Table 1 Chinese authorities' statements on the "dual-circulation" new development paradigm in 2020

Date	Main content	Source of Information
April 10	The economic advantage of a big country is the feasibility of internal circulation…. There is no contradiction between the effort to expand its domestic demand and that to further the opening-up drive. With a stronger domestic circulation, the country will have a greater appeal to global resources. This is conducive to the new development paradigm with the "domestic circulation" as the mainstay and the "domestic circulation" and "international circulation" promoting each other, constituting new advantages in participating in international competition and cooperation	"Certain Major Issues for Our National Medium- to Long-Term Economic and Social Development Strategy"
May 14	It is essential to deepen the supply-side structural reform and give full play to China's super-large-scale market and domestic demand potential, developing a new development paradigm whereby the "domestic circulation" and "international circulation" reinforce each other	Press release for the meeting of the Standing Committee of the Poliburo of the CPC Central Committee
May 23	As China has to seek development in an increasingly unstable world, it is a must to focus on our domestic needs, build a complete domestic-demand system, and promote our innovation especially scientific and technological innovation. Efforts shall be exerted to strengthen various links such as production, distribution, circulation, and consumption, gradually forming a new development paradigm with the "domestic circulation" as the mainstay and the "domestic circulation" and "international circulation" promoting each other. This will create new advantages for China's participation in the international cooperation and competition under the new situation	Speech by Xi Jinping when meeting with the CPPCC members from the economic sector

(continued)

Table 1 (continued)

Date	Main content	Source of Information
July 21	China plans to foster a new development paradigm whereby the "domestic circulation" serves as the mainstay, and the "domestic circulation" and "international circulation" reinforce each other. This plan is mainly based on the following considerations: Today's world is experiencing a great change that has never happened in a hundred years, and a new round of scientific and technological revolution emerges. The rising protectionism, weak global economy, and shrinking global market propel China to fully leverage its super-large domestic market. With domestic economic prosperity and a strong growth driver, i.e. the "domestic circulation," China will become better able to facilitate the recovery of the world economy. To realize these purposes, it is essential to modernize our industrial chains and supply chains, vigorously promote scientific and technological innovation, and accelerate progress on core technologies and our new advantages for future development. It must be realized that the focus on the "domestic circulation" does not mean that China will cut itself off from the world. Instead, it requires to exert the potential of our domestic demand for better connections between our domestic and international markets. With such connections, China will be better at adapting to markets and pooling resources at home and abroad, which is conducive to its vigorous and sustainable development. China will deepen its reforms, open wider to the outside world, and promote international cooperation in the field of science and technology. It will remain devoted to the building of an open world economy and that of a community with shared future for mankind	Speech by Xi Jinping at the symposium of entrepreneurs
August 20	In view of the shrinking global market, we must concentrate on doing our own things well. We shall give full play to our advantage, i.e. our super-large-scale domestic market, and accelerate the formation of a new development paradigm whereby the "domestic circulation" serves as the mainstay and the "domestic circulation" and "international circulation" promote each other. The Yangtze River Delta shall actively explore paths to realize this new paradigm by leveraging its advantages in talent, science and technology, manufacturing industries, industrial and supply chains, and market potential	Speech by Xi Jinping at the symposium on solidly promoting the integrated development of the Yangtze River Delta

(continued)

Table 1 (continued)

Date	Main content	Source of Information
August 24	The new development paradigm shall be aimed to smooth the barriers in our national economic cycle. It is proposed in accordance with the changes regarding our development stage, environment, and conditions. It is a strategic choice to create China's new advantages in the international cooperation and competition. Given changes in the external environment and various factors in recent years, the "international circulation" that depends on foreign markets and resources become increasingly weak in driving our economic growth. The "domestic circulation," however, is getting increasingly strong with the increased domestic demand. Since the 2008 Global Financial Crisis, China has become increasingly dependent of its "domestic circulation." This requires the country to grasp the strategic chance to expand its domestic demand while adhering to its supply-side structural reform. It shall see that its production, distribution, circulation, and consumption links rely more on the domestic market, and that its supply system better adapt to its domestic demand. It shall realize a high-level, dynamic balance between supplies and demands. It must be noted that the new development paradigm refers to a "dual-circulation" (domestic and international circulations) paradigm rather than a closed domestic circulation. As China's position in the economy continues to rise and its ties with the global economy become closer, the country will provide more market opportunities for other countries and attract more international resources and commodities. In this process, China must vigorously enhance its independent innovation capability and make breakthroughs regarding core technologies. This is a major issue related to the overall development of the country and the establishment of the new paradigm with the "domestic circulation" as the mainstay	Speech by Xi Jinping at the symposium of experts from the economic and social fields

(continued)

Table 1 (continued)

Date	Main content	Source of Information
September 1	Given changes regarding its development stage, environment, and conditions, China decides to accelerate the formation of a new development paradigm whereby the "domestic circulation" is the mainstay and the "domestic circulation" and "international circulation" promote each other. This is a systematic and in-depth change that concerns the overall situation. To realize this goal, it is a must to resolutely push forward our reforms and maintain our strategic determination. We must uphold and improve our socialist system with Chinese characteristics, promote the modernization of our national governance system and governance capacity, and carry out in-depth reforms, We shall upgrade our opening-up, providing a strong driving force for the establishment of our new development paradigm	Speech by Xi Jinping at the 15th meeting of the Central Committee for Deepening Overall Reform
September 11	The drive to establish a new development paradigm requires further efforts to boost scientific and technological innovation. When promoting the "domestic circulation," it is necessary to stick to the supply-side structural reform and improve our supply system and quality management. In this process, science and technology play a key role in fostering new demands through the creation of new supplies. In a word, the "dual circulations" shall be guaranteed by improved scientific and technological strength and stabilized industrial chains and supply chains	Speech by Xi Jinping at the symposium of scientists

(continued)

Table 1 (continued)

Date	Main content	Source of Information
October 14	So far, the main contradiction in our society has changed, and our people's requirements for a better life have grown stronger. Fortunately, long-term expectations for our economy are optimistic, which boasts large market space and high development resilience. At present, a new development paradigm in which the "domestic circulation" serves as the mainstay and the "domestic circulation" and "international circulation" complement each other is gradually taking shape. This paradigm does not only entails a closed domestic circulation. Instead, it highlights the "dual circulations", with open domestic and international markets. Under such circumstances, we must optimize and upgrade relevant systems regarding our production, distribution, circulation, and consumption; we shall consolidate our internal economic ties, expand our domestic market, and enhance the functions of our "domestic circulation" and those of our "international circulation;" we must promote institutional opening-up to optimize our standards and rules, and establish an economic system that adapts to a higher-level of openness	Speech by Xi Jinping at a grand gathering held to celebrate the 40th anniversary of Shenzhen Special Economic Zone

(continued)

Table 1 (continued)

Date	Main content	Source of Information
October 29	The Plenary Session requires to foster a strong domestic market and establish a new development paradigm. It emphasizes the strategic basis of expanding domestic demand and calls for rapid progress in the building of a complete domestic demand system. It points out that the efforts to expand domestic demand and those to deepen the supply-side structural reform shall be combined organically, and that innovation-driven, high-quality supplies can be utilized to creates demand. It clarifies that the new development paradigm entails efforts to promote consumption in an all-round way and create investment opportunities	The "Communique of the 5th Plenary Session of the 19th CPC Central Committee"
October 29	It is necessary to promote the "dual circulations" (domestic and international circulations). We must proceed from our "domestic circulation" and give full play to our comparative advantages, fostering a strong domestic market for China and developing it into a trade powerhouse. We shall increase our appeal to global resources with a strong "domestic circulation," make full use of markets and resources at home and abroad, actively expand domestic and foreign demands. We shall strive for increases in our imports and exports, utilize more foreign capitals, and make more investments overseas, ensuring that our international income and expenditure are basically balanced. Relevant rules and standards for domestic and international trade need to be be unified, such as relevant laws and regulations, supervision mechanisms, rules for business qualification and certification, quality standards, inspection and quarantine standards, and other accreditation standards. Domestic and international market layouts, commodity structures, and trade modes should be optimized. Export quality should be improved, and more high-quality products should be imported. Trade and investment activities should be integrated, and a modernized logistics system should be established	The "Proposals of the CPC Central Committee on the Formulation of the 14th Five-Year Plan (2021–2025) for National Economic and Social Development and the Long-Range Objectives Through the Year 2035 (Vision 2035)"

Source of Information Zhu (2020) and relevant supplementary information

Thirdly, some researchers clarify misunderstandings about the paradigm, refuting some people's views that the paradigm is to pursue closedness and sole domestic circulation. General Secretary Xi Jinping has also repeatedly emphasized that the paradigm is aimed to realize open "dual circulations." Liu, W. (2020), Liu, Y. (2020), and Zhang (2020) have further clarified that these two circulation are intended to promote rather compete against each other, and that a superficial equilibrium between the two shall be avoided.

Fourthly, some researchers focus on the issues of supplies and demands. Just as General Secretary Xi Jinping has remarked, it is necessary to improve the adaptability of the supply system to the domestic demands and create a high-level dynamic equilibrium between supplies and demands, seeing that demands drive supplies and that supplies stimulate demands. Liu, H. (2020) has summarized that the related research in this field usually focus on the supply-side structural reform and the strategy of expanding domestic demand.

Fifthly, some research papers such as Liu, Z. (2020) have invented new terms "home" and "away" globalization to explain the new development paradigm. Liu, Z. (2020) believe that the paradigm is aimed to transform China's "away" globalization, which features high reliance on the international market, into "home" globalization, which entails globalization underpinned by strong domestic needs and a super-large domestic market and domestic-demand-driven connections between the domestic and international markets.

Sixthly, some research such as Lu (2020) claim that the new development paradigm is about changes in the GDP composition: The future growth will be driven mainly by consumption and investment. Related research focuses on the relationship between consumption and investment, giving suggestions on increasing domestic demand and effective investment.

Seventhly, some scholars actively explore the new paradigm's connections to other development strategies. Liu, W. (2020) pointed out that the paradigm embodied the fundamental requirements of a modernized economic system. Liu, Y. (2020) considered it as a continuation of the supply-side reform because it relies on high-quality development and a successful supply-side reform.

The above seven perspectives are summarized in Table 2. According to these seven perspectives, we extract seven inherent requirements for the new development paradigm. First, it is essential to have the "domestic circulation" as the mainstay, and see that the "domestic circulation" and "international circulation" promote each other. Second, it is necessary to ensure smooth circulations. Third, openness must be embodied. Fourth, the dynamic balance between supplies and demands needs to be ensured. Fifth, the "home" globalization shall be pursued. Sixth, consumption and investment shall be further expanded. Seventh, the new paradigm shall be coordinated with existing development strategies. Given these seven requirements, we can clarify that the effort to build the new development paradigm shall be based on stronger industrial chains and "domestic circulation," expanded domestic demands, innovation-drive growth drivers, openness, and unleashed potential of domestic demand. These connotations, basis, foundations, and key links are summarized in the "internal logic."

Table 2 Summary of Chinese Academia's statements on the "dual-circulation" new development paradigm

Connotations	The primary and secondary circulations; the economic circulation; the closedness and openness; supplies and demands; "home" and "away" globalization; GDP composition; and existing strategic relationships
Foundation	The "domestic circulation"
Strategic basis	Expanded domestic demand
Central task	The strengthening of industrial chains
Key issues	The coordination between investment and consumption; the improvement in residents' consumption and effective investment (efforts to tap the demand potential of the domestic market); the innovation-driven development (the independent innovation of core technologies); and the openness (for both domestic and international markets)
Internal logic	Expanding domestic demands → attracting global resources → forming a modernized industrial system dominated by emerging industries → building a big economic cycle with the "domestic circulation" as the mainstay for upgraded primary industries and modernized industrial chains → establishing a new development paradigm whereby the "domestic circulation" and "international circulation" promote each other

Source of information The information in this Table is compiled based on relevant materials

1.2 Regional-Development Problems Regarding the"Dual-Circulation" Development Paradigm

The new "dual-circulation" development paradigm mentioned above has the "domestic circulation" as its mainstay: Only with a smooth "domestic circulation" can the paradigm wherein the "domestic circulation" and "international circulation" reinforce each other be established. Unfortunately, the "domestic circulation" is still troubled by many fundamental problems. From the perspective of regional development, these problems can be described as barriers in regional economic circulations.

A smooth "domestic circulation" has to be based on the integrated development of economic regions, which requires an integrated market. Nevertheless, just as Xia et al. (2018) has observed, the inter-regional economic ties in China are still heavily influenced by the division of administrative regions. This leads to separated markets for production factors (i.e. labor and capitals) and commodities, which is a phenomenon widely existing among provinces and cities of the same or different provinces. Huang (2020a, 2020b) studied this phenomenon of separated markets at the provincial level in the period from 1998 to 2016: There was a downward trend with fluctuations indicating the generally decreasing market fragmentation or market segmentation, but this problem still widely existed among provinces; The labor market suffered the most from such separation, followed by the commodity market; The capital market boasted the lowest separation degree. Lv and He (2020) analyzed the market fragmentation at the municipal level: There was a downward trend in the degree of separation from 2001 to 2015, but as the market fragmentation was more obvious at the provincial level than at the municipal level, some inter-city trade ties were adversely affected. This widely-existing problem of market fragmentation undoubtedly hinders the formation of a unified, open, competitive, and orderly domestic market system, the flow of factors and commodities, and the integrated development of economic regions.

As Huang (2020a, 2020b) has noted, the integrated development of economic regions constitutes the most essential prerequisite for a smooth "domestic circulation." Nonetheless, there are many problems hindering the integrated development such as the inter-regional economic inequality, the imbalance between urban and rural areas, and the insufficient development. In recent years, many new changes regarding these problems have taken place. Regarding the inter-regional economic inequality, the gap between the western and eastern regions has been narrowed, but that between the northern and southern regions has been expanded. Zhang (2019) studied the north–south gap in per-capita GDP, which started widening in 2015, attributing the growing gap to the increasing north–south difference in economic growth rates. As to the urban–rural gap, it has been narrowed gradually, but the absolute value of the gap is still large. From 2013 to 2018, the ratio of per-capita disposable income of urban residents to that of rural residents continuously decreased, but the absolute value of the urban–rural gap kept growing. The ratio of per-capita road area of urban residents to that of rural residents and the absolute value of the urban–rural gap in this regard have experience similar changes. As for books per 100 people, the urban–rural ratio and the absolute value of the urban–rural gap have both decreased, but the gap is still evident. In 2015, the urban data was 22 times as much as the rural data. In summary, the inter-regional and the urban–rural inequalities lead to the north–south and east–west gaps and the imbalance between urban and rural areas, which hinders the coordinated development of economic regions and results in the rural areas' low presence and participation in the "dual-circulation" paradigm.

Regional development policies are important tools for the Central Government to intervene in the regional economy, but the influence of the local governments reduces the effectiveness of these tools. The eight major national strategies, namely the BRI, the coordinated development of the BTH, the integration of the YRD, the development of the GBA, the effort to comprehensively deepen reform in Hainan, the development of the YREB, the ecological protection and high-quality development in the YRB, and the development of the CCEC, have already entailed arrangements that cover nearly all the territorial space of China. Nevertheless, the wide coverage and the need for cooperation among different localities have created obstacles for the implementation of the strategies. Sun and Yuan (2014) pointed out that the Chinese authorities began adjusting its regional development strategies in 2006 and adopted a series of plans for experimental areas, demonstration areas, and regional development projects, which amounted to 101 according to incomplete statistics. Sun observed that such wide coverage weakened the regional development strategies in terms of spatial arrangement and greatly reduced their effects. Therefore, these strategies can hardly smooth out the difficulties in the regional economic circulations, which hinders the establishment of the "dual-circulation" development paradigm.

2 Theoretical and Practical Characteristics of Metropolitan Areas

As of 2019, China's urbanization rate has reached 60.6%. According to the urbanization curve, first suggested by Northam (1979), this percentage suggests that China is in the second half of the acceleration phase.That is, its social and economic activities continue to gather in cities. Metropolitan areas are advanced urban forms. They are the product of our long-term urbanization trend and the new features and key regions of our current development stage. The growth of metropolitan areas has become a new trend of our regional economic development. Given their theoretical and practical characteristics, they play a significant role in smoothing out difficulties in the economic circulations at home and abroad, which is conducive to the "dual-circulation" paradigm.

2.1 Theoretical Characteristics

1. **Function of Gathering Externalities.** Referring to the concepts and connotations proposed by Zhang (2003) and Wang et al. (2019), a metropolitan area can be defined as a large urban area consisting of at least one central city and surrounding urban communities, which boast strong economic and social ties and convenient transportation and meet the requirements regarding commuting time. This definition highlights the following two distinctive characteristics of metropolitan areas. First, a metropolitan area features a central-peripheral spatial structure, consisting of usually one big central city and its adjacent small cities and towns. Second, the central city has a strong appeal to regional resources and a great driving force with regional influence, ensuring the integrated development of the metropolitan area. From the perspective of the spatial economics, the big central city usually boasts an urbanization economy and diversity, and the surrounding small cities and towns feature a localization economy. Their integrated development is conducive to Marshallian Externalities and Jocobs Externalities, and these two types of externalities are embodied in the spatial and temporal evolutions of the metropolitan area. In the development of a metropolitan area, most cities and towns play the role of "Nursery Cities," providing diversified environments for new enterprises to learn, explore, experiment, and create and reducing the pressure caused by the risks of innovative activities. Dong and Yuan (2014) has pointed out that Jocobs Externalities created through big industrial clusters enable emerging enterprises to develop rapidly and quickly reach maturity

and that when they become mature and need more Marshallian Externalities provided by specialized industrial clusters, they will move to suitable, nearby small cities. Such a dynamic process wherein enterprises emerge in big cities and then move to nearby small cities promotes the circulation of production factors and commodities and boosts the integration of big and small cities, i.e. the formation of metropolitan areas.

2. **Economic Resilience.** Sun and Sun (2017) has noted that economic resilience refers to a socio-economic system's ability to recover from shocks. Xia and Sun (2019) hold that from the perspective of spatial scale, a metropolitan area shall meet a specific scale standard: its agglomeration mechanism is mainly underpinned by technological innovation and technology spillover; its form of agglomeration is the coagglomeration of manufacturing and service industries. Song et al. (2020) discovered that in comparison with the national- or regional-level city clusters, metropolitan areas have stronger economic resilience because they are more market-sensitive and stable in space and they have a higher level of integration. Faced with external shocks, these metropolitan areas that boast solid foundations and clear division of labor can quickly mitigate external risks and effectively rebuild the smooth flows of people, logistics, and information for their economic-and-social-development networks, which promotes the restart of their economic circulations. Their strong economic resilience has been proved by their performance during the COVID-19 pandemic in 2020. In accordance with a news report from Xinhuanet in 2020, Beijing, Shanghai, Guangzhou, Shenzhen, and other 14 metropolitan areas have achieved positive results in epidemic prevention and the resumption of work and production through cooperation mechanisms such as the exchange of epidemic prevention information, the mutual recognition of quarantine results, the guarantee of material supplies, and the smooth flows of personnel and transportation.

2.2 Practical Characteristics

Practice has proved that compared with national- and regional-level economic belts and city clusters, metropolitan areas feature less internal administrative barriers, higher market sensitivity, and strong endogenous development momentum. Given this, they become increasingly favored by policies. The information provided in Table 3 show how their functions have been increasingly highlighted in the national and regional development strategies since the promulgation of the 13th Five-Year Plan (2016–2020). First, the development of metropolitan areas is considered conducive to the new-type urbanization. The "Guiding Opinions of the National Development and Reform Commission on Cultivating and Developing Metropolitan Areas," adapted in 2019, regards it as an important means to realize the new-type urbanization. Since 2018, it has been listed as a key task to promote the new-type urbanization for three consecutive years. Second, the development of metropolitan areas is believed to be beneficial to the integrated development of the economic regions. The "Outline for the YRD Integrated Development Plan" regards it as an important step in the drive to realize integrated development. In the process of formulating regional coordinated development strategies, metropolitan areas, as policy units with relatively smaller spatial scales, are often expected to play a role in building new regional-coordination mechanisms and regional-economic layouts that leverage complementary advantages and pursue high-quality development. The "Recommendations on the 14th Five-Year Plan (2021–2025)" even require metropolitan areas to play an important role in the drive to optimize China's territorial-space arrangement and promote its regional coordinated development. Various policies have enriched the functional orientation of metropolitan areas. The functions of boosting the development of city clusters and enhancing central cities' influences have been expanded to include the optimization of urban layouts and the promotion of regional coordinated development and the new-type urbanization. With the promulgation of the "Guiding Opinions of the National Development and Reform Commission on Cultivating and Developing Metropolitan Areas" and the "Notice on Accelerating the Development of Metropolitan-Area (Suburban) Railways" and the upgrading of the CCEC development plan into a national-level development strategy, the process of selecting key metropolitan areas out of city clusters has been basically completed.

The practical characteristics of metropolitan areas are closely related to their functional orientation, i.e. optimizing the territorial-space arrangement and promoting the coordinated regional development and the new-type urbanization, which facilitate smooth regional economic circulations. Will the focus of development policies placed on the metropolitan areas, the problems caused by the wide coverage, which are tricky issues regarding the regional economic circulations, are solved. Therefore, the development of metropolitan areas is a key method and foundation for the "dual-circulation" paradigm. The metropolitan areas serve as portals, nodes and platforms for the circulations.

At present, academic circles have come up with three methods to determine the scope of a metropolitan area: the radius method, the commuting-time method, and the comprehensive method. The effective application of each method needs to rely on the support of economic, social and geographical data. Among them, the comprehensive method is more accurate and scientific than the other two because it involves multi-dimensional indicators such as radius, commuting time, proportion of commuters, population, and economic activity. In the "Guiding Opinions of the National Development and Reform Commission on Cultivating and Developing Metropolitan Areas," a metropolitan area is defined as an urban area that takes a mega or super mega city as its central city and boasts an one-hour commuting circle. In the following analysis of metropolitan areas and their cities, we adopts the definition and classification made by Ren et al. (2019): There are 24 metropolitan areas with over 10 million people in China, namely the Shanghai metropolitan area, the Suzhou-Wuxi-Changzhou metropolitan area, the Shenzhen-Dongguan-Huizhou metropolitan area, the Guangzhou-Foshan-Zhaoqing metropolitan area, the Beijing metropolitan area, the Hangzhou metropolitan area, the Wuhan metropolitan area, the Nanjing metropolitan area, the Chengdu metropolitan area, the Zhengzhou metropolitan area, the Tianjian metropolitan area, the Qingdao metropolitan area, the

Table 3 The functional positioning of metropolitan areas in relevant policies and plans

Policies and plans	Time of issue	The functional positioning of metropolitan areas
The "National New-Type Urbanization Plan (2014–2020)"	2014	Enhancing the radiating and driving functions of central cities
The "Outline of the 13th Five-Year Plan (2016–2020)"	2016	Improving the transportation of city clusters and strengthening the radiating and driving functions of central cities
The "Notice on Implementing Key Tasks of Promoting New-Type Urbanization Construction in 2018"	2018	Improving the construction quality of city clusters
"Key Tasks for New-Type Urbanization in 2019"	2019	Optimizing urban layouts
The "Guiding Opinions on Fostering and Developing Modernized Metropolitan Areas"	2019	Supporting the high-quality development of city clusters and the economic transformation and upgrading
The "Outline for the YRD Integrated Development Plan"	2019	Facilitating the establishment of a new development paradigm featuring the coordinated regional development
"Key Tasks for New-Type Urbanization and Integrated Urban–Rural Development in 2020"	2020	Optimizing urban spatial layouts and improving the comprehensive carrying capacity of cities
The "Proposals of the CPC Central Committee on the Formulation of the 14th Five-Year Plan (2021–2025) for National Economic and Social Development and the Long-Range Objectives Through the Year 2035 (Vision 2035)"	2020	Optimizing territorial spacial planning, promoting the coordinated regional development, and advancing the new-type urbanization

Source of information The information in this Table is compiled based on relevant materials

Xiamen-Quanzhou-Zhangzhou metropolitan area, the Ningbo metropolitan area, the Changsha-Zhuzhou-Xiangtan metropolitan area, the Chongqing metropolitan area, the Hefei metropolitan area, the Jinan metropolitan area, the Shenyang metropolitan area, the Xi'an metropolitan area, the Changchun-Jilin metropolitan area, the Nanchang metropolitan area, the Harbin metropolitan area, and the Shijiazhuang metropolitan area.

1. **Metropolitan AreasPortals for the "Dual Circulations."** China's Special Economic Areas (SEAs) are often located within its metropolitan areas. These Areas serve specific purposes such as promoting economic development and enjoy preferential policies. According to Sun & Yuan (2014), by 2020, China has built 19 national-level new districts, 21 free trade zones, five comprehensive

supporting reform pilot zones, 10 pilot financial reform zones, 99 cross-border e-commerce comprehensive pilot zones, and 71 comprehensive bonded zones. Most of these SEAs are situated in China's 24 metropolitan areas: They promote economic development, openness, in-depth reform, and institutional innovation; They play a vital role in the establishment of the "dual-circulation" development paradigm.

2. **Metropolitan Areas: Nodes for the "Dual Circulations."** The central cities of metropolitan areas are often municipalities directly under the Central Government, cities specifically designated in the state plan, or provincial capitals. These cities boast high-level economic and social development and physical and institutional infrastructure, constituting regional- or national-level growth poles. Therefore, the metropolitan areas become "nodes" for the "dual circulations," i.e. hubs for the flows of people, logistics, information, technologies, and capitals, which pool resources from their surrounding areas and exert influence on them. In the initial stage of development, the siphon effect of central cities are stronger than their diffusion effect. With progress in their economic and social development, they will enter the high-quality development phase wherein their diffusion effect outshine the siphon effect because of the strong inter-city ties and the influences of the governments and markets. In this development phase, the central cities became importance distributors of people flows, logistics flows, information flows, technology flows, and capital flows; the peripheral cities benefit from the diffusion effect and narrow the gap between themselves and the central cities. Thanks to this diffusion effect, the metropolitan area will take the lead in completing the goal of regional integration. Among the 24 metropolitan areas with over 10 million people, the Shanghai metropolitan area has taken the lead in entering the high-quality development phase and pursuing regional integration. The central city, Shanghai, mainly exerts diffusion effect on its surrounding cities and towns, playing a significant role in promoting their development. There are three cities under its influence, namely Suzhou, Wuxi, and Ningbo, boasting trillion-level economies. Given its role in the "domestic circulation" and "international circulation", it is fair to say that the Shanghai metropolis area has already become a "node" for the "dual circulations."
3. **Metropolitan Areas: Platforms for the "Dual Circulations."** Metropolitan areas embody economies of scale: With the sharing, learning, and matching effects, they can create higher output with limited production factors and realize higher production efficiency. At present, metropolitan areas are important platforms for China's high-quality economic development. This point can be proved by their economic aggregate. Ren (2019) summarized according to the data released in 2018 that the 24 metropolitan areas, which occupied only 6.7% of the country's land, gathered about 33% of the country's population and created 54% of its GDP. From the perspective of industrial clustering, these areas also offer platforms for high-tech industrial clusters. Among the 66 strategic industry clusters selected by the the National Development and Reform Commission, 43 are located in the 24 metropolitan areas (Notice of the National Development and Reform Commission on Accelerating the Development of Strategic Emerging

Industry Clusters, 2019). In terms of innovative activities, the metropolitan areas are in the forefront of scientific and technological innovation. Ren (2019) count the number of patent applications at the municipal level: Among the more than 1.26 million applications, about 0.79 million applications came from the 24 metropolitan areas, accounting for 62.7% of the total number. As to science, education, culture, and health resources, the 24 metropolitan areas also have absolute advantages. According to China City Statistical Yearbook 2019, which contained data for 2018, these metropolitan boasted 1,935,727 hospital beds, which accounted for 31.4% of the national total number of urban hospital beds; they had about 447,620 books in their public libraries, accounting for 46.5% of the books collected by urban public libraries; They had about 1,300 higher-education institutions, accounting for 50.4% of the total number of urban higher-education institutions.[1,2]

To sum up, the theoretical characteristics of metropolitan areas are embodied in their economic resilience and their ability to gather economic externalities; their practical characteristics are reflected in their position as the portals, nodes, and platforms for the "dual circulations." These theoretical and practical characteristics help to solve regional problems of China. To begin with, the high market sensitivity of the metropolitan areas is conducive to the integration of markets and that of an economic region, which helps to smooth out difficulties in our regional economic circulations. Furthermore, the metropolitan areas' strong appeal to economic activities and population will gather more enterprises and population in China's southeast coast and its urban areas. In this process, China's per-capita development level will be further improved, its regional coordinated development and urban–rural integrated development will be promoted, and the barriers in its regional circulation will be removed. Last but not least, as spatial units smaller than city clusters, the metropolitan areas can serve as good portals, nodes, and platforms for the "dual circulations," avoiding the efficiency problems caused by wide coverage of policies. They facilitate the implementations of regional policies, smooth regional economic circulations. Given these functions, the development of metropolitan areas are important methods and foundation for the establishment of the "dual-circulation" paradigm.

3 The Pathz that Metropolitan Areas Can Take to Promote the "Dual Circulations"

As shown in Table 2, the key to building the "dual-circulation" paradigm lies in tapping the potential of domestic demand and promoting innovation-driven progress and openness. As an important method and foundation for the realization of the

[1] The Chinese Research Data Services Platform (CNRDS).

[2] China City Statistical Yearbook 2019.

paradigm, metropolitan areas shall take the following paths to give full play to their due roles.

3.1 Path to Increased Domestic Demand

GDP consists of three parts: investment, consumption, and net export. Among them, consumption and investment are collectively referred to as domestic demand. Metropolitan areas promote consumption and effective investment through the new-urbanization drive. First, the development of the metropolitan areas is conducive to progress in urbanization, and such progress boost consumption. As shown in Fig. 4, the proportion of rural per-capita consumption expenditure to urban per-capita consumption expenditure has steadily increased from 40.5% in 2013 to 47.5% in 2019, but the proportion is still below 50%. This suggests that further urbanization can further tap the consumption potential. There are two ways in which metropolitan areas promote progress in urbanization: the rural–urban migration and the spatial expansion of core cities. As to the rural–urban migration, this way is usually realized by core cities, especially those with large scales. These cities usually boast strong economies, good social governance, abundant employment opportunities, high-quality public services, and large market scales. Therefore, they can attract a large number of rural surplus laborers. As for the spatial expansion of core cities, it is a natural result of the development of metropolitan areas: With strong inter-city ties, a metropolis area can actualize the integrated development; In this integration process, counties will be upgraded or merged into municipal districts. Second, the development of metropolitan areas create more employment opportunities, which in conducive to the increase in consumption. The integrated development of central cities and peripheral cities will gather more social and economic activities. Under such circumstances, industrial clusters will create a large amount of jobs. With increase in employment, there will be increase in income and consumption. Third, the development of modernized metropolitan areas can effectively boost investment: The investment in infrastructure projects such as those regarding highway networks, rail transit networks, logistics networks, and information networks are a must for the development of modernized metropolitan areas; The continuous inflow of population will attract investments in fields closely-related to people's livelihood to satisfy people's needs for education, medical care, old-age care, and unemployment relief, because the public services can hardly be improved within a short period of time to satisfy all these needs.

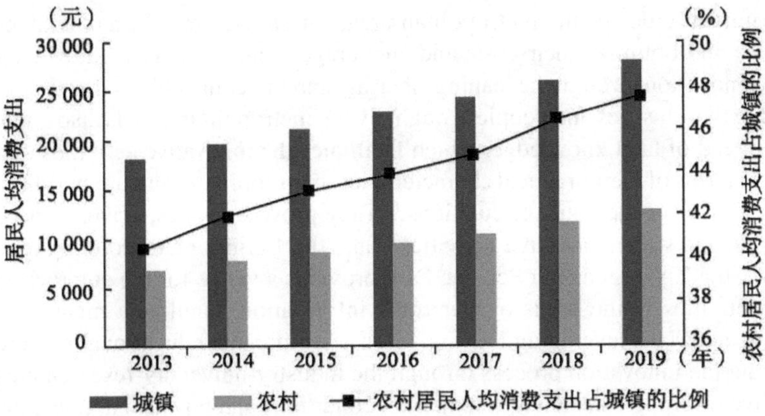

Fig. 4 Per-capita consumption expenditures of urban and rural residents 2013–2019 and rural–urban consumption ratio (*Data source* The data of this Figure comes from the the EPS database and estimates based on data from this database)

城镇	Urban areas
农村	Rural areas
农村居民人均消费支出占城镇的比例	Rural–urban consumption ratio
(年)	(year)
居民人均消费支出	Per-capita consumption expenditure of residents
农村居民人均消费支出占城镇的比例	Rural–urban consumption ratio
(元)	(yuan)

3.2 Path to Innovation-Driven Growth

The innovation process is about the creation of new knowledge and entails the accumulation, sharing, acquisition, transfer, and application of knowledge. In terms of their theoretical characteristics, a metropolitan area often feature the co-existence of an urbanization economy and a localization economy: Both the grouping of various industries and that of enterprises belonging to the same industry can generate positive externalities. In this case, enterprises in different life cycles can benefit from the metropolitan area's innovation activities. For start-up and growing enterprises, the industrial diversity provided by the central city facilitates their development by allowing them to learn rapidly, and accumulate knowledge. Its advantages in production factors and market size give these enterprises sufficient motivation to innovate. As to the innovation risks the enterprises have to face, they are decreased by the central city's mature market mechanism, business environment, and funding conditions. For mature enterprises, they usually benefit from specific industry clusters in the peripheral cities of the metropolitan area. These industry clusters enable them to reduce and optimize their costs and encourage them to increase knowledge and pursue innovation through the leaning, sharing, and matching effects. In addition, the gathering of industries and people brought by the metropolitan area is also conducive to the spread of tacit knowledge, which facilitates the innovative activities of enterprises. In terms of their practical characteristics, metropolitan areas are portals, nodes and platforms for the "dual circulations." They provide venues, drivers, and distribution channels for innovative activities. Take the Beijing metropolitan area as an example, the Zhongguancun Science Park provides a venue for the enterprise innovation activities in the fields of electronic information, intelligent manufacturing, and new materials, leveraging Beijing's rich scientific and educational resources to accelerate the innovation process through the industry-university-research cooperation. Given Beijing's advantages in market conditions and physical and institutional infrastructure, the innovation achievements can be transformed, applied, and diffused in a timely manner.

3.3 Path to Promote Openness

Liu, Z. (2020) held that openness, for both domestic and international markets, was the key factor influencing the "dual-circulation" paradigm and that the openness of the domestic market, or internal openness, was a more urgent need in comparison with the opening-up to the world, or external openness. Metropolitan areas boost our internal openness and facilitate our external openness from the following two aspects. First, they promote the integration of our domestic market. A super-large-scale domestic market is the distinctive feature and key advantage of the Chinese economy, but the existing "administrative-region economy" makes it difficult to utilize the full potential of this market. Bigger than cities and smaller than city clusters, metropolitan areas are natural results of the development of markets: They are an existence beyond the administrative divisions. With the development of a metropolitan areas, its internal inter-city economic-and-social ties grow increasingly strong, and so do its inter-city coordination mechanisms. In the process, administrative barriers are eliminated, and a unified, open, competitive, and orderly market is built. This is conducive to the unification of our super-large-scale domestic market. Second, metropolitan areas promotes our opening-up to the world, which pushes forward our efforts to realize our internal openness through the institutional innovation. The actualization of a higher-level opening-up to the world depends on metropolitan areas' functions of serving as portals, nodes, and platforms for the "dual circulations," and the establishment of Special Economic Areas, usually in our metropolitan areas, embodies our institutional innovation. Various economic regions in our metropolitan areas greatly promote China's opening to the outside world. Free trade zones enjoy preferential policies on the freedom of trade, investment, capital flow, and transportation, expanding and upgrading our opening-up endeavor. Adopting measures to realize tax exemption without tickets, income-tax collection on a deemed basis, and fast customs clearance, cross-border e-commerce comprehensive pilot zones facilitate the building of a complete industrial chain and ecological chain in cross-border electronic commerce and help to open more sectors and carry out in-depth opening-up.

4 Suggestions on Fostering and Developing Modernized Metropolitan Areas Under the New Paradigm

Metropolitan areas have spatial and temporal features: Temporally, their development is part of the urbanization process, embodying the trend of regional integration and boasting strong gathering and diffusing effects; Spatially, a metropolitan area often features a layered structure: the combination of a big central city and its peripheral small cities. With these two features, China's metropolitan areas have experienced fast growth and developed into important platforms for our high-quality development. Metropolitan areas also play a significant role in the establishing of the "dual-circulation" paradigm. Leveraging their theoretical and practical characteristics, they help to smooth out barriers in our regional economic circulation, thus promoting our domestic and international circulations, or the "dual circulations." They offer an effective method and a solid foundation for the establishment of the "dual-circulation" paradigm, which is embodied in their efforts to increase domestic demand, and promote innovation and openness. At present, China's metropolitan areas still rely on the market-driven spontaneous development and urgently need guidance and support, which are delivered through policies. In this regard, we believe that these policies shall entail two aspects, namely the inter-metropolitan-area and intra-metropolitan-area aspects. They shall call for accelerated efforts to push forward relevant urban planning projects, improve inter-metropolitan-area physical and institutional infrastructure, and foster endogenous development drivers.

4.1 Accelerated Efforts to Push Forward Urban Planning Projects

At present, China's metropolitan areas are still in the early stage of their development. Only a limited number of development plans have been formulated for them, and most of the available plans concentrate solely on provinces along the eastern coast. Such plans constitute predictive intervention and guidance for the development of modernized metropolitan areas, and thus it is a must to accelerated our efforts to promulgate such plans. First, it is essential to coordinate our development efforts and adopt policies according to local conditions. As the metropolitan areas in the eastern region have been planned earlier and developed faster, it is necessary to adopt plans to speed up the development of metropolitan areas in the central and western regions. When formulating these plans, it is a must to recognize the fact that the central and western metropolitan areas are different from their eastern counterparts in terms of development stage, economy, population size, traffic smoothness, and commuting needs. Therefore, these development plans for metropolitan areas must contain measures adapted to local conditions and avoid applying fixed standards to all metropolitan areas. To be more specific, metropolitan areas in different regions can be divided into three types according to corresponding economic-and-social-development indicators, namely starting-up type, developing type, and mature type, and construction standards shall be formulated based on this classification. Second, it is important to coordinate our planning efforts. At present, the problem of uncoordinated planning is not prominent because the number of our metropolitan-area development plans are still limited. Nevertheless, the inconsistency of statements in some planning documents reminds us of this problem. For instance, "the Suzhou-Wuxi-Changshu Metropolian Area Development Plan (2001–2020)" and "the YRD Metropolian Area Development Plan (2001–2020)" clearly define the Suzhou-Wuxi-Changshu Metropolian Area, but "the Shanghai Metropolitan Area Overall Development Plan (2017–2035)" include the three cities, i.e. Suzhou, Wuxi, and Changshu, in the Shanghai Metropolitan Area. Future plans for metropolitan areas must exclude such overlapping to prevent conflicts among themselves. Third, it is necessary to learn from the advanced experience of foreign countries. The term 'metropolitan area" originated in Japan. It is otherwise known as Metropolitan Statistical Area (MSA) and Travel to Work Area (TTWA), respectively in the United States and the UK. These countries have accumulated decades of experience in the development of metropolitan areas, which can be used for our reference. For example, we can use their multi-indicator model to determine the size of a metropolitan area, which involves indicators such as population size, proportion of commuters, age range of commuters, and light density at night.

4.2 Accelerated Efforts to Improve Inter-Metropolitan-Area Physical and Institutional Infrastructure

Liu (2020a, 2020b, 2020c, 2020d, 2020e) emphasized that the "dual-circulation" was a development strategy for the nation instead of a regional or municipal strategy. That is, the realization of the "dual circulations" entails the cooperation among multiple closely-related metropolitan areas instead of relying on a single metropolitan area. To ensure close relations and smooth circulations among metropolitan areas, it is a must to enhance the physical and institutional infrastructure. The institutional infrastructure, i.e. cooperation mechanisms and sharing platforms for metropolitan areas, shall be improved through the policy coordination and cooperation among metropolitan areas. Nevertheless, due to the influence of the so-called "administrative-region economy" and the competition among central cities, metropolitan areas are often in a competitive relationship. To solve this problem, the Central Government or provincial governments can try to establish cross-regional or inter-provincial quasi government organizations for the communication and coordination among metropolitan areas and build sharing platforms to promote progress information exchange, technology, tourism, and trade. The physical infrastructure refers to transportation, energy, telecommunications, and environmental protection facilities. Among all these types, infrastructure facilities of transportation and telecommunications need to be given priority. Governments at all levels shall encourage all types of private investment in the special funds for infrastructure development in metropolitan area. They must try their best to make up for the existing defects in the infrastructure networks such as those of roads, railways, and telecommunications and enhance key infrastructure projects such as ports and airports, ensuring smooth flows of people, goods, information, and capitals.

4.3 Accelerated Efforts to Foster Endogenous Development Drivers

Metropolitan areas are the natural product of the development of market. This development process entails increased importance of people and technology factors in comparison with that of land and capital factors, improved factor flows, and enhanced entrepreneurship promoting production activities based on the combination of factors. In a word, to foster endogenous development drivers for metropolitan areas, the emphasis shall be placed on the efforts to promote factor flows and strengthen the entrepreneurial spirit. As to factor flows, improvement in this regard depends on the integration process occurring in metropolitan areas. To promote the integration of a metropolitan area, it is a must to do well in the following three aspects: the integration of its infrastructure, i.e. the overall enhancement of its rail-transit, highway, information, and logistics networks; the integration of its public services, namely the overall improvement of its basic service, social-security system, intelligence level, especially the institutional guarantee for the free flow of people; the integration of its market, which is realized through reforms in the household-registration and land systems, the financial sector, and the field of science and technology and aimed to better facilitate the flows of various factors. As for the entrepreneurship, progress in this regard depends on the "streamlining of administration, delegating powers, and improving regulation and services" reform. Administrative examination and approval centers of metropolitan areas shall be built to simplify the examination and approval process, unify relevant standards, and remove institutional obstacles for enterprises to enter the market and participate in the market competition. Market supervision mechanisms should be optimized in accordance with the Anti-Monopoly Law of the PRC and the Anti-Unfair Competition Law of the PRC, and the law enforcement ability should be strengthened to implement the anti-monopoly and anti-unfair competition laws. This can create a fair market environment for enterprises to compete.

References

Dong, X., & Yuan, Y. (2014). Firm innovation, life cycle and agglomeration economies. *China Economic Quarterly, 13*(02), 767–792.

Dong, Z., & Li, C. (2020). China's double-circulation new development pattern: source, implications and policy orientation. *Journal of the Party School of the Central Committee of the CPC (Chinese Academy of Governance), 24*(05), 47–55.

Dushiquan xietong fangyi kangyi lianhe fugong fuchan shuang jianxiao (Effective inter-city cooperation for epidemic contral and the resumption of work and production). *Xinhuanet*. http://www.xinhuanet.com/info/2020-02/28/c_138825022.htm

Huang, Q. (2020, October 27). *Yi gengshen cengci gaige tuidong goujian wanzheng neixu tixi* (Construct a complete domestic demand system through in-depth reforms). *Economic Daily*, 001.

Huang, Q. (2020b). Efforts should be made to create new development pattern whereby domestic and foreign markets can boost each other, with the domestic market as the mainstay. *China Economic Weekly, 2020*(09), 74–81.

Huang, Z., & Yao, T. (2020). Market segmentation and regional productivity: Theoretical mechanism and empirical evidence. *Journal of Finance and Economics, 46*(01), 96–110.

Liu, H. (2020, November 25). *Jiakuai goujian yi guonei da xuenhua wei zhuti, guonei guoji shuang xunhuan xianghu cujin de xin fazhan geju* (Accelerate the construction of a new development paradigm with the domestic circulation as the mainstay and the domestic and international circulations reinforcing each other). *People's Daily*, 006.

Lu, M. (2020, Sptember 4). *Guonei da xunhua zenme gao? Xian zhaochu neixu chichi buneng tizhen de yuanyin* (How to promote the domestic circulation? First find out the reasons for the weak domestic demand). https://www.163.com/dy/article/FLLQ0LBQ051481US.html

Liu, W. (2020, September 24). *Yi xin fazhan geju chongsu woguo jingji xin youshi* (Reshape China's economic advantages through the new development paradigm). *Economic Daily*, 001.

Liu, W. (2020, October 19). *Zenyang dudong "shuang xunhua" xin fazhan geju* (How to understand the new development paradigm featuring "dual circulations"). *Henan Daily*. http://newpaper.dahe.cn/hnrb/html/2020-10/19/content_449858.htm.

Liu, Y. (2020, September 7). Guonei guoji shuang xunhuan shi xianghu cujin de (Domestic and international circulations promotes each other). *Beijing Daily*, 013.

Liu, Z. (2020). Seeking the new logic of reshaping Chinese economic cycle in domestic and foreign markets. *Exploration and Free View*, 2020(07), 42–49 and 157–158.

Lv, B., & He, Y. (2020). On the trend of regional market integration: Measurement and analysis on Chinese commodity market segmentation based on city data. *Economic Theory and Business Management, 2020*(04), 13–25.

Quan, H. (2020, October 21). *Xin fazhan geju shi kaifang de guonei guoji shuang xunhuan* (The new development paradigm boasting open domestic and international circulations). *Study Times*, 003.

Ren, Z., Xiong, C., & Yan, K., (2019, December 28). *Zhongguo shida zuiju qianli dushi quan: 2019* (China's Top 10 Promising Metropolitan Areas: 2019). https://www.sohu.com/a/363257340_467568

Song, Z., Sun, J., & Xia, T. (2020). The scale, mechanism and system of metropolitan area under the strategy of urban agglomeration—Based on the comparative perspective of planning and development. *Academic Research, 2020*(09), 92–99.

Sun, J., & Sun, X. (2017). Research progress of regional economic resilience and exploration of its application in China. *Economic Geography, 37*(10), 1–9.

The National Development and Reform Commission of the PRC (2019), Notice of the National Development and Reform Commission on Accelerating the Development of Strategic Emerging Industry Clusters (document number: fagaigaoji [2019] 1473). https://www.ndrc.gov.cn/

Wang, G., Ye, Q., Li, F., & Gao, Y. (2019). Peiyu xiandai hua dushi quan de ruogan sikao (Some thoughts on developing modernized metropolitan areas). *Urban Planning Forum, 2019*(05), 14–23.

Xia, T., Sun, J., & Lin, W. (2018). A review of the development of China's administrative region economy and regional economy—comment on the development direction of regional economics in China. *Economist, 2018*(08), 94–104.

Xia, T., & Sun, J. (2019). *Jiyu quyu jingji lilun de xin shidai kongjian chidu chonggou yanjiu* (Research on spatial scale reconstruction in the new era based on regional economic theory). *Urban Studies, 26*(06), 13–20.

Zhang, H., Li, S., & Xu, Z. (2019). Evolution of regional disparity in China since the reform and opening-up. *Reform, 2019*(04), 78–87.

Zhu, H. (2020). New development pattern of mutual promotion of domestic and international dual circulation:Its structure and misunderstanding clarification. *Journal of Dongbei University of Finance and Economics, 2020*(06), 3–11.

Zhang, J. (2020, November 29). *Lizu guonei da xunhuan ruhe yingxiang zhongguo jingji de fazhan?* (How does the domestic circulation influence the development of the Chinese economy?). *Xinhuanet*. https://politics.gmw.cn/2020-11/29/content_34411899.htm

Zhang, W. (2003). The basic concept, characteristics and planning of metropolitan regions in Jiangsu. *City Planning Review, 2003*(06), 47–50.

Chapter 9
China's Industrial Layout Under the "Dual-Circulation" Paradigm

1 Theme of the "Dual-Circulation" Development Paradigm: Security and Smoothness

For a big developing economy, the increase in its internal demand is of great significance: It underpins the economy's ability to deal with external risks and challenges, promote economic development, and enhance economic security. In recent years, the traditional international economic circulations have been weakened by the anti-globalization movement, the COVID-19 pandemic, and the rising unilateralism and protectionism, which has posed a threat to the economic security of China.

1.1 Domestic and International Conditions for China's Industrial Layout

The "Proposals of the CPC Central Committee on the Formulation of the 14th Five-Year Plan (2021–2025) for National Economic and Social Development and the Long-Range Objectives Through the Year 2035 (Vision 2035)," adopted at the 5th Plenary Session of the 19th CPC Central Committee, contains the following statements: "Accelerated efforts shall be made to build a modernized economic system and a new development paradigm with the domestic circulation as the mainstay and the domestic and international circulations promoting each other. Progress shall be realized in the modernization of our national governance system and governance capacities. Sustainable and steady economic growth and social stability and harmony shall be pursued to facilitate our drive to build a modern socialist country in an all-round way." The "dual-circulation" paradigm is a strategic arrangement made for our 14th Five-Year Period (2021–2025) and the follow-up periods in the future. The term"dual-circulation" reflects an objective law governing the development of a major economy of the world: The development of such an economy cannot rely

J. Sun, *Regional Economic Development in China in 2020*,
https://doi.org/10.1007/978-981-95-3604-7_9

solely on its external market. Historically, the United States has become the largest economy of the world depending on its strong domestic demands and economic circulation. Similarly, China's economy has to rely on its domestic demands to realize further development. The "dual-circulation" paradigm can safeguard China's economic security, enhance its economic resilience, and broaden its development space. Now that efficiency and quality have become two key factors determining China's economic development, it is a must for the country to realize high-quality development. It is essential to improve efficiency, product quality, and the competitiveness of industrial chains. It is important to carry out structural adjustment, expand domestic demand, and accelerate the transformation, upgrading, and cross-sector integration of the industrial chains. Given the opportunities provided by the establishment of the "dual-circulation" paradigm, China's industrial clusters have been improved, and the digital-technology-driven upgrading of its traditional industries has been sped up.

The 5th Plenary Session of the 19th CPC Central Committee has proposed to coordinate development and security issues and reminded all sectors to always pay close attention to security problems and resolve all kinds of risks affecting China's modernization process. In 2020, the COVID-19 pandemic plunged the global economy into the worst recession since World War II. By the end of 2020, the United States and the euro zone have seen their economy decline by 4.3% and 8.2%, respectively. Among the major economies in the world, China is the only one that is expected to achieve positive growth: Its economy experienced a brief decline this year, but it quickly recovered. As shown in Fig. 1, the growth rates of China's GDP in the four quarters of 2020 are −6.8%, −1.6%, 0.7%, and 2.3%. China's remarkable achievements in fighting the COVID-19 epidemic highlight the superiority of the socialist system with Chinese characteristics and the unified leadership of the CPC. Given the new domestic and international conditions, China timely grasps the new changes and takes the initiative of development by proposing the goal of building a "dual-circulation" paradigm. It has placed its economic work focus on ensuring its economic security and the smoothness of its economic circulations, preventing itself from being under the control of others in the future.

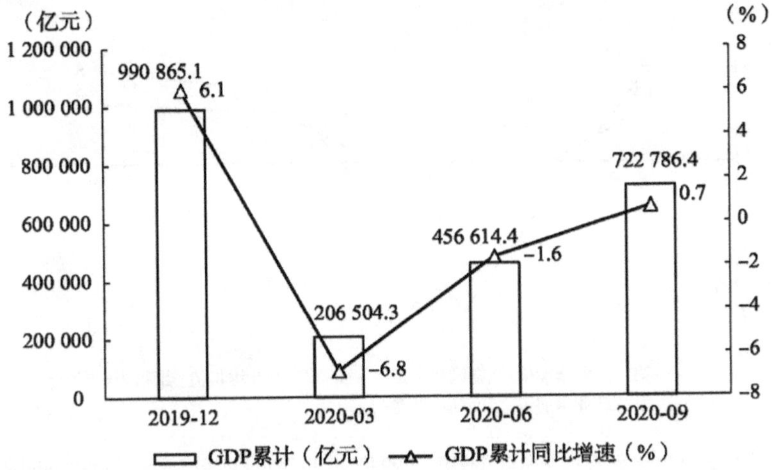

Fig. 1 China's GDP growth (*Data source* the National Bureau of Statistics of China)

(亿元)	(100 million yuan)
GDP累计	GDP
GDP累计同比增速	

The COVID-19 pandemic caused different degrees of impact on different Chinese industrial sectors. As shown in Fig. 2, its secondary industry suffered the greatest impact. In the first quarter of 2020, the added value of the secondary industry decreased by 9.6%, but the primary and tertiary sectors saw only 3.2% and 5.2% decreases, respectively. In the second quarter, with the resumption of work and production activities, the added value of the secondary industry grew by 4.7%, which was higher than those of the primary and tertiary industries. This trend continued into the third quarter, when the added value of the secondary industry increased by 6%.

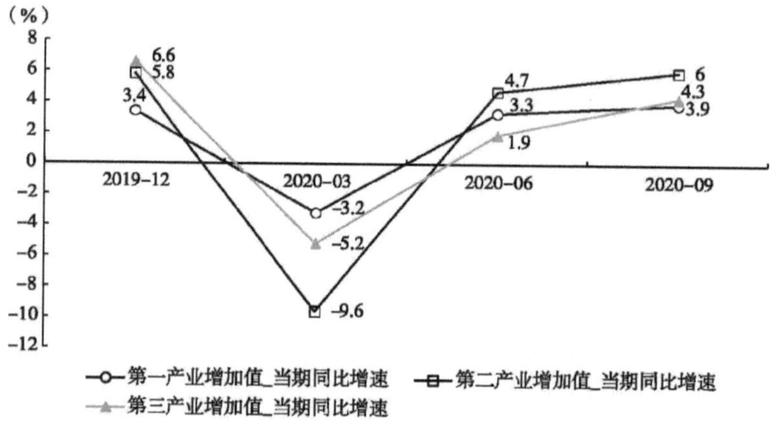

Fig. 2 Changes in the growth rate of added value in the national primary, secondary, and tertiary industries (*Data source* the National Bureau of Statistics of China)

第一产业增加值-当期同比增速	Year-on-year growth rate of added value of the primary industry
第二产业增加值-当期同比增速	Year-on-year growth rate of added value of the secondary industry
第三产业增加值-当期同比增速	Year-on-year growth rate of added value of the tertiary industry

1.2 Efforts to Safeguard the Security of Domestic Industrial Chains and Supply Chains

In view of the drastic changes in the international environment, China needs to build a new development paradigm with the domestic circulation as the mainstay and the domestic and international circulations promoting each other, which can better guarantee China's economic security and expand its development space. To ensure that the domestic and international circulations promote each other, China must actively promote the new-type globalization, the BRI projects, and global circulation based on the smooth connection between domestic and international circulations. It has to focus on the main contradiction and make up the defects of its economy.

As the only country that has all the industrial categories listed in the UN industrial classification, China has to avoid mistakes made by Japan and the United States, namely industrial hollowing-out and de-industrialization. In the meantime, it needs to pay attention to the "de-Chinaization" of global industrial chain. This is especially important at present given the on-going anti-globalization movement: Some countries, such as the United States and Japan are accelerating their efforts to revive their manufacturing industries. For its economic security and national development interests, China will have to prioritize the needs to ensure the safety of its industrial chains and supply chains when adjusting its industrial layout. It shall also actively make up for the shortcomings of its industrial chains and supply chains and increase investment in these sectors. At the 3rd China International Import Expo (CIIE), General Secretary Xi Jinping remarked, "Among China's 14 billion people, there are 400 million belong to the middle-income group. That is, China has the greatest market potential in the world. The cumulative value of its imported goods in the next 10 years is expected to exceed $22 trillion. Leveraging our super-large market and complete industrial system, we shall promote the large-scale application and timely upgrading of new technologies and accelerate the transformation of scientific and technological achievements into real productive forces. We must strengthen our industrial chains and ensure their safety.

The smooth operation of our industrial chains and that of our supply chains underpin our national economic circulations. The realization of such operation counts on the upgrading of our industrial foundations and industrial chains. First, barriers in every link of our industrial chains shall be removed. The operation of enterprises that have important status in our industrial chains and supply chains shall be given priority. Circulation difficulties in key sectors such as raw materials, crucial parts, capitals, and logistics shall be solved. The synergistic development of upstream and downstream enterprises and that of production, supply and sales enterprises shall be promoted regardless of the sizes of the enterprises. The normalized operation of shopping malls, market places, and other daily-life service providers shall be realized, ensuring smooth industrial flows, market flows, and economic circulation. Second, our underlying industrial abilities and control over key links of the industrial chains shall be improved, and the shortcomings that influence the safety of our industrial chains shall be overcome. Efforts must be exerted to solve the problems that choke our development, seeing that our industries and industrial chains can develop independently. Our domestic demands and future industrial competitiveness in the world shall be prioritized when we promote the restructuring of our industrial system, industrial chains, and supply chains through the new-type infrastructure projects.

Since adopting the policy of reform and opening-up, China has made remarkable economic achievements, especially in the manufacturing sector. It has become a major manufacturing country with strong industrial production capacity, which is a unique advantage of our economic development. As shown in Fig. 3, the added value of China's manufacturing industry in 1996 was only 2,952.891 billion yuan. In 2020, however, the scale of China's manufacturing industry has reached 31,307.1 billion yuan. With the rising proportion of service industry, China's manufacturing industry still maintains a certain scale and keeps increasing, which makes strong industrial production capacity a new comparative advantage of China. Our complete industrial system is of great strategic significance for our upgrading from a big manufacturing country to a strong manufacturing country and our drive to move up the global value chain and enhance our manufacturing competitiveness in the world. By promoting inter-regional industrial transfer within China, we can effectively enhance our high-end manufacturing industries while keeping the low-end ones.

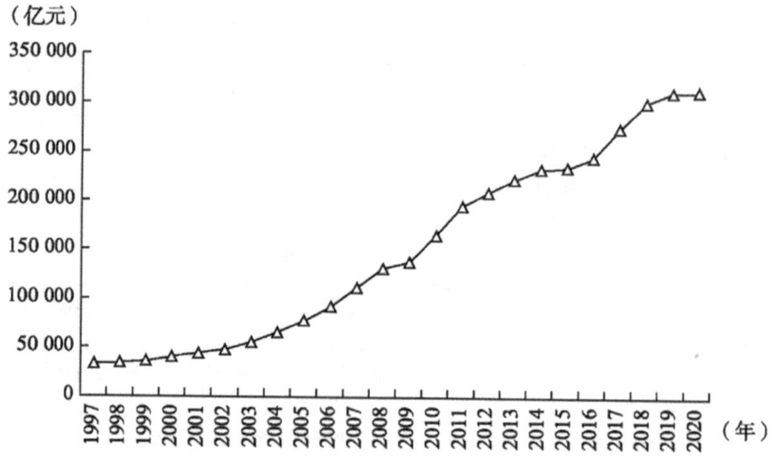

Fig. 3 Changes in the manufacturing value added in China from 1996 to 2020 (*Data source* China Statistical Yearbooks)

(亿元)	(100 million yuan)
(年)	(year)

Figure 4 is about the GDP contribution by industrial sectors. It shows that with the development of economy, the contribution of the tertiary industry to China's GDP kept increasing; that of the primary industry remained low and kept decreasing; that of the secondary industry and that of all industries started falling after a fluctuating rising trend. These changes in the industries conform with the general law of economic development that has been proved in other countries. Nevertheless, different from the situation in the United States, the UK, Japan and other developed countries, China has maintained a strong industrial productivity in this development stage and avoided industrial hollowing-out. Despite that the country's service industry continuously grows and that its proportion in GDP continues to rise, China still manages to maintain a strong manufacturing capacity and regards it as a new comparative advantage and a strong support for its domestic circulation. When establishing its new development paradigm featuring the "dual circulations," China places the focus of its industrial-planning-and-development work on the efforts to upgrade its industrial structure, strengthen its industrial production capacity, maintains its manufacturing industry scale, and conduct the innovative development of high-end manufacturing industries.

Figure 5 shows changes in the industry added value of China's four major economic regions. The industry added value created by the eastern region constituted the largest proportion, remaining around 50%, but its proportion decreased with the increase in the western region's proportion, which was an achievement of the western-region development campaign. China also plans to further narrow the east–west gap by further increasing the proportions of the industry added value created by the western and central regions and encouraging manufacturing industries

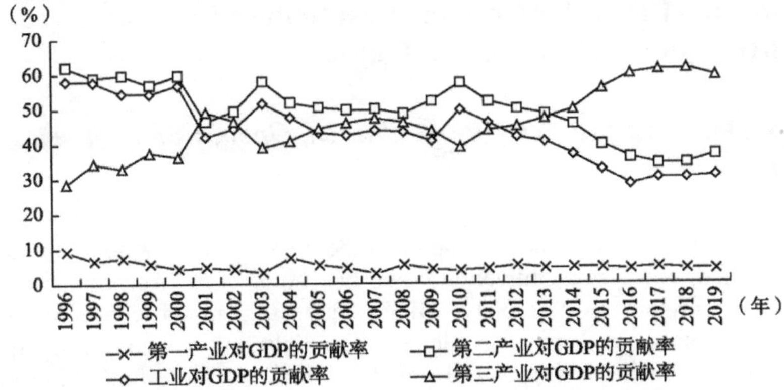

Fig. 4 Contributions to GDP by industry from 1996 to 2019 (the primary/secondary/tertiary industry's contribution to GDP, All industries' contribution to GDP) (*Data source* China Statistical Yearbooks)

to move from the eastern region to the western and central regions. This is because the "dual-circulation" paradigm requires to eliminate the existing wide inter-regional development gaps for the smooth operation of the domestic circulation. It also calls for an optimized industrial spatial layout that facilitates the domestic circulation.

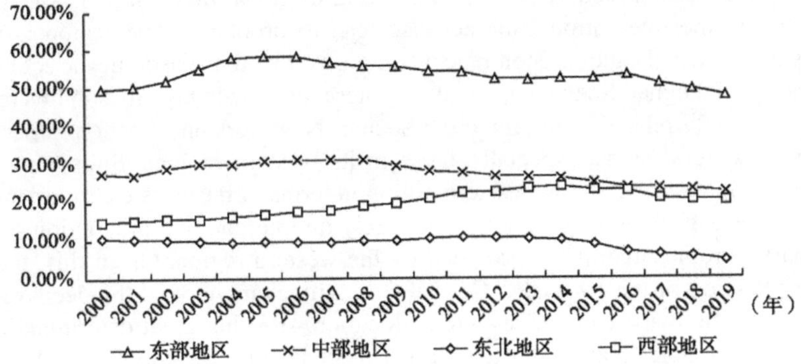

Fig. 5 Changes in the proportion of industry added value created by China's four major sectors from 2000 to 2019 (the eastern region, central, northeastern region, western) (*Data source* The data in this figure are extracted from china statistical yearbooks)

东部地区	The eastern region
中部地区	The central region
东北地区	The northeastern region
西部地区	The western region
(年)	(year)

2 Industrial Transfer from the East to the West: Narrowed Inter-regional Development Gaps

2.1 A Domestic Market That Is Unified, Open, Competitive, and Orderly

A smooth domestic economic circulation is the prerequisite to a domestic market that is unified, open, competitive, and orderly. The essence of economic circulation is the reproduction of society, which require the smooth connection of all links in the national economy. To connect these links and realize market equilibrium, or balance between supply and demand, it is important to further modernize our industrial chains supply chains, improve our development quality and competitiveness. So far, China has not unified its domestic market due to the so-called administrative-region economy, administrative barriers, and insufficient credit and openness. This is a problem needs to be solved fairly quickly, and the problem solving needs a smooth domestic circulation that helps to aggregate China's fragmented administrative-region economies into open regional economies, and condense the separated markets into a unified and strong domestic market.

Available studies have revealed that the rise of the eastern region is driven by market factors. The region has realized rapid economic development by taking the lead in marketization and actively participating in global division of labor, but its close ties to the international market also lead to problems. The region is more deeply involved in the global economic integration rather than the domestic economic integration. Shanghai, Shenzhen, and other internationalized big cities in this region are more closely related to foreign cities such as New York and London rather than China's vast rural areas, especially those in its western regions. Simply put, the western region lags behind the eastern region in terms of business environment and marketization level, and the eastern region rely on international production factors and markets rather than those provided by the western region. Given this, the 3rd Plenary Session of the 19th CPC Central Committee emphasized the decisive role of the domestic market in our resource allocation, showing great determination to remove all obstacles that hinder the market mechanism from playing its role. The integration of the domestic market realized through in-depth economic-institution reforms and the removal of market barriers will ensure that the central and western regions benefit from the spill-over effects of the economic development in the eastern region.

Industrial transfer depends on the flow of production factors such as capital, technology, labor, and information. That is, the free, inter-regional flow of production factors constitutes a prerequisite for industrial transfer. The absence of a unified market leads to a series of problems, such as increased inter-regional transaction costs, overlapped industrial structures, high industrial similarity, excessive competition, and increased competition costs and protection costs. To avoid these problems, inter-regional industrial transfers must be based on an open, fair market competition mechanism, which smooths out difficulties including regional blockades, regional segregation, local protectionism, industry monopolies, and other institutional barriers. With expanded market space and capacity, each region can realize high-level, large-size, and sustainable economic development.

To ensure the decisive role of the domestic market in our resource allocation, it is necessary to further clear away the administrative monopoly and regional blockades and promote the openness and integration of the market and the modernized circulation methods. Finance, land, technology, and labor factor markets shall be further developed, and all kinds of intermediary organizations shall be further regulated. The price formation mechanism of commodities and factors shall be further improved, and the inter-regional and inter-sector factor allocation efficiencies shall be increased. Market order needs to be further standardized, and strict measures must be taken against the production and sales of counterfeited goods, commercial fraud, tax evasion,and the infringement of intellectual property rights.

Without a unified national market, our national economy will consist of several separated economic circulations, and cannot enjoy the advantage brought by the super-large-scale domestic market. The old economic development model has led to a long-existing problem, i.e. the "administrative-region economy," which has caused the fragmentation of our domestic markets. Under this development model, local governments can directly intervene in the market through administrative means and influence market access requirements and enterprise competition. Now that the new development paradigm with domestic circulation as the mainstay and domestic and international circulations promoting each other is being established step by step, the western region shall give priority to domestic economic integration rather than international economic integration and improve the openness of its domestic market while promoting its opening-up along the border. Local governments in the region must not split the domestic market under the pretext of pursuing opening-up to the world. Some of these localities attempt to increase their status in the regional division of labor through this method, trying to protect their disadvantageous industries or even turn them into advantageous industries that can compete with their counterparts in the developed areas.

The integration of the domestic market and the improvement of the resource-allocation efficiency are the endogenous requirements of China's economic development. To meet these requirements, it is a must to advance our reform and opening-up and accelerate the formation of a unified, transparent, orderly, and standardized market environment. The western region lags behind in terms of development philosophy and economic-system reform. These weak points cause the erosion, leakage, and repulsion of factors from other regions. Given this, it is essential to remove the regional blockades, cancel the administrative restrictions on access, and reduce the local governments' intervention into the market. The fragmented economic circulation must be integrated and upgraded for a unified, open, competitive, and orderly market. It is worth noting that the policy about the integration of market is not targeted at a specific region or locality. It aims to coordinate our regional development and eliminate barriers in the flow of production factors, which are caused by institutions and policies. The elimination of local protectionism is the institutional basis for integrating the fragmented economic circulation.

In view of the unsmoothness and fragmentation of the domestic economic circulation, China's international competitiveness can hardly be improved rapidly. Its endeavor to further its opening-up also calls for the support of a strong domestic circulation. This judgement can be fully supported by economic theories. For instance, Adam Smith, the father of classical economic theory, wrote about specialization and division of labor with regard to market scale in *The Wealth of Nations*, and regarded specialization and division of labor as the fundamental driver of economic growth (Smith's theory). The east–west development gap in China is partly caused by geographical barriers, but the top cause is institutional reason, especially the "administrative-region economy." Due to China's fiscal decentralization, different localities of the country care about only their own economic interests rather than economic activities across administrative regions, which leads to the fragmentation of our domestic market and unbalanced development. Under such circumstances, it is necessary to accelerate the integration of factor and commodity markets and reduce fragmented economic circulations.

It is worth noting that our regional development policies are essentially consistent with the requirement for a unified domestic market, and that in the process of realizing such integration, incentives will be provided to encourage local governments to develop their local economies. Such efforts to maintain the localities' enthusiasm for economic development underpin the economic layout that pursue complementarities and high-quality development. The new development campaign of the western region shall not be an endeavor driven solely by the Central Government. Local governments shall also be encouraged to actively participate in the regional economic division of labor and industrial development. Their participation constitutes an essential of our national economic circulation. The weak east–west ties are a result of multiple causes: the inconvenience of transportation, caused by geographical conditions, administrative divisions, and outdated transportation facilities; and the eastern region's neglect of the market potential of the western region, which results from the western region's protectionism and the eastern region's export-oriented economic model. In the future, the western region needs to promote its openness and integrate its fragmented regional economic circulation into the unified national circulation. It shall give full play to its comparative advantages in market potential, natural resources, and labor force, and strengthen its transportation infrastructure and economic reform, effectively cooperation with the eastern region and promoting the establishment of a unified domestic market. Similar efforts shall be adopted widely to eliminate the fragmentation of our market and build a unified and open national market system that is conducive to the coordinated development of our economic regions.

While adjusting its industrial layout, China is facing drastic changes unseen in a century. Fortunately, it is still in an important period of strategic opportunities, but the opportunities and challenges it faces have changed. Given the complex internal and external changes, it decides to pursue a "dual-circulation" development paradigm to rebuild its development advantages and achieve high-quality development. Since adopting the policy of reform and opening-up, especially since joining the WTO, China with its manufacturing system which is the largest and most comprehensive in the world, has deeply participated in the international division of labor and the global industrial system. So far, it has occupied an important position in the global industrial chain and supply chain.

2.2 *Expanded Space for Economic Development: More Strategic Growth Poles in the Central and Western Regions*

From a regional perspective, China's coastal areas are facing new challenges when participating in the international circulation. In recent years, their foreign trade has been suffering from the increased costs of factors and Sino-US economic and trade frictions. For their in-depth participation in the international division of labor, they become vulnerable to fluctuations in the international market. To effectively alleviate the impact caused by changes in the international economic situation, China has to expand its economic development space and enhance its economic resilience. To this end, it must accelerate the economic development of its central and western regions.

From the perspective of the supply side, China has the world's most complete and largest industrial system and complete supporting capabilities, with 130 million registered market entities and more than 170 million highly-educated workers with various professional skills. Nevertheless, it has adhered to its traditional development model for years, which relies on imported advanced foreign technologies and equipment. Under the traditional model, Chinese enterprises increase their competitiveness by utilizing their superior factor combination abilities and institutional environment to improve their product quality and continuously reduce their product costs. This model played an active role in the past as it adapted to the environment and conditions. At present, with the changes in our development stage and those in the international market, this model faces many challenges. On one hand, the traditional ways used to upgrade technologies and conduct foreign cooperation must be changed. On the other hand, the pressure for expanding exports increases continuously.

The "Guiding Opinions of the CPC Central Committee and the PRC State Council on Promoting the Development Campaign of the Western Region in the New Era and Starting a New Chapter" (hereinafter the "Guiding Opinions"), released on May 17, 2020, makes it clear that the new development campaign of the western region is a major arrangement that meets the needs of China's endeavor to advance its socialism with Chinese characteristics in the new era, upgrade its coordinated regional development, and coordinate the situations at home and abroad. Despite the remarkable economic achievements made since the beginning of reform and opening-up, China still faces obvious development problems, i.e. insufficient and imbalance development and the wide east–west development gap. The "Guiding

Opinions" clearly expresses China's determination to narrow the east–west development gap by describing the development goals of the new development campaign of the western region as follows: The western region shall catch up with the eastern region in terms of basic public services, accessibility to basic infrastructure, and people's living standards. The CPC Central Committee has also clarified the significance of this new development campaign to the new "dual-circulation" paradigm by launching this campaign when promoting the establishment of the new paradigm.

The western region must make full use of its advantages of backwardness to attract industrial transfer from the eastern region. When implementing its new development campaign, it shall take into consideration the comparative advantages of different regions from the perspective of macroeconomics and start a new development paradigm that is based on the rational division of labor and allocation of resources throughout the country. The key to the new paradigm is new changes and new forms regarding the spatial structure of our economic activities: The development campaign of the western region is aimed to adjust our regional economic structure and narrow inter-region development gaps, especially the one between the eastern and western region. Our governments at all levels must learn advanced international experience, i.e. how a government functions in a mature market economy. When pushing forward the development campaign of the western region, they shall better play its strategic leading role besides regulating the market. Apart from fulfilling their general economic functions, they shall give full play to their special functions For instance, they shall coordinate the economic management systems of various localities, which have formed because of the "administrative-region economy," to remove administrative barriers that hinder the flows of production factors and commodities.

Since being put forward 20 years ago, the development campaign of the western region has made little progress in changing our industrial layout with industries, especially high-end industries, concentrated in coastal areas. This is because of the path dependence and technology lock-in created by our export-oriented economy. At present, the weak global economy and the rising counter-globalization trend make our export-oriented development model unsustainable, but this situation creates opportunities for the economic development of the western region. Due to the rising cost of land and labor, some industries in the eastern region of China have to be moved to the central and western regions or even overseas. That is, the western region of our country must compete with some foreign countries with low-cost labor and land for the industrial transfer from the eastern region. This requires the western region to effectively increase its appeal to the transferred industries by adopting reasonable regional development policies and promoting its openness. In this way, the western region can establish a new development paradigm and prevent our excessive industrial transfer to Southeast Asia and South Asia, which helps to maintain the competitive advantage of our manufacturing industries. Given the frequent fluctuations in international economy and trade and the increasingly enhanced domestic circulation of China, the western region boast obvious competitive advantages over their overseas competitors. In the new round of the western-region development campaign, the western region must give full play to its advantages of backwardness and strengthen its participation in the inter-regional division of labor and inter-regional trade, attracting more industries transferred from the eastern region and developing itself into a national-level industrial base.

Over the past 20 years, the implementation of the development campaign of the western region has greatly enhanced the region's infrastructure, especially its transportation, hydro-power, and energy infrastructure. This has greatly improved its conditions for industrial development. Nevertheless, there is still a big gap between the western region and the eastern region in terms of industrial development conditions. In comparison with the western region, the eastern region is stronger in science, technologies, and innovation ability. The YRD and the PRD have already realized the transformation and upgrading of their industrial structures and embarked on the path of innovation-driven development. Fortunately, the western region can leverage its advantage of backwardness. It can achieve faster growth by introducing advanced and mature technologies from the developed eastern region and gradually catch up with the developed region. The western region also needs to enhance its institutional development, optimize its business environment, and increase its investment in public infrastructure, which will facilitate the transfer and spread of technologies from the east to the west. The transfer of industries from the east to the west is of important significance to the new development paradigm of the western region, and the development of the western region constitutes an important part of the drive to expand the space for our regional development and unify our domestic market. In this sense, the east–west industrial transfer can effectively improve the allocational efficiency of resources across the country.

The key to the new development paradigm of the western region lies in the east–west industrial transfer, especially the transfer of the industrial center to the central and western regions. This will promote our diversified division of labor and the strategic depth and integration of our economy. This effort to change the current situation, i.e. the excessive concentration of manufacturing industries in the overseas areas also helps to the effectively narrow our inter-regional development gaps. Nonetheless, since the manufacturing industries in the eastern region have not developed to generate their endogenous needs for industrial transfer, this industrial transfer has to rely more on the government's regional balanced development strategy. A possible solution is to reduce the circulation costs of domestic trade by strengthening transportation infrastructure and reforming the economic institutions and establish a relatively self-sufficient and independent national economic system wherein the eastern, central, and western regions boast complementary advantages and jointly pursue high-quality development.

2.3 *Accelerated Industrial Transfer From the Eastern Region to the Western Region*

The central and western regions cover vast areas, accounting for 81.7% of the total land area of China. This hinders the industrial transfer from the eastern region. First, the great geographic distance leads to high transportation costs and barriers to technology exchanges. Second, the population distribution in the central and western regions is very uneven: The central region, with 10.7% of the country's land, is home to 26.51% of the country's population; the western region occupying up to 71% of the country's land, but its population accounts for only 28% of the national total population. In view of the above-mentioned characteristics, the central and western regions shall take a centralized approach to receive industrial transfer. Such an approach is conducive not only to high-quality urbanization in the central and western regions, but also to higher economic benefits of industrial transfer. To this end, governments at all levels must strengthen guidance, and enterprises need to pay special attention to their transfer destinations' industrial foundations, infrastructure, institutional costs, and market size. The central cities in the central and western regions shall give full play to their roles in receiving industrial transfer in a centralized way, developing themselves into desirable transfer destinations of the eastern enterprises.

Industrial transfer is considered the primary way for China to realize its industrial transformation and upgrading since it is of great significance to both the transfer-out and the transfer-in places. For the eastern coastal areas, they can transfer out the following industries: mature industries that are in the late stage of their industrial development and product-life cycle; declining industries; industries that no longer have comparative advantages due to resource bottlenecks and environment constraints; and outdated industries that no longer meet the needs of high-quality development in the future. Such industrial transfer is conducive to the smooth realization of industrial transformation and upgrading in the eastern region: It allows the region to effectively allocate its resources and concentrate more resources on their advantageous and emerging industries. It enable the region to optimize its industrial structure, accelerate the replacement of its old growth drivers by the new ones, and improve the quality of its economic development, propelling the region's economic development to reach a new height. For the central and western regions, the east–west industrial transfer enables them to absorb advanced technologies and equipment, leverage their advantages in resources, and boost their local economic development. Such an industrial transfer can also effectively promote the development of emerging industries in the central and western regions and facilitate their progress in production technologies, management ability, and development philosophy, which is conducive to their drive to optimize their own industrial structures. With the transferred-in industries, more employment opportunities can be created in the central and western regions. Therefore, these regions' surplus labor force can be transferred to better-paying industrial sectors, realizing the optimal allocation of resources. In view of the great significance of the east–west industrial transfer, the Central Government needs to attach great importance to the industrial transfer and fully inspire the two parities' enthusiasm. In the top-level design, it is necessary to give this industrial transfer human-resource, finance, and material support through various policies and also optimize the industrial layout at the national and regional levels, helping the two parties to improve their coordination mechanism for the transfer. The eastern region should be encouraged to actively participate in the industrial transfer to realize its own industrial transformation and upgrading. The central and western regions should be required to continuously strengthen their own physical and institutional conditions for investment and speed up the enhancement of their infrastructure and industrial supporting facilities, readily receiving the transferred industries.

The east–west industrial transfer of can prevent the problem of industrial hollowing-out. Now that China's economic development has entered a new stage, its labor costs has been on the rise, and its comparative advantages have changed. These changes are in line with the general law of economic development. The question is how we grasp and utilize this law in our endeavor to cultivate new comparative advantages, accelerate the upgrading and transfer of our domestic industries, and create a favorable environment for our manufacturing industries, which features low costs, investment convenience, and fair competition. China has to do well in this regard to ensure the sustainable development of its local manufacturing industries and prevent the problem of industrial hollowing-out. Boasting vast development space and a huge advantage in cost, the central and western regions of China shall speed up the improvement of its physical and institutional conditions, actively absorb industries transferred from the east, and develop their own leading industrial clusters based on their local characteristics.

Emphasis should be placed on industrial parks and the establishment of a complete policy system in the drive to optimize the central and western regions' model of receiving transferred industries. Industrial parks can effectively gather industries, share resources, and overcome the negative external effects. They also drive the development of related industries, effectively promoting the formation of industry clusters. By 2020, there are 626 national-level industrial parks and 1,166 provincial-level industrial parks in China. Through the proper guidance of the government, the inter-regional cooperation among industrial parks can be utilized to promote the orderly east–west industrial transfer. This is also one of the major methods used to realize industrial transfer in China.

According to the theory of spatial economics, the spatial distribution of industries is the result of the joint action of three production-factor mechanisms, i.e. mechanisms of concentration, selection, and classification. Leveraging their preferential policies, industrial parks can accelerate the concentration of various production factors and promote industrial transfer through the construction of new industrial parks. In China's regional economic development, such industrial clustering has played an important role: They have adjusted the regional distribution of industries based on the needs of the market and accelerated the inter-regional industrial transfer for the formation of economic centers in the central and western regions. Attracting investments based on industry clusters has also been proved to be an effective model to conduct industrial transfer: This form of investment solicitation is based on the segmentation of market and division of labor and the influence of specific industrial clusters and their supporting environment; it is targeted at enterprises related to the specific industry clusters and the relevant supporting projects. The key to this form of investment solicitation lies in introducing leading enterprises and giving full play to their leading roles. Industrial parks can cooperate with listed companies or leading enterprises located inside or nearby and leverage their high-quality assets to attract well-known investment institutions at home and abroad for the establishment of specific industrial development funds that serve the enterprises' needs for technological innovation and industrial upgrading and transformation. While helping these leading enterprises to realize further growth, industrial parks shall enhance their functions of giving guidance and industrial clustering and give full play to their advantages in attracting capitals and talents.

The investment-attraction function of industrial parks in the central and western regions should be strengthened. Industrial parks in these regions shall improve their attractiveness and reduce the cost of enterprises by adopting preferential policies regarding taxation, land use, and water and electricity subsidies. They shall provide transferred enterprises with relocation subsidies to reduce the cost of industrial transfer and ensure that these transferred industries can maintain their competitive advantages in the local markets. The development of an industry is often the result of the synergistic development of related industries. As the eastern, central and western regions are in different stages of industrialization, the east–west industrial transfer faces many practical problems. In order to promote the balance between our spatial structure and economic structure, i.e. a relatively balance of industrial distribution, the eastern industries should be transferred to the central and western regions in accordance with their industrialization stages, ensuring that the transferred-in place can readily absorb the transferred industries.

In conclusion, since the beginning of the reform and opening-up, the eastern region has outshone the other regions in terms of economic development because of the international industrial transfer and the support of preferential policies. Nevertheless, with the continuous economic progress, the region faces increasingly obvious resource bottleneck and rising labor cost, which deprives its traditional industries of their production advantages and the ability to boost sustainable regional economic growth. In this case, the eastern region must carry out industrial transformation and upgrading. First, the eastern region no longer has advantages in resources and labor, so it has to transfer its resource-intensive and labor-intensive traditional industries to the central and western regions. For such industrial transfer, the transfer threshold shall be appropriately lowered, relevant procedures shall be streamlined, and the follow-up guarantee for transferred industries shall be enhanced. This calls for the support of relevant policies. Second, the eastern region must leverage its solid industrial foundations and relatively mature market mechanism to actively absorb the international transfer of high-end industries. To this end, it shall further improve its business environment and step up its efforts to attract foreign investments, developing itself into China's pioneer in advanced manufacturing industries and economic globalization. In the meantime, it shall utilize its strong appeal to talents and resources and enhance its universities and research institutes, building itself into a national-level scientific-and-technological-innovation base.

Located in the hinterland of China and troubled by inconvenient transportation and high transportation costs, the western region has lagged behind in terms of economic development over the years. Nonetheless, the region has obvious advantages in many aspects: It boasts abundant natural resources and labor force, and its labor cost is low; It is located at the forefront of our westward opening-up and benefited by multiple preferential policies. The western region is an important destination for the industrial transfer from the eastern region. To attract transferred industries, the leading role of regional key cities such as Chongqing, Chengdu, and Xi'an should be brought into full play. They shall leverage their advantages in science, technology, and human resources to attract high-end processing and manufacturing industries and information industry from the eastern region and even developed countries, building themselves into regional-level high-tech and advanced manufacturing bases. Besides, the westward opening-up endeavor shall be pushed forward. When the problem of high transportation costs is solved, which is the most important factor reducing the western region's appeal to transferred industries, and when the foreign markets to the west are opened up, the western region can complete production and export activities conveniently in places near the target markets, turning its development disadvantages into advantages.

2.4 *Industrial Development System Wherein Supplies Create Demands, and Demands Guide Supplies*

The 15th Plenary Session of the CPC Central Committee emphasized the strategic importance of expanding domestic demand and the necessity of relying on the domestic market for a healthy economic circulation. This persistence in expanding domestic demand satisfy multiple needs: the need to cope with the impact of the COVID-19 epidemic; that to maintain the long-term, sustainable, and healthy development of China's economy; and that to meet people's growing needs for a better life.

In the past, the eastern region seized opportunities brought by the in-depth development of economic globalization and utilized China's abundant labor resources and low production costs to absorb labor-intensive industries transferred out of developed countries and built an export-oriented economy, which relied on foreign markets and resources. This economic development model aggravated the existing wide inter-regional development gaps in China. Nowadays, given the surging anti-globalization trend, the uncertainty of the external environment supporting China's economic development is increasing. Under the circumstances, China has to make its domestic circulation the mainstay of its national economic circulation to replace its economic reliance on external demands with that on domestic demands, and the drive to expand its domestic demand becomes the key to a smooth domestic economic circulation. Over the years, the implementation of the west-region development campaign has greatly improved the consumption capacity in the region. As shown in Fig. 6, the western region's proportion in the national consumption kept declining before the implementation of the western-region development campaign, but its proportion started increasing after 2003, reaching its peak 21.96% in 2017. During this period, the proportion of its population maintained about 27.88%. That is, the consumption in the western region has been lower than the national level for a long time. The backward economic situation in the western region is an important reason for the weak domestic demand of China. The excessive inter-regional development gaps have restrained our domestic consumption.

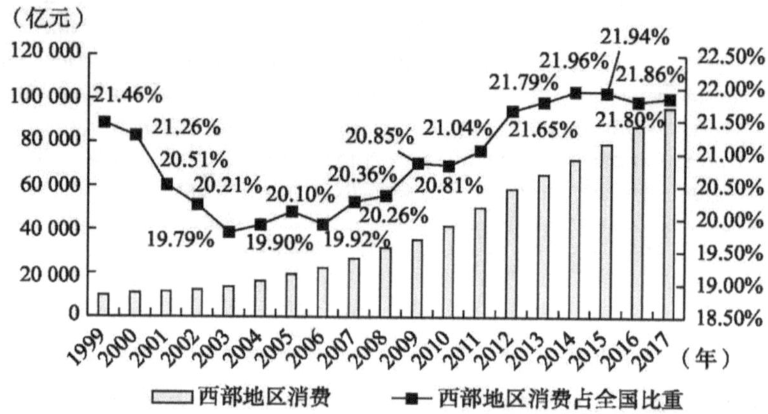

Fig. 6 Changes in consumption contribution made by the western region (*Data source* The data in this figure is extracted from china statistical yearbooks for regional economy released from 2000 to 2018)

(亿元)	(100 million yuan)
西部地区消费	Consumption in the western region
西部地区消费占全国比重	Its proportion in the national consumption
(年)	(year)

The shortage of effective demand leads to excessive supplies in the national economic circulation, which hinder social reproduction, i.e. further economic activities and cause economic fluctuations. Only a smooth domestic economic circulation can make our production, distribution, circulation, and consumption links more dependent on the domestic market and improve the adaptability of our supply system to our domestic demands, creating a high-level dynamic balance between supplies and demands: demands stimulating supplies, and supplies create demands. When discussing social reproduction, Marxist political economy theory holds that production and consumption are mutually determined and that the core of social reproduction is the realization of the total output of a social group. Insufficient consumption will affect the realization of the total output, and thus the cycle of social reproduction cannot be carried out regularly. To be more specific, goods and services produced by a social group must enter the consumption link through circulation channels. They can be reproduced only after their values are compensated through consumption. Without sufficient consumption demands, the value compensation for the capital consumed by production is sufficient. In this case, the cycle of social reproduction will be hindered, thus affecting the normal economic circulation. At present, China is troubled by overproduction and insufficient domestic demand. In essence, this is a problem caused by the mismatch between supply and demand and embodied in domestic-circulation difficulties and the lagging-behind of the western region. The state has increased its financial and material input into the targeted poverty-alleviation campaign to increase the purchasing power of the poor people in the western region. As to the eastern region, the emergence of a large number of "zombie enterprises", which are characterized by low operational efficiency and production and excessive reliance on government subsidies and bank loans for survival, reflects the problem of overcapacity.

One of the main reasons for China's weak domestic demand is its wide inter-regional development gaps. Given the low consumption ability of its western region and the weak external demand, its eastern region's strong supply capacity does damage instead of remedy to its economy, causing the contradiction between over-supply and insufficient domestic demand. To adjust the export-oriented development strategy, China proposes the "dual-circulation" development paradigm, planning to expand its domestic demand through its supply-side structural reform and increase the consumption capacity of the underdeveloped areas in its western region by boosting their economic growth. An important aspect of its supply-side structural reform is adjustments made to the spatial layout of its economic activities. In the new era, it is a must to make full use of the backwardness advantages of the western region, i.e. its rich resources and huge market potential. Among the economic regions, the western region has the largest proportion of low-income groups. That is, there is huge potential for its consumption upgrading. At present, China's industries, especially high-end industries, are excessively concentrated in its eastern coastal areas, which is a unreasonable industrial layout. Given this, it is essential to encourage eastern industries to move to the central and western regions by strengthening our inter-regional division of labor and inter-regional trade.

3 Enhanced Independent-Innovation Capability for the Accelerated Development of Strategic and Emerging Industries

3.1 The New System for Mobilizing Resources Nationwide to Make Key Technological Breakthroughs, Solving "Bottleneck" Problems

The 5th Plenary Session of the 19th CPC Central Committee adhered to the new development philosophy that highlighted innovation, coordination, green development, openness, sharing, and security while highlighting the core position of innovation in China's modernization drive. Apart from making a series of major arrangements that were aimed to promote the innovation-driven development, make the country strong in science and technology, and develop a modernized industrial system, the Session emphasized the scientific and technological self-reliance as the strategic support for China's development and called for efforts to optimize and upgrade the country's economic system.

China makes the strategic decision to pursue a "dual-circulation" paradigm for internal and external reasons. Internationally, China is faced with all-round containment and suppression from some Western countries, which increases the challenges for and uncertainties about China's economic growth. Domestically, China's total GDP and per-capita GDP have reached new heights in 2019, but its economic problems cannot be ignored, such as its declining economy growth rate, relatively low-quality growth, and reliance on crucial technologies from foreign countries. In view of the above-mentioned problems, China must strengthen its basic innovation capacity and give full play to its new system that mobilizes resources nationwide to make breakthroughs in the fields of high technologies and basic sciences and solve "bottleneck" problems.

The 5th Plenary Session of the 19th CPC Central Committee proposed that "realizing key technological breakthroughs and becoming an innovative country" should be included in China's 2035 Vision. The Session highlighted the core position of innovation in China's modernization drive in the 14th Five-Year Plan period (2021–2025) and the strategic support of scientific and technological self-reliance. It called for efforts to strengthen the nation's scientific and technological strength, enhance the technological innovation capability of enterprises, stimulate the innovation vitality of talents, and improve relevant systems and mechanisms for scientific and technological innovation.

Sciences are the major source of technological innovation. China's weakness in basic sciences explains its lack of crucial technologies. That is, to fundamentally solve the "bottleneck" problems and create its own crucial technologies, China must strengthen its basic science research. The Western countries' success in the Scientific Revolution and the Industrial Revolution is based on their progress in basic science research. To turn itself form a follower into a leader in the field of science and technology, China has to step up its basic science research. Only in this way can we provide a solid guarantee and lasting driving force for China's technology application.

Innovation underpins the progress of a nation and the prosperity of a country, and the Chinese nation is well-endowed in this regard. China has the institutional advantage of concentrating its resources for great endeavors. It shall give full play to its new system that mobilizes resources nationwide to make key technological breakthroughs and strengthen the guarantee for crucial links, areas, and products. It shall formulate an action plan to strengthen the country's sciences and technologies, aiming at frontier fields such as artificial intelligence, quantum information, and integrated circuits and initiating a number of national-level science-and-technology projects. It shall strengthen its guidance for innovation activities and promote industry-city, industry-education, and industry-research integrated development. It shall strengthen its basic innovation platform, make up its shortcomings in scientific and technological innovation ability, and increase its investment in relevant research to meet the requirements for the buildings of a innovative country. So far, China has not yet solved its deep-seated institutional problems that hinder its scientific and technological innovation or gotten rid of its reliance on imported crucial technologies.

The integrity of China's industrial chains and supply chains depends on reliable imports, which have been impacted by the COVID-19 pandemic and the complicated international environment. Under such circumstances, China must replace some imported parts, especially high-tech products, with domestically-made ones. This requires the strong support from its science-and-technology sectors and the self-reliance for crucial technologies. China must arrange its innovation projects based on its industrial chains and allow these projects to influence the industrial chains, to boost the high-quality development of its economy. The country shall enhance its scientific-and-technological-innovation and emergency-response capabilities through its institutions and mechanisms and accelerate the establishment of the system that mobilizes resources nationwide to make key technological breakthroughs. It shall make up for its shortcomings, improve its weaknesses, and fix relevant loopholes, enhancing the systematic ability of scientific and technological innovation. It shall further concentrate its national scientific research resources on key areas, key projects, and key institutes. It must strengthen the transformation mechanism of scientific and technological achievements, connecting the innovation links in the industry-university-research institute cooperation and upgrading the relevant industry and value chains. It must ensure the stability and development of its domestic and international industrial chains.

3.2 *The "Dual-Circulation" Paradigm, Bringing Opportunities for Digital Transformation*

Since the beginning of the twenty-first century, the new-generation information technologies represented by cloud computing, big data, and mobile Internet have advanced by leaps and bounds, bringing about subversive changes in our daily lives and production methods. Data resources has been growing exponentially and explosively, starting a new era: the digital age. At present, the digital-technology sector boasts the strongest innovation vitality: Big data, artificial intelligence, block-chain, quantum computing, and other digital technologies continue to evolve, providing an inexhaustible driving force for the innovation-driven development of digital China. With 940 million Internet users, China owns the world's largest consumer Internet and manufacturing capacities. Its digital economy boasts great potential and innovation vitality, and its "dual-circulation" development paradigm that is aimed to tap the potential of its super-large-scale domestic market is conductive to the increase of its digital productivity and the release of its digital dividends. Its digital transformation driven by the accelerated economic activities also influences its regional development and reshapes the regional structure. Its digital economy promotes the coordination between and even the integration of the Internet and various industrial sectors in the national economy, reshaping the upstream and downstream ecology of the industrial chains.

In March 2020, the CPC Central Committee and the PRC State Council issued the "Opinions on Building a More Complete System and Mechanism for Market-Oriented Allocation of Factors." The top authorities called for accelerated efforts to develop a data factor market and new industries, new businesses, and new models of the digital economy while emphasizing the necessity of establishing a mechanism that evaluates and remunerates production factors in accordance to their contribution. General Secretary Xi Jinping pointed out that in the Internet-empowered economy, data is a new production factor, a basic and strategic resource, and a productive force. According to the report released by the World Bank in 2020, the development of digital economy will have a fundamental impact on the structure and process of the current economic system. In the United States, for example, the development of digital technology has brought about great changes in recent years, and the intercontinental digital-information flows have also increased remarkably.

Due to the impact of the COVID-19 pandemic, the public's recognition of digital transformation has been further enhanced, which determines that the focus of the next-stage industrial upgrading shall be placed on the digital infrastructure, cloud services, the Internet of Things, and the Internet of remote things. During the epidemic period, various online-consumption and new-consumption models represented by online retail business developed rapidly, which greatly promoted consumption recovery and consumption upgrading and provided an important support for our domestic circulation and the post-epidemic consumption recovery. The advantages of online retailing were particularly prominent when the authorities implemented strict epidemic-prevention-and-control measures. Back then, some local governments and e-commerce platforms jointly distributed electronic coupons, stimulating online and offline consumption through subsidies. The PRC Ministry of Commerce mobilized e-commerce platforms to launch a Shuangpin Online Shopping Festival ("Shuangpin" refers to brand and quality), realizing nationwide online retail sales of 430 billion yuan. During the 618 mid-year shopping festival of 2020, the transaction volumes of Tmall and JDCOM reached 698.2 billion yuan and 269.2 billion yuan respectively, unleashing the consumer demands suppressed by the epidemic situation. Emerging technologies represented by new-generation information technologies such as big data, artificial intelligence, 5G helped to track close contacts and carry out medical image detection, making a significant contribution to the epidemic-prevention-and-control campaign. So far, big-data analyses have been adopted to help enterprises simplify their supplier-selection process; cloud computing has been increasingly used to promote and manage their supplier relationships; automation and Internet of Things have been utilized to optimize logistics and transportation processes. China is accelerating its digital infrastructure development, i.e. 5G research and development, the development of its data centers, and that of Internet of Things construction. On this basis, the country is actively reshaping its industrial chains and supply chains, and developing its digital economy, intelligent manufacturing, life and health industry, new material industry, and other strategic, emerging industries through its concerted efforts to promote the development of its industrial Internet. It encourages the development of digital commerce, which can support its digital supply chains. It comprehensively promotes the Internet + model, encouraging enterprises to create new advantages of digital economy and pursuing digital transformation based on their own needs.

To accelerate the growth of new business and models spawned by the epidemic-control situation is a requirement put forward by General Secretary Xi Jinping. This requirement is well-founded because our digitalization process has been greatly accelerated during the epidemic, which leads to the emergence of diversified innovative business models such as platform economy, serves economy, and sharing economy. This has injected new kinetic energy into China's economic development. During the epidemic, the traditional,off-line service industries have suffered but seen new development opportunities brought by online channels: Online take-out orders, online education, online diagnosis and treatment, and delivering services, which are provided by individuals or enterprises, have developed vigorously, helping small and medium-sized enterprises get rid of the economic difficulties caused by the epidemic and boosting the digitalization of our service industries. The PRC Ministry of Industry and Information Technology issued "Notice on Promoting the Accelerated Development of the Industrial Internet," planning to promote the integration and innovation of the industrial Internet in a wider scope, further depth, and higher intensity and offer stronger growth drivers for the high-quality economic development.

While experiencing vigorous development, China's digital economy also faces some difficulties, especially its weakness in basic scientific research and reliance on imported crucial technologies. Crucial technologies are of great significance for the country, especially for its high-quality development and national security. At present, the sci-tech innovation competition is increasingly focused on systems and ecology, which puts forward higher requirements for our innovation system of digital technology. It must be acknowledged that the overall efficiency of China's innovation system is not high enough, and the scattered distribution of our scientific and technological innovation resources, the meaningless repetition, and the lower efficiency still hinder the transformation of our innovation achievements. As a result, China has not yet succeeded in getting rid of its reliance on imported crucial technologies.

Besides, China's R&D investment has long been focused on hardware instead of software: The limited progress in its software is also a cause of China's technological weakness.

Regarding the development of digital economy, China is at the forefront of the world and is likely to lead a new round of information-technology revolution: The added value of its digital economy has been ranked second in the world for many years. The "White Paper on China's Digital Economy Development 2020," released by the China Academy of Information and Communications Technology (CAICT), disclosed the following data: The scale of the country's digital economy reached 35.8 trillion yuan in 2019, accounting for 36.2% of its annual GDP and contributing 67.7% of its GDP growth; The country's digital economy structure continued to be optimized and upgraded, and the added value created through the industrial digitalization accounted for 80.2% of its digital economy. The innovative activities of new business models have increasingly prominent influence on our employment and economy, and it is necessary to create an innovative atmosphere and fine business environment for the development of our digital economy.

3.3 *Advanced Modern Service Industry, Optimizing the Spatial Layout*

The rationality of industrial layout underpins the effects of the regional advantages and the sustainable growth of the national economy. According to the universal law of economic development of human society, an agricultural economy will be developed into an industrial economy and then a service economy. With the development of information technology, various services related to the production and processing of information can be provided not only at a long distance but also at a very low cost. These services have new characteristics: They can be stored and traded. The emergence of these new services has greatly broadened the space of service trade. A large amount of across-border services provided through cyberspace have greatly improved the production efficiency of service industries, especially cyberspace service industries. Advanced transportation and communication technologies have reduced the transportation cost and the cost of providing services at a long distance, which has changed the spatial layout of service industries. Intermediate services or producer services have increased, expanding the scale of trade service: Intermediary services such as information, finance, logistics, and business services now account for more than 2/3 of the world's total service trade. In the Internet age, service industries have realized higher productivity and globalization level, and in some fields, they do better than manufacturing industries in these two aspects. For instance, videos and text messages on the Internet can be copied countless times at a very low cost, enjoying almost unlimited benefits brought by the economies of scale. No manufactured products can compete with them in this regard.

Today, with the progress of sciences and technologies, services and their providers can be separated. That is, the way that services are supplied are changed: the supply and consumption of services are no longer limited by space and time. The traditional factor endowment theory holds that the commodity price differences among different regions constitute the basis of commodity trade, and these differences mainly come from the comparative advantages of production factors in different regions. This theory faces challenges now because of the fast-developing new service industries driven by the information revolution. At present, many services can be transmitted remotely on a large scale without large-scale labor mobility. This greatly reduces the cost of service transactions and rapidly narrows service price gaps, allowing areas with service advantages to easily deliver their services to the entire country or even the whole world. From this point of view, the development of contemporary service industries can enhance the clustering of economic activities and improve their professionalism. For example, providers of high-quality music or video contents can gather together to provide low-cost and high-quality cultural services for both domestic and foreign audiences. For such service provides who can deliver their services through cyberspace, the explanatory power of traditional economic theory has declined. Some facts have proved the enhanced clustering of economic activities. Take retail business for example: JDCOM, Tmall, and other large online supermarkets can allocate resources nationwide, suppressing the living space of small off-line supermarkets. With the help of information technology, many service industries have realized high-level clustering and professionalism.

Today's service industries are also accelerating the optimization of their spatial layout. Guided by the government's plan to develop the BTH region, the GBA, Hainan, and the YRD, high-end service industries gather in international metropolises such as Beijing and Shanghai, turning them into national and international ton-notch service-industry centers. China's first-tier cities such as Beijing and Shanghai also have great comparative advantages in the development of service industries. In 2018, Beijing's service industry accounted for more than 80% of its GDP, boasting a strong service-industry foundation; Shanghai, whose services sector accounted for more than 70% of its GDP, ranked second only to Beijing and had obvious advantages in service industries. With service-industry proportions much higher than the national level, these two cities' advantages in service industries are further enhanced by the information revolution, and high-end service industries continuously flow there. As a response to this trend, it is a must to promote the equalization of public services and the integration of factor markets in our city clusters and metropolitan areas. We shall build a service network for our city clusters and metropolitan areas, promoting their synergistic progress and joint innovation. Our first-tier cities such as Beijing and Shanghai shall be developed into new regional hubs for the development of service industries. The service functions of our small and medium-sized cities shall be strengthened, creating a number of characteristic service towns that serve as new fulcrums to serve their surrounding urban and rural areas. Our maritime infrastructure shall be improved to develop service industries regarding marine logistics, tourism, information, engineering, finance, business, etc., creating a marine-service system with international competitiveness.

The development of service industry has a good penetration ability: Its development is not isolated but deeply involved and with agricultural and industrial development, boosting the growth of many related sectors. The progress of network and information technologies has increased the efficiency of traditional service industries and made services tradeable nationwide without being limited by locations. The most typical example is the online retail sales, which have maintained a high growth rate in recent years. According to available statistics, the nationwide online retail sales reached 3,877.32 billion yuan in 2015, with a growth rate of 33.3%. After that, this sector maintained an annual growth rate of over 20%, which was much higher than the growth rate of regional GDP in the same period. In 2019, the sector realized a total sales of 10,632.42 billion yuan. The rapid growth of online retail sales stand as proof that the role of contemporary service industries in allocating resources in a larger space is becoming prominent: They effectively allocate resources across the country and thus greatly expand the scope of economic activities.

3.4 The Central Role of Innovation, Improving the Quality of Industrial Development

The 5th Plenary Session of the 19th CPC Central Committee emphasized the central role of innovation in our modernization drive, which reflected that the CPC Central Committee with General Secretary Xi Jinping at the core attached great importance to the innovation-driven development strategy. The Session admitted the serious problem of unbalanced and insufficient development in China and pointed out that the reform tasks in crucial areas and key links were still arduous and that the current innovation ability still could not meet the requirements of the high-quality development. Given the set long-term goal of basically realizing socialist modernization by 2035, it required persistence in innovation for our modernization projects and self-reliance in sciences and technologies for our national development, calling for accelerated efforts to thoroughly implement the strategy of rejuvenating the country through science and education and with talents and the innovation-driven development, improve the national innovation system, and develop the country into a world leader in science and technology.

During the 13th Five-Year Plan period (2016–2020), China increased the proportion of its R&D investment from 2.06% to 2.23% and doubled its input into the basic scientific research, improving its ranking in the Global Innovation Index from 29 to 14th. During this period, major innovations emerged in China, and significant progress was made in building an innovative country. Nevertheless, the country's innovation ability still could not meet the requirements of its high-quality development, and there are still shortcomings and weaknesses in its industrial system. In its endeavor to realize socialist modernization in an all-round way, it has to seize the great opportunities brought about by the new round of scientific-and-technological revolution and industrial transformation and pursue a new development paradigm.

It must and reshape its development advantages and enhance its innovation ability. For this new endeavor in the new era, China must put innovation at the core and take self-reliance in science and technology as the strategic support for its national development. As to the new development paradigm, it refers to the model that takes the domestic circulation as the mainstay and propels the domestic and international circulations to promote each other. Adopting this paradigm is a strategic decision made by China: It is aimed to coordinate the domestic and foreign situations, focus on China's domestic affairs, and turn crisis into opportunities. For the goal of developing China into a world leader in science and technology, China must do well in the following aspects: combining scientific-and-technological innovation with institutional innovation; paying close attention to cutting-edge technologies, economic situations, major national needs, and people's lives and health conditions; pushing forward the strategy of rejuvenating the country through science and education and with talents and innovation-driven development; advancing the institutional reform in the scientific and technological sectors; enhancing the national innovation system.

Given the new trends of our times, China must enhance its ability to conduct independent innovation, especially that related to crucial technologies. It has to get rid of its reliance on imported crucial technologies, high-end equipment, and key parts. On the basis of fully absorbing and utilizing international resources, China needs to pursue "open innovation": It shall leverage its resource endowment and industrial advantages to gain advantages in some important sectors and core technologies, trying to form a system of mutual constraints and dependencies with other technologically advanced countries. Influenced by the COVID-19 epidemic, China has realized the necessity of adjusting its thinking on the development of smart cities, which is a term referring to the integrated application of digital technologies in urban management and public services. In recent years, China has invested a lot of financial and material resources in the development of smart cities, and what it has finished has played an irreplaceable role in the epidemic-prevention-and-control work. Nonetheless, the epidemic situation has also exposed problems and deficiencies of its current smart-city model, such as the lack of inter-regional and inter-sectoral cooperation, unrealistic projects, and obstacles to data sharing. In this sense, the epidemic situation has examined the effectiveness of our smart-city model and put forward higher requirements for our public governance capacity. With summaries and reflections on relevant situations, China will adjust the model of smart cities for their future development.

In the "dual-circulation" paradigm, it is a must to give full play the government's role in the top-level design and overall organization and coordination while encouraging enterprises to play the major role in basic innovation activities. It is necessary to consolidate our basic research capacities while accelerating the transformation of our scientific and technological achievements into real productivity. It is important to vigorously improve our independent-innovation ability while persisting in "open innovation" and strengthening our international scientific-and-technological exchanges and cooperation. Large metropolitan areas and city clusters shall be regarded as hubs and strategic fulcrums for the "dual-circulation" paradigm. With their rising positions in the domestic circulation, the YRD and the GBA are growing into international city clusters. Intra-cluster urbanization is advancing rapidly, narrowing the gap between central cities and their peripheral cities and towns. As the most innovative areas in China, metropolitan areas and city clusters provide an excellent environment for innovation activities. Therefore the vigorous development of these areas is conducive to the improvement of China's innovation ability.

The huge size and current characteristics of the Chinese economy determine that it must rely on innovation in frontier fields for high-quality development and encourage innovation in each of its sector and industry. It must smooth out the existing institutional barriers through market-oriented reforms, such as the reform of the commercial system, to create a first-class business environment. It shall promote interconnections among the new-type infrastructure facilities to create a good environment for innovation and entrepreneurship activities, and guide the public to respect knowledge, sciences, innovation, and talents and encourage dedication to scientific exploration, laying solid foundations for its endeavor to become a world leader in science and technology. It must uphold the core position of innovation in its modernization drive and seize the great opportunities brought about by the new round of scientific-and-technological revolution and industrial transformation. It must pursue the "dual-circulation" development paradigm and gain new development advantages. Only in this way can we survive the crisis by locating emerging opportunities in the crisis and actualize new growth during times of change.

References

Cai, Z. (2020). The formation path of regional economic layout with high-quality development: Based on the perspective of regional complementary advantages. *Reform, 2020*(08), 132–146.

Fan, J., & Yang, B. (2002). Concentration transmission of manufacturing industry in early America and its enlightenment for China's western development. *Economic Research Journal, 2002*(08), 66–73.

Guan, W., & Yao, S. (2002). Some pivotal problems about the urbanization in the process of western development. *Economic Geography, 2002*(05), 579–584.

Li, Y., & Fu, R. (2010). Weishenme dongbu chanye bu xiang xibu zhuanyi: Jiyu kongjian jingji lilun de jieshi (Why do industries in the eastern region refuse to move to the western region: An explanation based on a spatial economic analysis). *The Journal of World Economy, 33*(08), 59–71.

Liu, J., & Shu, Q. (1996). Zhongguo quyu jingji de xin shijiao: Xingzheng qu jingji (A new perspective on China's regional economy: Administrative-region economy). *Reformation & Strategy, 1996*(05), 1–4.

Lin, Y., & Zhang, P. (2005). The advantage of latter comers, technology imports, and economic growth of developing countries. *China Economic Quarterly, 2005*(04), 53–74.

Research group of the economic center at the Party School of the CPC Central Committee. (2000). Xibu da kaifa de jingji xue sikao (Economic thinking on the development campaign of the western region). *Economic Research Journal, 2000*(06), 33–40.

Sun, J. (2020). *Tanxun xin shidai xibu da kaifa 2.0 de xin fangwei* (Exploration into the orientation of the 2.0-version development campaign of the western region). *People's Tribune, 2020*(26), 42–45.

Wei, Q., Wang, A., & Wang, J. (2014). The rise of China's coastal areas: Power of market. *Economic Research Journal, 49*(08), 170–183.

Yang, Q. (2016). The track and effect of China's western development policy. *Reform, 2016*(05), 6–24.

Zhu, Z., & Liu, H. (2020). Market segmentation, regional heterogeneity and the quality of economic growth. *Reform, 2020*(04), 86–99.

Part IV
Summary and Prospect

Chapter 10
Summary of China's 13th Five-Year Plan Period and Prospect for Its 14th Five-Year Plan Period

Regional economic development is a classic topic that always welcomes new ideas. Over the past 40-plus years, China's coastal areas have gradually become the backbone of the country's economy with the launch of the reform and opening-up campaign and the promulgation of relevant development strategies. This period has also witnessed increasingly prominent imbalance among economic regions, especially that between the coastal and hinterland areas. Targeting these problems, the Central Government of China has carried out theoretical and practical explorations for the coordination development of economic regions. Since 1999, it has successively adopted strategies for the development of the western region, the revitalization of the northeast, and the rise of central China as supplements to the high-speed development strategy for the eastern region, establishing a framework for its regional economy. Since the 18th National Congress of the CPC, convened in 2012, the Central Government furthered its regional-development policies. It has carried out new endeavors to pursue coordinated development in the BTH region, rapid progress in the YREB and the GBA, integration of the YRD, and high-quality development and ecological conservation in the YRB. The CPC Central Committe's report to the 19th National Congress of the Party, held in 2017, first proposed the coordinated regional development as a overall strategic plan of the country and a path that must be followed to solve the contradiction between the growing needs of people and the inadequate and imbalanced development in the new era.

During the 13th Five-Year Plan period (2016–2020), China faces increasing uncertainties: the downward pressure from macroeconomic fragility and the complexity of the international political and economic environment. Fortunately, China's regional economic development is still characterized by overall stability, which has become an important prerequisite for the national economy to move from a high-speed growth stage to a high-quality development stage. In the 14th Five-Year Plan period (2021–2025), China will still have to face challenges brought by the rising unilateralism and the COVID-19 pandemic. Given the drastic change unseen in the past century, China will urgently needs to build a new development paradigm with the domestic

J. Sun, *Regional Economic Development in China in 2020*,
https://doi.org/10.1007/978-981-95-3604-7_10

circulation as the mainstay and the domestic and international circulations promoting each other. This new development paradigm will also be a key breakthrough point in the innovative practice of regional coordinated development after 2020. This chapter reviews the development of China's regional economy during the 13th Five-Year Plan period and summarizes its main characteristics, and on this basis, provides prospects for its development trend in the 14th Five-Year Plan period and the path to basically realize socialist modernization in China.

1 The Development of China's Regional Economy in the 13th Five-Year Plan Period

The 13th Five-Year Plan period (2016–2020) witnesses the change regarding the main contradiction in our society: The current contradiction is between the people's growing demand for a better life and the imbalanced, inadequate development. China wants to achieve new progress in the changing situation and find new opportunities in the crisis. It has to establish a new "dual-circulation" development paradigm and realize the high-quality regional development. For these goals, it must change its model of development, optimize its economic structure, and upgrade its growth drivers. The following paragraphs analyze the 2016–2019 economic situations in China's eastern, central, southwestern, northwestern, and northeastern regions from the following aspects: economic scale, financial strength, total consumption, total investment, and foreign trade.[1,2]

1.1 The Steadily-Expanding Regional Economies

All the five regions have seen economic growths in recent years, which offers a solid material basis for the "dual-circulation" paradigm. Figure 1 shows data about their regional GDPs and their growth rates from 2016 to 2020. In 2016, the regional GDPs of the eastern, central, southwestern, northwestern, and northeastern regions were 41.02 trillion, 16.06 trillion, 9.67 trillion, 6.01 trillion, and 5.24 trillion. In 2020, the

[1] Considering the data accessibility, this analysis only includes Q2 economic-scale data for the period from 2016 to 2020, and other included economic data are annual data for the period from 2016 to 2019.

[2] In the analysis, the 31 provinces in the country (excluding Hong Kong, Macao, and Taiwan) are divided into five economic regions. The eastern region includes 10 provinces: Beijing, Tianjin, Hebei, Shandong, Jiangsu, Shanghai, Zhejiang, Fujian, Guangdong, and Hainan. The central region has six provinces, namely Shanxi, Henan, Anhui, Hubei, Hunan, and Jiangxi. The southwestern and the northeastern regions are also composed of six provinces. The former has Chongqing, Sichuan, Guizhou, Yunnan, Guangxi, and Tibet, and the latter has Inner Mongolia, Shaanxi, Gansu, Ningxia, Qinghai, and Xinjiang. The three northeastern provinces, i.e. Heilongjiang, Jilin and Liaoning, constitute the northeastern region.

above data were changed into 52.58 trillion, 22.22 trillion, 14 trillion, 7.33 trillion, and 5.11 trillion: Except for the northeastern region, all the regions realized positive growth.

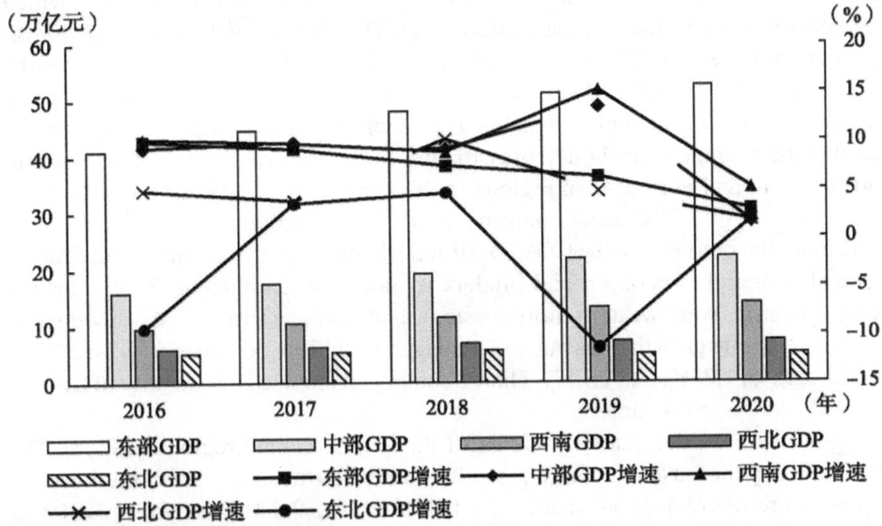

Fig. 1 The five economic regions' GDPs and their growth rates 2016–2020 (*Data Source* the National Bureau of Statistics of China)

东部GDP	GDP of the eastern region
中部GDP	GDP of the central region
西南GDP	GDP of the southwestern region
西北GDP	GDP of the northwestern region
东北GDP	GDP of the northeastern region
东部GDP增速	GDP growth rate of the eastern region
中部GDP增速	GDP growth rate of the central region
西南GDP增速	GDP growth rate of the southwestern region
西北GDP增速	GDP growth rate of the northwestern region
东北GDP增速	GDP growth rate of the northeastern region
(万亿元)	(trillion yuan)

The economic growth rate of the eastern region has been on the decline. Its annual GDP growth rates from 2016 to 2019 were 9.97%, 9.18%, 7.4% and 6.27%. The central region, however, has seen its growth rate on the rise. Its annual GDP growth rates from 2016 to 2019 were 9.32%, 9.86%, 9.16%, and 13.54%. Its success in realizing an over 10% increase in 2019 effectively remedied its economic backwardness. From 2016 to 2019, the annual GDP growth rates of the southwestern region were 10.25%, 9.77%, 8.86%, and 15.22%. The region's growth pole, the Chengdu-Chongqing city cluster, has been growing into the economic backbone of the region. During this period, the northwestern region's annual GDP growth rates were 4.92%, 3.8%, 10.16%, and 4.79%. Except in 2018, its growth rate remained under 5%, which was generally lower than the national average. The GDP growth rates of the northeastern region were -9.35%, 3.52%, 4.6%, and -11.46%. Despite the positive growths in 2017 and 2018, it suffered serious declines during this period. This suggests that the revitalization of the region remains to be a challenging task. By calculation, we find that the average annual GDP growth rates of the eastern, central, southwestern, northwestern, and northeastern regions in this period are 7.61%, 10.84%, 11.25%, 6.21%, and -1.39%. The result confirms the above judgments.

The sudden outbreak of the COVID-19 at the beginning of 2020 impacted China's regional economic operation and hinders its domestic circulation. That year, the eastern, central, southwestern, northwestern, and northeastern regions saw decline in their economic growth rates but still managed to achieve positive growths: 2.85%, 1.60%, 5.15%, 1.73%, and 1.74%. This effectively guaranteed the stability of China's regional economic fundamentals.

Figure 2 shows the quarterly GDPs of the five economic regions in 2020. The Q1 GDPs of the eastern, central, southwestern, northwestern, and northeastern regions were 10.73 trillion yuan, 4.35 trillion yuan, 2.9 trillion yuan, 1.54 trillion yuan, and 0.99 trillion yuan, decreased by 21.26%, 35.58%, 34.48%, 22.36% and 7.72% quarter-on-quarter. In the second quarter, when the epidemic was under control, the eastern, central, southwestern, northwestern, and northeastern regions saw their quarterly GDPs increase qaurter-on-quarter by 20.57%, 27.93%, 18.76%, 15.17%, and 19.72%. This strong economic recovery underpinned China's success in realizing positive economic growth and bringing its national GDP across the threshold of one million yuan in 2020. Thanks to effective epidemic-prevention-and-control measures, this growth trend continued into the third and fourth quarters. The eastern, central, southwestern, northwestern, and northeastern regions achieved 6.49%, 5.65%, 6.20%, 7.38%, and 11.83% quarter-on-quarter increases in the third quarter; and 9.92%, 9.25%, 8.99%, 10.42%, and 20.31% quarter-on-quarter increases in the fourth quarter. Generally speaking, the development of the regional economies were influenced by two factors in 2020: the severity of the regional epidemic situation and the regional economic resilience, i.e. the regional capacity to stand and recover from adverse shocks. Wuhan City of Hubei Province, the locality hit the hardest by the epidemic in 2020, was a top-notch city of the central region, and the intersection of the Beijing-Guangzhou and Shanghai-Chengdu railways. During the Chinese New Year travel rush, as much as 30 million travelers passed through this place, facilitating the spread of the virus across the country. Among all provinces of China, Hubei and its neighboring provinces in the central region, such as Hunan, Jiangxi, Anhui, and Henan, saw more confirmed COVID-19 cases. To cut the spread of the virus and safeguard people's health and safety, provinces throughout the country successively raised their emergency response for public health emergency to level I: Manufacturing enterprises delayed their start dates after the Spring Festival; Tourist destinations, hotels, shopping malls, and cinemas suspended their operation during the holiday. Such measures successfully controlled the virus spread but seriously impacted sectors such as manufacturing, tourism, catering, commerce, culture and entertainment, and education. Under such circumstances, all the five regions, especially the central region, saw stark decreases in the first quarter of 2020. Fortunately, with regularized epidemic-prevention-and-control measures and the resumption of production activities, their regional economies rebounded in the second quarter. Among them, the central region realized the strongest recovery, boasting a high economic resilience. In February 2021, according to the preliminary data from the National Bureau of Statistics of China, the national GDP for 2020 reached 101,598.6 billion yuan, achieving a positive growth of 2.3% at comparable prices. China was the only major economy of the world that achieved positive growth under the impact of the COVID-19 pandemic in 2020.

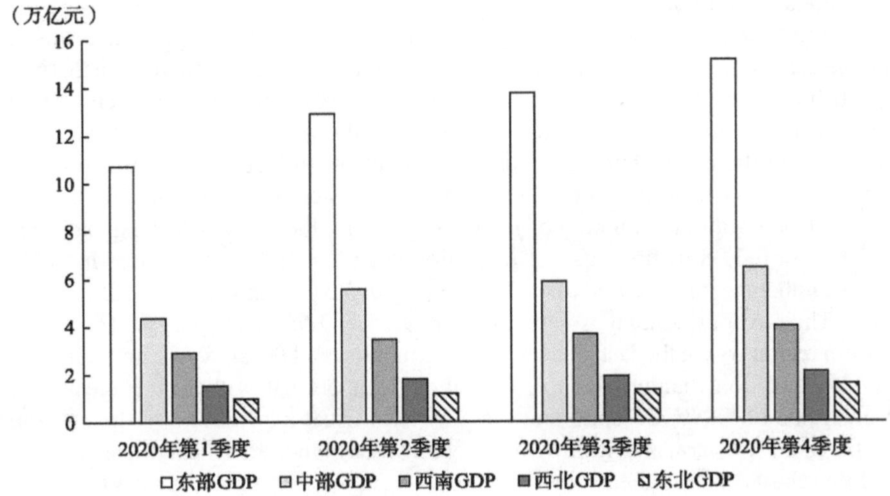

Fig. 2 The five economic regions' quarterly GDPs in 2020 (*Data Source* the National Bureau of Statistics of China)

东部GDP	GDP of the eastern region
中部GDP	GDP of the central region
西南GDP	GDP of the southwestern region
西北GDP	GDP of the northwestern region
东北GDP	GDP of the northeastern region
2020年第1季度	Q1 2020
2020年第2季度	Q2 2020
2020年第3季度	Q3 2020
2020年第4季度	Q4 2020
(万亿)	(trillion yuan)

1.2 The Growing Regional Financial Strengths

The success of the domestic circulation needs the support of public finance. Government agencies have to formulate and implement feasible fiscal policies to overcome any spontaneity, blindness, and lag within the market for the sustainable economic development. Since the 2008 Global Financial Crisis, limitations brought by the endogeneity of monetary policies propelled governments across the globe to re-attach importance to the regulatory role of fiscal policies and realize economic recoveries through a series of expansionary fiscal policies such as increasing government expenditure, reducing taxes and fees, and issuing bonds. Over these years, China has also pursued proactive fiscal policies. It has increased its fiscal expenditure and optimized its structure, facilitating the adjustment of its domestic demand and the optimization of its domestic circulation.[3]

As shown in Fig. 3, the eastern region outshone the other four regions in terms of total fiscal expenditure, increasing its expenditure from 6.79 trillion yuan in 2016 to 8.51 trillion yuan in 2019 and with an average annual growth rate of 7.81%. The region invested heavily in areas with strong positive externalities, such as advanced manufacturing, scientific and technological innovation, talent cultivation, and infrastructure construction. For central, southwestern, and northwestern regions, which ranked the 2nd, 3rd, and 4th in the ranking of fiscal expenditure, their total fiscal expenditures in 2016 were 3.38 trillion yuan, 2.73 trillion yuan, and 1.9 trillion yuan. In 2019, these numbers were increased to 4.47 trillion yuan, 3.6 trillion yuan, and 2.34 trillion yuan. Their average annual growth rate were 9.75%, 9.58%, and 7.23%. The northeastern region was at the last place in this ranking. Its total fiscal expenditure climbed from 1.24 trillion yuan in 2016 to 1.47 trillion yuan in 2019, with an average annual growth rate of 5.84%. Despite that the growth in fiscal expenditure in the eastern, central, southwestern, and northwestern regions was higher than the national average and that the growth in the northeast region's fiscal expenditure was far lower than the national average, there was not big difference in their annual growth rates. Their public financial expenditure helps to alleviate economic fluctuations and stimulate the vitality of the domestic circulation.

[3] The limitations brought by the endogeneity of monetary policies mainly include the following three aspects. First, In times of economic recession, the effects of expansionary monetary policies are not obvious. Second, the effect of time lag for monetary policies is strong, i.e. its incapability to realize economic recoveries within in a short time. Third, given the economic globalization, the effect of monetary policies is prone to the influences of the international balance of payments.

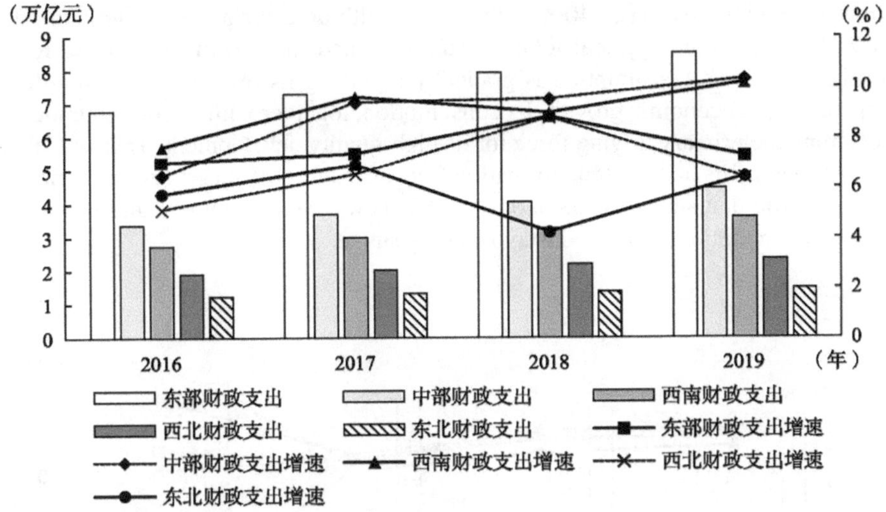

Fig. 3 The five economic regions' fiscal expenditures and their growth rates 2016–2019 (*Data Source* the National Bureau of Statistics of China)

东部财政支出	Fiscal expenditure of the eastern region
中部财政支出	Fiscal expenditure of the central region
西南财政支出	Fiscal expenditure of the southwestern region
西北财政支出	Fiscal expenditure of the northwestern region
东北财政支出	Fiscal expenditure of the northwestern region
东部财政支出增速	Fiscal-expenditure growth rate of the eastern region
中部财政支出增速	Fiscal-expenditure growth rate of the central region
西南财政支出增速	Fiscal-expenditure growth rate of the southwestern region
西北财政支出增速	Fiscal-expenditure growth rate of the northwestern region
东北财政支出增速	Fiscal-expenditure growth rate of the northeastern region

1.3 The Constantly Increasing Regional Consumption

As the last link in the four links of the domestic circulation chain, i.e. production, distribution, exchange, and consumption, consumption is a direct embodiment of the Chinese people's living standards. During the 13th Five-Year Plan period (2016–2020), every year's government work report has fully affirmed the fundamental role of consumption in our high-quality development. This clarifies the goal of the transformation of our regional economic drivers. From 2016 to 2019, the total retail sales of consumer goods (hereinafter "total consumption") in China climbed from 33.16 trillion yuan to 40.8 trillion yuan, with an average annual growth rate of 7.16%. As to the regional total consumption and its growth rate, please refer to Fig. 4. China's consumption is gradually replacing its investment in providing impetus for its economic growth. Its consumption, together with its foreign trade, is becoming the primary driving force for its high-quality development. That is, China is experiencing its historic transformation from an external-demand-driven economy to a domestic-demand-driven economy, and its new development paradigm with the domestic circulation as the mainstay is taking shape.

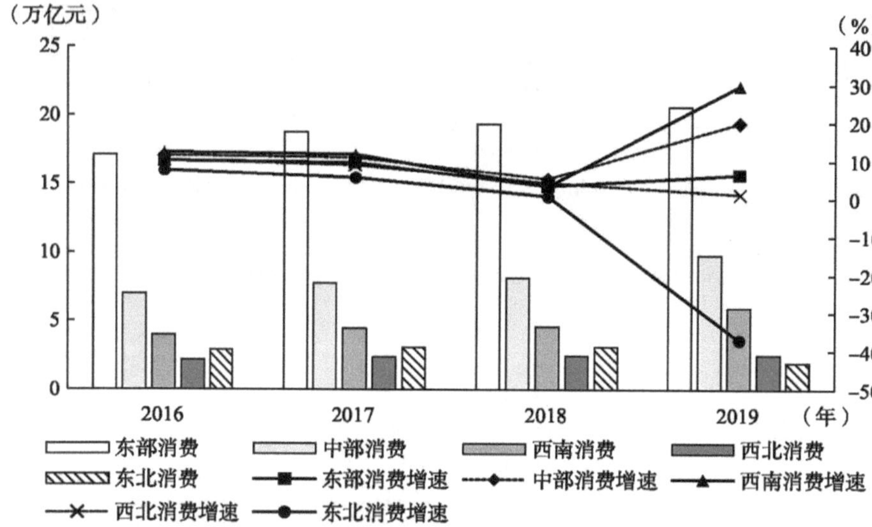

Fig. 4 Total retail sales of social consumer goods in the five economic regions and their growth rates 2016–2019 (*Data Source* the National Bureau of Statistics of China)

东部消费	Consumption in the eastern region
中部消费	Consumption in the central region
西南消费	Consumption in the southwestern region
西北消费	Consumption in the northwestern region
东北消费	Consumption in the northeastern region
东部消费增速	Consumption growth rate of the eastern region
中部消费增速	Consumption growth rate of the central region
西南消费增速	Consumption growth rate of the southwestern region
西北消费增速	Consumption growth rate of the northwestern region
东北消费增速	Consumption growth rate of the northeastern region

According to Fig. 4, the eastern region's total consumption increased from 17.11 trillion yuan to 20.63 trillion yuan from 2016 to 2019, with an average annual rate of 6.41%. In terms of total consumption, it has surpassed all the other four regions. This phenomenon is mainly caused by the following reasons: The per-capita income of many provinces and cities in the eastern region has been on par with that of moderately developed countries; The region's consumption of durable consumer goods, such as housing, automobiles, and health care products, has been on the rise; The region 's consumers have shown clear preference for high-end, personalized, and green products; The region has gained preliminary conditions for the development of a domestic circulation that relies on strong domestic consumption. Although the total consumption in the central and southwestern regions was less than that in the eastern region: The former increased from 6.98 trillion yuan in 2016 to 9.78 trillion yuan in 2019, and the latter from 3.98 trillion yuan to 5.96 trillion yuan. They outshone the eastern region in terms of consumption growth rates. Their growth rates were 11.91% and 14.4% respectively, 5.5 and 7.99 percentage points higher than the growth rate of the eastern region. It is fair to say that these two region are also increasingly dependent on domestic consumption. Also under the influence of the western-region development campaign, the northwestern region lagged far behind the southwestern region in boosting consumption. Its total consumption slowly climbed from 2.17 trillion yuan to 2.49 trillion yuan from 2016 to 2019, and its average annual growth rate was 4.74%, below the national average. It still has a long way to go in this regard. Nevertheless, the situation in the northwestern region was even worse. The regional total consumption shrunk from 2.91 trillion yuan to 1.95 trillion yuan from 2016 to 2019, with an average annual decrease rate of 12.54%. Such a cliff-style decline suggest that the weak regional consumption has become the crux of the campaign to revitalize the region in the new era. It is also a critical task that must be solved through policies when promoting the domestic circulation.

An analysis based on time periods finds that the consumption rates of the five regions fluctuated around 10% from 2016 to 2017, without obvious inter-regional differences, but they decreased year-on-year by 6.24, 5.68, 8.46, 4.72, and 4.9 percentage points in 2018., which did not prevent the five regions from realizing positive growths in that year. This reduction in their growth rates in 2018 was related to the decrease in the consumption of high-end durable consumer goods represented by automobiles. To solve this problem, China introduced a series of preferential measures to reduce relevant taxes and fees, which boosted the total consumption in the eastern, central, and southwestern regions in 2019. That year, the consumption rebounds in the central and southwestern regions were particularly strong. This greatly improved the efficiency of the domestic circulation. Unfortunately, these preferential measures failed to achieve any results in the northwestern and northeaster regions: The former was limited by its weak economy, and the latter, trapped in an economic depression. In 2019, the total consumption of the northeastern region dropped by 37.1%, an unprecedented huge decline.

1.4 Differentiation in Regional Investment

While enhancing the fundamental role of consumption, the key role of effective investment shall also be brought into full play. In the 13th Five-Year Plan period (2016–2020), China's top authorities have faithfully implemented the new development philosophy, improved the quality of the national economy, and transformed its growth drivers. They have launched seven new-type-infrastructure projects to develop the country's 5G base stations, ultra-high voltage transmission, inter-city high-speed railways and urban-rail-transit systems, new-energy-vehicles charging stations, big data centers, artificial intelligence, and industrial Internet. The above-mentioned investment in fixed assets has played a positive role in counteracting the negative effects brought by the macroeconomic downturn and increasing the efficiency of the domestic circulation.

As shown in Fig. 5, the total investment in fixed assets in the eastern, central and southwestern regions increased year by year from 2016 to 2019: rising from 25.29 trillion yuan to 29.57 trillion yuan in the eastern region; from 15.97 trillion yuan to 19.98 trillion yuan in the central region; and from 9.4 trillion yuan to 12.59 trillion yuan in the southwestern region. Their average annual growth rates were 5.35%, 7.75%, and 10.22%. The eastern region boasted the highest investment, but its growth rate was lower than those of the other two regions. Among the three, only the southwestern region achieved a double-digit growth rate. For the northwestern region, there was a mild rise in its regional investment from 2016 to 2017, but the decline in 2018 brought its average annual growth rate to -1.65%. As for the northeastern region, its investment growth was sluggish during this period, and the largest decline in its investment occurred in 2016, which was 23.38%. The downward trend slowed down between 2017 and 2018, but in 2019, the region's investment growth rate suddenly fell to -4.77%. This is an urgent problem and a weak point that must be solved for the success of the domestic circulation.

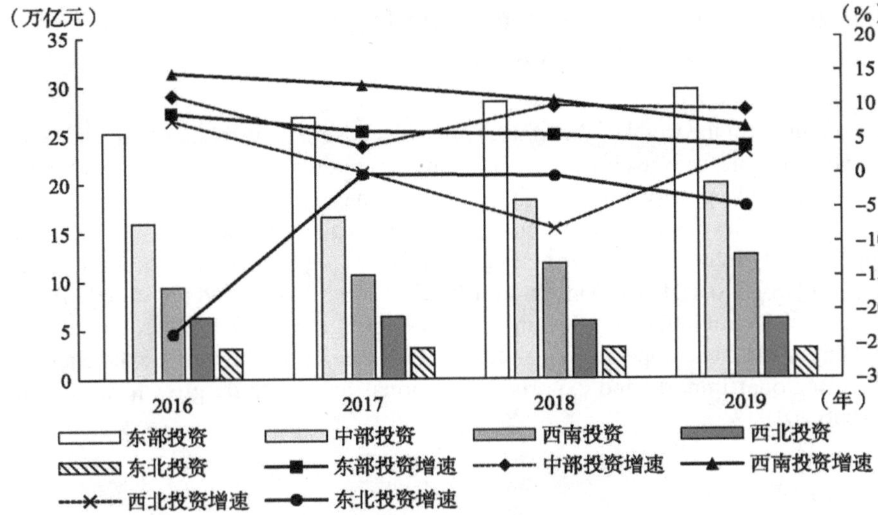

Fig. 5 Total investments in fixed assets in the five economic regions and their growth rates 2016–2019[4] (*Data Source* the National Bureau of Statistics of China)

东部投资	Investment in the eastern region
中部投资	Investment in the central region
西南投资	Investment in the southwestern region
西北投资	Investment in the northwestern region
东北投资	Investment in the northeastern region
东部投资增速	Investment growth rate of the eastern region
中部投资增速	Investment growth rate of the central region
西南投资增速	Investment growth rate of the southwestern region
西北投资增速	Investment growth rate of the northwestern region
东北投资增速	Investment growth rate of the northeastern region
(万亿)	(trillion yuan)

[4] Since 2019, China Statistical Yearbooks have stopped reporting on the localities' total investments in fixed assets. By considering the availability of data, we determine that the total investment in fixed assets in 2018 = the total urban investment in fixed assets in 2018 × (1 + the growth rate of the total urban investment in fixed assets in 2018) + the total rural investment in fixed assets in 2018 × (1 + the growth rate of the total rural investment in fixed assets in 2018) and that the total investment in fixed assets in 2019 = the total investment in fixed assets in 2018 × (1 + the growth rate of the total investment in fixed assets in 2018).

1.5 *Encouraging and Discouraging Differences in Regional Foreign-Trade Situations*

To keep up with the trend of the times, China has to step up its opening-up endeavor and actively integrate itself into the international circulation. Since General Secretary Xi Jinping put forward the BRI in 2013, China's export-oriented economy has entered a new development stage, enabling China to play an indispensable role in the international circulation. Nevertheless, given the rising unilateralism trend and the frequent US-Sino trade frictions in the 13th Five-Year Plan period (2016–2020), the international circulation has become unstable, bringing unprecedented systematic challenges to China's opening-up endeavor. Please refer to Fig. 6 to see the changes in the regional imports and exports and regional foreign-trade growth rates from 2016 to 2019.

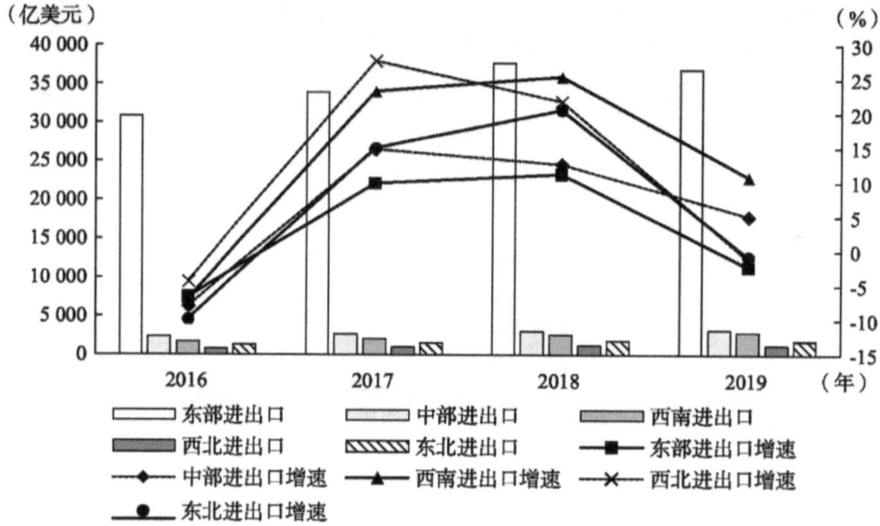

Fig. 6 Total import and export values in the five economic regions and their growth rates 2016–2019 (*Data Source* the National Bureau of Statistics of China)

东部进出口	Imports and exports of the eastern region
中部进出口	Imports and exports of the central region
西南进出口	Imports and exports of the southwestern region
西北进出口	Imports and exports of the northwestern region
东北进出口	Imports and exports of the northeastern region
东部进出口增速	Growth rate of the imports and exports in the eastern region
中部进出口增速	Growth rate of the imports and exports in the central region
西南进出口增速	Growth rate of the imports and exports in the southwestern region

(continued)

(continued)

西北进出口增速	Growth rate of the imports and exports in the northwestern region
东北进出口增速	Growth rate of the imports and exports in the northeastern region
(亿美元)	(100 million USD)
(年)	(year)

According to their foreign trade scales, the five economic regions can be ranked from biggest to smallest as follows: the eastern region, the central region, the southwestern region, the northeastern region, and the northwestern region. From the 1980s to the 1990s, the eastern region enjoyed preferential policies of the Central Government and established a batch of Special Economic Zones and Coastal Open Cities, creating four major coastal opening-up areas, namely the Bohai Rim, the southeastern area of Fujian, the YRD, and the PRD. Given this advantage, the region has outshone the other four regions in developing an open economy over the years. In the beginning of the twenty-first century, it reinforced its leading position in this regard by first implementing the free-trade-zone policy in Shanghai and actively participating in the cooperation of the international circulation. Statistics show that the total imports and exports of the region increased from $3,081.984 billion to $3,682.276 billion from 2016 to 2019 and that its total imports and exports accounted for an average of 82.02% of the national total, occupy an absolute leading position in the country. While vigorously promoting the opening of its eastern coastal areas, China also stepped up the opening-up of its hinterland and border areas, especially accelerating the opening of its hinterland provincial capitals, cities along the Yangtze river, and border cities. Thanks to such efforts, the total imports and exports of the central, southwestern, northwestern, and northeastern regions increased significantly during the period. Their total imports and exports in 2016 were $231.454 billion, $167.023 billion, $75.786 billion, and $129.311 billion, and in 2019, these numbers jumped to $314.852 billion, $286.269 billion, $116.536 billion, and $177.957 billion. This suggests that China's opening-up framework has been established and that the country has already become an important hub in the international circulation.

In 2016, a total of 27 countries and regions across the globe tried to impose 119 trade remedies against China, and emerging markets as well as developed markets suffered a widespread declined. As a result, all the five economic regions of China saw decreases in their total imports and exports. The decrease rates of the eastern, central, southwestern, northwestern, and northeastern regions were 6.6%, 7.94%, 6.87%, 4.3%, and 9.81%. China faced severe difficulties in integrating itself into the international circulation. Fortunately, benefited by the growth of the pilot Free Trade Zones, the regions saw strong rebounds in 2017 and 2018. The southwestern and northwestern regions even realized growth rates of more than 20%. In June, 2018, the Trump administration of the United States repeatedly imposed tariffs on the imported Chinese products and constantly escalated the trade frictions. Because of this, the five regions' total imports and exports declined again. The eastern region, which relied the most on the foreign trade, even suffered a negative growth of 2.26%. In this period, the total imports and exports of the northwestern and northeastern regions also shrank by 1.17% and 0.78%. Surprisingly, the central and southwestern regions managed to achieve positive growth under such difficult circumstances, and the growth rate in the southwestern region even reached as high as 10.82%.

In the first quarter of 2021, China's imports and exports declined because of the interruption in its production activities, which was caused by its Central Government's strict epidemic-control measures aimed to curb the spread of the COVID-19. In the second quarter, when the world suffered from the COVID-19 pandemic and when the global economic recovery was facing significant headwinds, China continued its good performance in curbing the spread of the virus and tried its best to restart the operation of China–Europe freight trains. What it did not only promoted the rapid recovery of its foreign trade but also facilitated the recovery of the international circulation. By doing so, China fulfilled its responsibilities as a major country of the world.

While vigorously promoting the development of its foreign trade, China has extensively implemented its negative list policy to continuously lower the entry threshold for foreign investors and actively participated in the international economic and technological cooperation. By the end of 2016, all 10 provinces in the eastern region has been included in the pilot free trade zones. The negative list policy implemented in Shanghai, Tianjin, Fujian, and Guangdong Free Trade Zones in 2015 has been introduced into other eastern provinces during the 13th Five-Year Plan period (2016–2020). Under the influence of the policy, the total foreign investment in the region jumped from $4,004.569 billion in 2016 to $6,629.215 billion in 2019, accounting for 76.98% of the national total and realizing an average annual growth rate of 18.3%. As shown in Fig. 7, with the expansion of pilot Free Trade Zones during the 13th Five-Year Plan period, the central, southwestern, northwestern, and northeastern regions also saw increases in the total amount of foreign investment from 2016 to 2019. The total foreign investment in the central region increased from $426.723 billion to $823.635 billion, with an average annual growth rate of 24.51%. Southwest region saw an increase from $284.967 billion to $610.256 billion and an annual growth rate of 28.9%. For the northwestern and northeastern regions, the total amounts of foreign investments they attracted respectively rose from $130.59 billion $263.747 billion and from $277.165 billion to $513.173 billion. Their average annual growth rates were 26.4% and 22.79%, respectively. During this period, China's economic ties with other countries in the world became stronger, and it gradually gained the initiative to promote the international circulation.

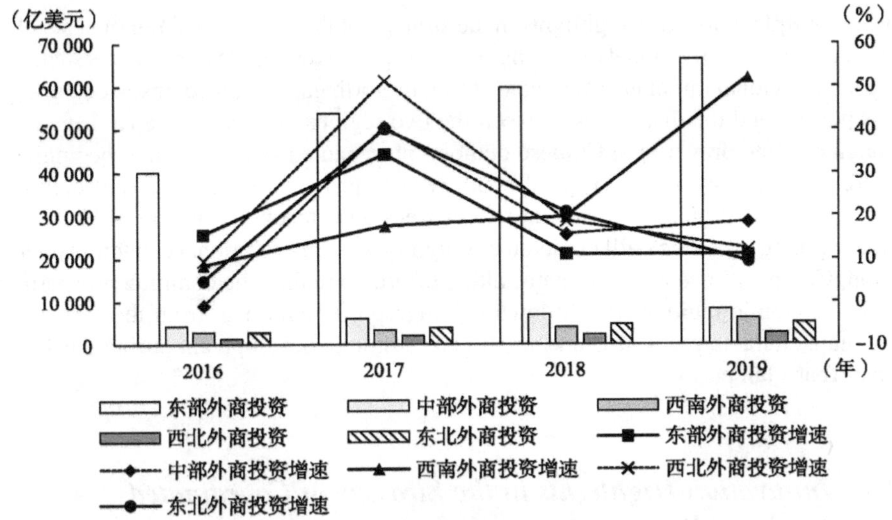

Fig. 7 Total foreign investments in the five economic regions and their growth rates 2016–2019 (*Data Source* the National Bureau of Statistics of China)

东部外商投资	Foreign investment in the eastern region
中部外商投资	Foreign investment in the central region
西南外商投资	Foreign investment in the southwestern region
西北外商投资	Foreign investment in the northwestern region
东北外商投资	Foreign investment in the northeastern region
东部外商投资增速	Foreign investment growth rate of the eastern region
中部外商投资增速	Foreign investment growth rate of the central region
西南外商投资增速	Foreign investment growth rate of the southwestern region
西北外商投资增速	Foreign investment growth rate of the northwestern region
东北外商投资增速	Foreign investment growth rate of the northeastern region

2 The Main Characteristics of China's Regional Economy in the 13th Five-Year Plan Period

During the 13th Five-Year Plan period (2016–2020), the Chinese government upheld the five new development concepts, namely innovative development, coordinated development, green development, open development, and sharing development, and kept pushing forward the supply-side structural reform in the established direction. This provided great opportunities for China's regional economic development: They were multiple innovation highlights in the strategy of the coordinated regional development, and new functional platforms kept emerging. Nevertheless, problems such as the inter-regional imbalance and the wide south-north gap remained unsolved during this period, and the disparities between the two regions divided by the Hu Line (an imaginary line drawn by a Chinese demographer named Hu Huanyong, beginning in Tengchong City in the tropical southwestern province of Yunnan and continuing northeast at an angle of around 45 degrees until reaching the city of Heihe in Heilongjiang Province) still hindered our high-quality inter-regional economic operation. China still needs to take painstaking efforts to realize the common prosperity of its economic regions. The following paragraphs summarize, from thee aspects, the main characteristics of China's regional economic development during the 13th Five-Year Plan period.

2.1 Innovation Highlights in the Strategy of Coordinated Regional Development

The establishment of a new, more effective mechanism for the coordinated regional development depends on the innovation of the supply-side strategy. During the 13th Five-Year Plan period (2016–2020), the Central Government kept optimizing its regional development plans, creating a new regional development paradigm featuring coordination, orderly competition,green development, and win–win cooperation. According to the sequence of their issuing times, the Central Government implemented the following four development plans during this period: the development of the YREB, the development of the GBA, the ecological conservationand high-quality development in the YRB, and the integration of the YRD.

1. **The Development of the Yangtze River Economic Belt (YREB).** As one of China's principal waterways, the Yangtze River, connects the country's three major geographical regions at different altitudes. The basin of this river is a region with great development potential. As early as the early 1990s, the Yangtze River Economic Belt (YREB) and the coastal region were planned as key development areas, jointly forming the "River-and-Sea" development paradigm. To ensure the constant growth of the YREB, the Central Committee of the CPC adopted the "Development Plan of the Yangtze River Economic Belt" in March 2016, which emphasized the necessity of protecting the environment and that of avoiding over-exploitation and discussed detailed measures for the YREB's high-quality economic development from the following aspects, namely the development goals, the spatial arrangement across the country, the ecological conservation, the shipping waterway construction, the industrial development, the opening-up, the unification of the domestic market, and the equal access to basic public services.
2. **The Development of the Guangdong-Hong Kong-Macau Greater Bay Area (Greater Bay Area or GBA).** The idea of the GBA was first proposed by China's top authorities in the Government Work Report released in March 2017.

The report called for in-depth cooperation among Guangdong, Hong Kong, and Macao, and announced the start of the strategic planning for the region. With the strong support of the Central Government, the development of the GBA became a national strategy on July 1, 2017, when the NDRC of the PRC, the Guangdong Provincial Government, and the governments of Hong Kong and Macao Special Administrative Regions, witnessed by President Xi Jinping, jointly signed the "Framework Agreement on Deepening Guangdong-Hong Kong-Macao Cooperation in the Development of the Greater Bay Area." The Agreement clarified the objectives and principles of cooperation for all parties and listed seven key cooperation areas, aimed to develop the GBA into a world-class bay area and city cluster. The Boao Forum for Asia, held in April 2018, organized a special sub-forum for the GBA. Representatives from Guangdong, Hong Kong, and Macao had a heated discussion in this sub-forum, and reached agreement on the following issues: As a financial center, the GBA could compete with New York Bay, and as a manufacturing base and a innovation hub, it could contend respectively against Tokyo Bay and San Francisco Bay; The further growth of the GBA had to rely on the in-depth cooperation among Guangdong, Hong Kong, and Macau. After nearly a year, the "Outline Development Plan for the Guangdong-Hong Kong-Macao Greater Bay Area" was adopted in February 2019, which covered the region's development in the following eight aspects: spatial arrangement, innovation ability, infrastructure development, industrial system, ecological conservation, people's livelihoods, opening-up, and inter-regional cooperation, pointing its regional development in the right direction.

3. **The Ecological Protection and High-Quality Development of the Yellow River Basin (YRB).** Similar to the Yangtze River, the Yellow River also flows across China from west to east. The river basin boasts rich energy resources and outstanding ecological importance. It is also a key area in China's targeted poverty alleviation campaign and its BRI cooperation. In September, 2019, General Secretary Xi Jinping announced the ecological conservation and high-quality development of the YRB as a national strategy. Different from the YREB, the YRB has a limited number of port cities and hub cities due to the Yellow River's natural defects that hinder its waterway function. It does not have the foundations to serve as a national-level economic belt. To boost its development, General Secretary Xi Jinping made clear arrangements at the 6th meeting of the Central Committee for Financial and Economic Affairs in January 2020. According to him, the YRB has to do well in the following two aspects. First, efforts must be exerted to curb water, air, and soil pollution and push forward major ecological management projects such as those intended to conserve fresh water, solve soil erosion, restore the wetland ecosystem in the Yellow River Delta, or improve the river channel and its beach areas. Second, the development of the central cities and city clusters in this region shall be boosted. To be more specific, the Lanzhou-Xining city cluster shall accelerate its development; the metropolitan areas in the Yellow River Ji-Shaped Bend shall pursue coordinated development; national-level central cities there, i.e. Xi'an and Zhengzhou, shall give full play

to their promoting role; and the city cluster in Shandong Peninsula shall play its leading role.

4. **The Integration of the Yangtze River Delta (YRD).** The Yangtze River Delta encompasses Shanghai Municipality and Zhejiang, Jiangsu and Anhui provinces. It is one of the regions with the highest economic and population densities in China. In November 2018, at the 1st China International Import Expo, General Secretary Xi Jinping showed his support for upgrading the integration of the YRD into a national strategy. To ensure the efficient, high-speed, and free flow of various high-quality factors such as capitals, talents, and technologies in the YRD, the "Outline for the Development of Regional Integration in the Yangtze River Delta" was promulgated in December 2019. It made a plan for the region's development in the next 15 years from the following aspects: innovation in industrial systems, infrastructure interconnection, joint ecological protection efforts, sharing of public services, and higher-level opening-up cooperation. To highlight the theme of high-quality development, General Secretary Xi Jinping presided over a symposium on solidly promoting the integrated development of the YRD in Hefei, Anhui, in August 2020. He emphasized that the development of the YRD needs to focus on two keywords: "integration" and "high quality." He called for concerted efforts to realize these two goals, which embodied the YRD's great significance for China's grand plan to basically realize socialist modernization from 2020 through 2035.

2.2 *Emerging Functional Platforms*

The economic development of various regions needs the strategic support of the new functional platforms. In the 13th Five-Year Plan period, the Chinese government adopted more than 80 national- and regional-level plans and approved a number of functional platforms such as free trade zones, national-level new districts, city clusters, and metropolitan areas, building a framework for China's endeavor to reshape its economic landscape.

1. **Free Trade Zones.** In the new era, to fully utilize and integrate the resources of the domestic and international markets, the Central Government established China's first free trade zone in Shanghai in 2013 as a landmark step toward its goal of furthering its opening-up endeavor. When this trade zone could provide other localities with mature experience for reference, the policy was introduced to other places across China, and the most rapid expansion of the policy happened in the 13th Five-Year Plan period (2016–2020). In four batches, the following 17 provinces adopted this policy: Liaoning, Zhejiang, Henan, Hubei, Chongqing, Sichuan and Shaanxi adopted this policy in March 2017; Hainan established its free trade zone in April 2018; Shandong, Jiangsu, Guangxi, Hebei, and Yunnan started similar practice in July 2019; and Beijing, Hunan, and Anhui were included in the fourth batch in September 2020. Pilot projects under this

policy covers a total of 35,818.45 km^2, and the following Table 1 contains the basic information about the 21 free trade zones.

2. **National-Level New Districts.** Established for major national development goals and strategic tasks, national-level new districts are naturally growth drivers for their surrounding areas. During the 13th Five-Year Plan period (2016–2020),

Table 1 Overview of China's 21 Free Trade Zones

Region	Founding time (month-year)	Area (km^2)	Areas covered
Shanghai	September 2013	120.72	Waigaoqiao Free Trade Zone, Waigaoqiao Bonded Logistics Zone, Yangshan Bonded Zone, Pudong Airport Comprehensive Bonded Zone, Jinqiao Export Processing Zone, Zhangjiang Science City, and Lujiazui Finance and Trade Zone
Guangdong	April 2015	116.2	Guangzhou area, Shenzhen area, and Zhuhai area
Tianjin	April 2015	119.9	Tianjin Port Area, Tianjin Airport Area, and Binhai New Area
Fujian	April 2015	118.04	Fuzhou area, Xiamen area, and Pingtan area
Liaoning	March 2017	119.89	Shenyang area, Dalian area, and Yingkou area
Zhejiang	March 2017	119.95	The northern and southern regions of Zhoushan Islands and the off-island region of Zhoushan
Henan	March 2017	119.77	Zhengzhou area, Kaifeng area, and Luoyang area
Hubei	March 2017	120	Wuhan area, Yichang area, and Xiangyang area
Chongqing	March 2017	119.98	Liangjiang area, Xiyong area, and Guoyuangang area
Sichuan	March 2017	119.99	Chengdu Tianfu New Area, Chengdu Qingbaijiang Railway Port Area, and the South Sichuan Lingang Area
Shaanxi	March 2017	119.95	Xi'an Central Area, Xi'an International Port Area, and Yangling Demonstration Zone
Hainan	October 2018	33,900	Hainan Island
Shandong	August 2019	119.98	Jinan area, Qingdao area, and Yantai area
Jiangsu	August 2019	119.97	Nanjing area, Suzhou area, and Lianyungang area
Guangxi	August 2019	119.99	Nanning area, Qinzhou Port area, and Chongzuo area
Hebei	August 2019	119.97	Xiong'an area, Zhengding area, Caofeidian area, and Daxing Airport area
Yunnan	August 2019	119.86	Kunming area, Honghe area, and Dehong area
Heilongjiang	August 2019	119.85	Harbin area, Heihei area, and Suifenhe area
Beijing	September 2020	119.68	Science and technology innovation area, international business service area, and high-end industry area
Hunan	September 2020	119.76	Changsha area, Yueyang area, and Chenzhou area
Anhui	September 2020	119.86	Hefei area, Wuhu area, and Bengbu area

three national-level new districts were created: Changchun New Area, founded in February 2016; Ganjiang New Area, set up in June 2016; Xiong'an New Area, established in April 2017. Together with the national-level new districts founded from 2010 to 2015, they become fulcrums for the high-quality development of our national economy. The three national-level new districts are located in the northeaster, central, and eastern regions, echoing the national strategies for the revitalization of the northeastern region, the development of the YREB, and the coordinated development of the BTH region. They become landmark development projects for the 13th Five-Year Plan period, and among them, Xiong'an New Area is the most representative one. As a major strategic decision made under the BTH coordinated development strategy, Xiong'an was built for a very realistic purpose: the decentralization of Beijing's non-capital functions. Different from the YRD and the GBA, the BTH region has few well-developed peripheral urban communities around its central city. Therefore, Beijing has concentrated a lot of non-capital functions, which induced a series of "metropolitan diseases." To solve this problem, General Secretary Xi Jinping proposed to decentralize its non-capital functions at the 9th meeting of the Central Leading Group for Financial and Economic Affairs in February 2015, highlighting it as the key to promoting the coordinated development of the BTH region. In April 2017, the CPC Central Committee and the PRC State Council formally decided to establish Xiongan New Area in Hebei Province, aiming to continue the success of Shenzhen Special Economic Zone and Shanghai Pudong New Area. As a major part of the development of a long-term plan, For the national goal of developing Xiong'an into a city of the future and a world-class new city with Chinese characteristics, China's top authorities adopted the "Outline of Xiong'an New Area Development Plan" and the "Xiong'an New Area Mater Development Plan (2018–2035)" respectively in April and December of 2018. These two planning documents systematically discuss the development of Xiong'an from the following aspects: urban–rural integration, environmental protection, comprehensive transportation networks, and high-end industries. It echoes the spatial arrangements made by the "Outline of the Coordinated Development Plan for the Beijing-Tianjin-Hebei Region," which highlights the central role of Beijing, the close ties between Beijing and Tianjin, the development of three industrial belts (Beijing-Tianjin, Beijing-Baoding-Shijianzhuang, and Beijing-Tangshan-Qinhuangdao) and the four functional areas (the central area, the coastal area, the northwestern ecological area, and the southern function-expansion area), and the multi-fulcrum structure consisting of regional-level central cities such as Shijiazhuang, Tangshan, Baoding, and Handan and major cities such as Zhangjiakou, Chengde, Langfang, Qinhuangdao, Changzhou, Xingtai. Hengshui, etc.

3. **City Clusters and Metropolitan Areas.** During the 13th Five-Year Plan period (2016–2020), China's national-level city clusters were developed in an orderly way: The Harbin-Changchun city cluster, the Chengdu-Chongqing city cluster, the YRD city cluster, the Central Plains city cluster, the Beibu Gulf city cluster, the Hohhot-Baotou-Ordos-Yulin city cluster, the Lanzhou-Xi'an city cluster, and the GBA city cluster were approved and set up successively. Statistics show that

these national-level city cluster effectively aggregated economic activities and population. (Please see Table 2 for more details.) Among these city clusters, the YRD and the GBA city clusters each boasted regional GDPs exceeding 10 trillion yuan; the YRD, the Central Plains, and the BTH city clusters each had their resident populations surpassing 100 million; and the YRD, the GBA, and the Hohhot-Baotou-Ordos-Yulin city clusters had their per-capita GDPs exceeded 100,000 yuan. With strong economic ties and high-quality transportation networks, the integration trend in their development had become increasingly prominent. While building a city cluster system featuring the clear division of labor and the orderly development manner, China also enhanced the influence of central cities (often super-mega, mega, or big cities) on their peripheral cities and towns and boosted the development of metropolitan areas, each of them consisting of a central city and featuring a "one-hour commute circle." To this end, the National Development and Reform Commission (NDRC) promulgated the "Guiding Opinions of the NDCR on Cultivating and Developing Metropolitan Areas," which discussed how to improve our metropolitan areas in terms of spatial structure, production-factor flows, and industrial coordination and division of labor from following aspects: infrastructure integration, unified and open market, sharing of high-quality public-service resources, and joint efforts to protect the environment.

Table 2 Development Status of National-Level City Clusters

City cluster name	Founding time	GDP(trillion yuan)	Permanent population	Per-capita GDP (yuan)
The middle reaches of the Yangtze River	March 2015	8.53	132,786,200	64,274.5
Harbin-Changchun	February 2016	2.61	46,188,900	56,591.90
Chengdu-Chongqing	April 2016	5.55	95,161,700	58,371.12
The Yangtze River Delta	May 2016	17.86	146,732,400	121,746.84
The Central Plains	December 2016	7.17	165,605,500	43,283.13
Beibu Gulf	January 2017	1.99	42,251,000	47,166.61
Guanzhong Plain	January 2018	1.93	39,276,400	49,162.55
Hohhot-Baotou-Ordos-Yulin	February 2018	1.35	11,511,300	116,990.44
Lanzhou-Xi'an	March 2018	0.59	12,046,500	49,149.69
Guangdong-Hong Kong-Macao	February 2019	10.87	71,119,900	152,798.41
Beijing-Tianjin-Hebei	N/A	8.51	112,700,000	75,545.60
The West Coast of the Taiwan Strait	N/A	5.85	95,051,400	61,570.7

2.3 *The Inter-Regional Imbalance and the South-North Gap*

With the theoretical and practical progress in the coordinated regional development, China's economic regions realized positive growths in 13th Five-Year Plan period (2016–2020). Nevertheless, problems such as the inter-regional imbalance and the wide south-north gap remained unsolved.

1. **The Inter-Regional Imbalance.** The analysis in Section I reveals the following facts: The central, southwestern, and northeastern regions' GDPs grew faster than that of the eastern region, and their rapidly-growing economic strengths gradually narrowed the west–east development gap. Nonetheless, the inter-regional imbalance remained to be a problem. Figure 8 below compares the five major economic regions' contributions to the national GDP: The shares of the eastern, central, southwestern, northwestern, and northeastern regions in the 2016 national GDP were 52.58%, 20.59%, 12.4%, 7.71%, and 6.72%; In 2020, their percentages were 51.93%, 21.95%, 13.83%, 7.24%, and 5.05%. During this period, the GDP shares of the eastern and southwestern regions increased by 1.36 and 1.43 percentage points, respectively; those of the eastern, northwestern, and northeastern regions decreased by 0.65, 0.47, and 1.67 percentage points. Nevertheless, the increment of the eastern region was far greater than those of the other regions. This suggests that the overall structure of China's regional development has remained basically unchanged and that the imbalance between the two major regions divided up by the Hu Line has hinder overall prosperity across the country.

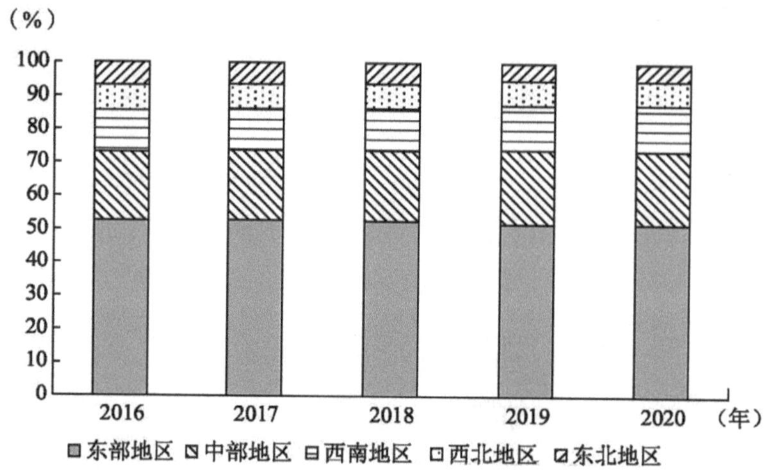

Fig. 8 GDP share of the five economic regions 2016–2020 (*Data Source* the National Bureau of Statistics of China)

东部地区	The eastern region

(continued)

(continued)

中部地区	The central region
西南地区	The southwestern region
西北地区	The northwestern region
东北地区	The northeastern region

While paying attention to the expansion of its economic scale, China also tries its best to ensure that its progress benefits as much people as possible to bring happiness and sense of fulfillment to its people. As shown in Fig. 9, from 2016 to 2019, the per-capita GDP growth rates of the eastern central, southwestern, northwestern, and northeastern regions were 6.8%, 10.3%, 10.53%, 5.44%, and -1.04%. The central region realized the highest per-capita GDP growth rate, followed by the southwestern, eastern, northwestern, and northeastern regions. Except for the northeastern region, all the regions saw positive growths in their per-capita GDPs. The central and southwestern regions even had double-digit increases, which narrowed the gap between their per-capita GDP and that of the eastern region. In terms of absolute values, the eastern region still outshine the other regions in terms of per-capita GDP: In this regard, it is equivalent to moderately developed countries. Its per-capita GDP is 1.71 times that of the central region, 1.92 times that of the southwestern region, 1.67 times that of the northwestern region, and 1.75 times that of the northeastern region. China still has a long way to go to reduce such persistent inter-regional gaps and realize common prosperity on both side of the Hu Line.

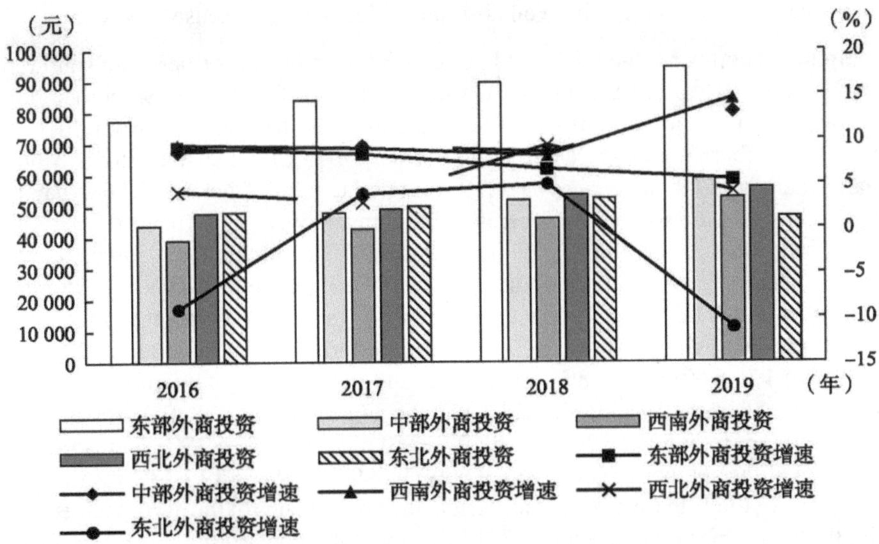

Fig. 9 Per-capita GDP and its growth rate in the five economic regions 2016–2019 (*Data Source* the National Bureau of Statistics of China)

东部外商投资	Foreign investment in the eastern region
中部外商投资	Foreign investment in the central region
西南外商投资	Foreign investment in the southwestern region
西北外商投资	Foreign investment in the northwestern region
东北外商投资	Foreign investment in the northeastern region
东部外商投资增速	Foreign investment growth rate of the eastern region
中部外商投资增速	Foreign investment growth rate of the central region
西南外商投资增速	Foreign investment growth rate of the southwestern region
西北外商投资增速	Foreign investment growth rate of the northwestern region
东北外商投资增速	Foreign investment growth rate of the northeastern region

2. **The South-North Gap.** In 1908, Zhang Xiangwen, the 1st president of the Geographical Society of China, first put forward the Qinling–Huaihe Line at 35°N as a reference line to distinguish between northern and southern China. According to this division standard, the northern China include Beijing and Tianjin municipalities and 13 provinces, namely, Hebei, Shanxi, Inner Mongolia, Liaoning, Jilin, Heilongjiang, Shandong, Henan, Shaanxi, Gansu, Qinghai, Ningxia, and Xinjiang, covering a total area of about 5,775,300 km^2, almost 59.7% of China's total land area. The rest place are defined as the southern China. This north–south division method is also widely used in the analyses of the Chinese economy. During the 13th Five-Year Plan period (2016–2020), the south-north gap was a more obvious problem than the inter-regional imbalance: The north–south disparities were stark in both an absolute and in a relative sense.

Figure 10 compares the GDPs and per-capita GDPs of the northern and southern regions from 2016 to 2020. In an absolute sense, when the GDP of the northern region rose with fluctuation from 30.97 trillion yuan to 35.66 trillion yuan from 2016 to 2019, the southern region increased its GDP from 47.04 trillion yuan to 65.58 trillion yuan; when the northern per-capita GDP struggled to climb from 53,502.68 yuan to 59,767.93 yuan, the southern per-capita GDP jumped 59,723.54 yuan to 77,617.48 yuan. During this period, the ratio of the northern GDP to the southern GDP increased from 1: 1.52 to 1: 1.84. As to the ratio between their per-capita GDPs, it rose from 1:1.1 to 1:1.3. All this suggests that the south-north gap is still widening. In a relative sense, the annual GPD growth rates of the northern region from 2016 to 2020 were 4.19%, 6.54%, 6.67%, −0.78%, and 2.21%, and those of the southern region were 10.54%, 9.95%, 8.81%, 13.04%, and 3.1%; the annual per-capita GPD growth rates of the northern region from 2016 to 2019 were 3.68%, 6.14%, 6.33%, and −1.02%, and those of the southern region were 9.69%, 9.09%, 7.96% and 12.23%. In the first three years of the 13th Five-Year Plan period, the growth rates of the north accelerated slightly but the negative growth of 2019 disrupted the upward trend. The southern region, however, maintained a relatively stable and high-speed growth. The biggest south-north disparities in GDP and per-capita GDP appeared in 2019: The southern GDP and per-capita GDP growth rates were 13.83 and 13.25 percentage points higher than those of the northern region, which further widened the south-north gap. So far, this gap has attracted wide attention from political and academic circles.

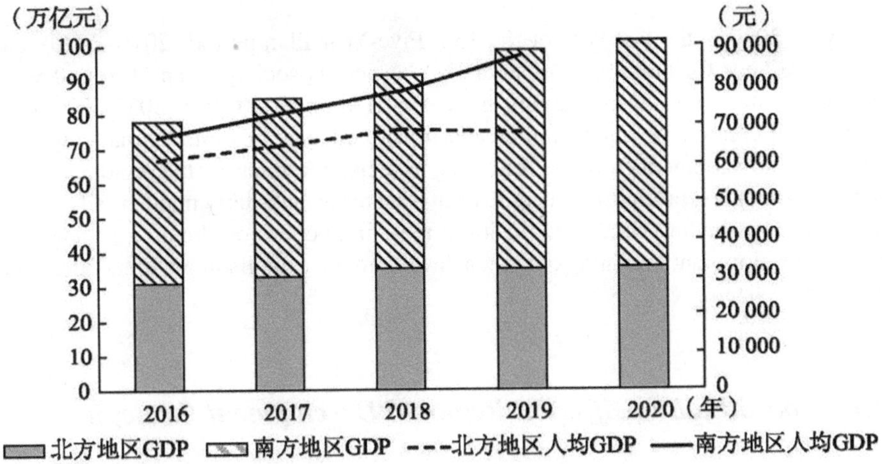

Fig. 10 GDP and per-capita GDP in North and South China 2016–2020 (*Data Source* the National Bureau of Statistics of China)

北方地区GDP	GDP of the northern region
南方地区GDP	GDP of the southern region
北方地区人均GDP	Per-capita GDP of the northern region
南方地区人均GDP	Per-capita GDP of the southern region
(万亿元)	(trillion yuan)
(元)	(yuan)

The south-north gap is also reflected in the numbers of their cities whose GDP exceed 1 trillion. By the end of 2020, China has seen 23 cities' GDPs exceeded the trillion-yuan mark, 17 of them being southern cities, accounting for 73.91% of the total. When cities with high administrative levels, i.e. municipalities directly under the Central Government and special administrative regions, are excluded in this list, the proportion of southern cities will be further increased to 78.95%, and only four norther cities, namely Qingdao, Zhengzhou, Jinan and Xi'an, will remain on the list. The northern region lags behind the southern region in terms of economic development for multiple reasons: insufficient investment in high-quality factors such as talents, technologies, and information; weak endogenous growth resulting from the overcapacity in traditional economic sectors; delayed institutional reforms; and changes in international political and economic environments.

3 Prospects for China's Regional Economy in the 14th Five-Year Plan Period

The year 2020 is the last year of the 13th Five-Year Plan period (2016–2020) and a decisive year for building a moderately prosperous society in an all-round way, starting a transition toward the 14th Five-Year Plan period (2021–2025). The new period will create new opportunities and inject new energies into China's regional economic development from the following aspects: coordinating the regional development strategies; promoting the industrial transfer in an orderly manner; advancing new urbanization and rural revitalization; realizing better coordination of land and marine development; pursuing the green development; establishing a sound and open economic system.

3.1 Coordinating Different Regional Development Strategies

Since the beginning of the twenty-first century, a number of regional development strategies have been formulated, represented by those made for the overall regional development; the establishment of functional ares; the coordinated development of the BTH region; the development of the YREB; the development of the GBA; the integration of the YRD; the ecological conservation and high-quality development of the YRB; and the development of city clusters, metropolitan areas, and national-level central cities, high-tech zones, and economic and technological development zones. They lay a solid institutional foundation for the drive to solve the contradiction between people's growing demands for a better life and the imbalanced, inadequate development in the new era.

The term "regional scale analysis" is a core concept of Regional Geography. It is used to analyze a place by dividing it into multiple regions with different features. Given this, it naturally implies classification. When this concept is applied to the analysis of regional development strategies, the national- and regional-level strategies, designed as an overall measure of dispersion, can be described as a neutral; the municipal-level strategies, formulated for the concentration of advantages, are more locally based. The two types of strategies complement each other, promoting the common progress of all economic regions and facilitating their economic recovery during the 14th Five-Year Plan period (2021–2025). The national- and regional-level strategies for China's regional development the following: the overall strategy of regional development, the development of the major functional regions, the coordinated development of the BTH region, the development of the YREB, the development of the GBA, the integration of the YRD, the ecological protection and high-quality development of the YRB, the development of city clusters and metropolitan areas. As to the municipal-level strategies, they refer to those regarding the development of certain national-level central cities, high-tech zones, and economic and technological development zones. The strategies at different levels are mutually connected and influenced. They redefine the inter-regional division of labor for the 14th Five-Year Plan period and blur the administrative boundaries. Apart from thoroughly implementing the available development plans, efforts shall also be exerted to develop new-type economic geographic areas such as the Grand Canal Cultural Belt, Hexi Corridor, Sichuan-Shanxi-Gansu Golden Triangle, and regional-level city clusters and central cities in the 14th Five-Year Plan period. More work shall be done to renew our development plans from administrative and spatial-arrangement perspectives, constantly improving the paradigm of our regional economic development.

3.2 Promoting the Industrial Transfer in an Orderly Manner

In the post-2018 Financial Crisis era, the original international industrial transfer trend has profoundly changed. On the one hand, some of our manufacturing industries has been relocated foreign countries. On the other hand, the transfer of our labor-intensive and capital-intensive manufacturing industries from the eastern coastal areas to the central and western regions has been accelerated.

Faced with a change unseen in the past 100 years, China will have to handle more complicated domestic and international conditions for its industrial transfer during the 14th Five-Year Plan period. First, its industrial transfer will become bi-directional: Its labor-intensive manufacturing industries such as textile and clothing industries and capital-intensive manufacturing industries such as steel, petrochemical and non-ferrous-metal productions will tend to move to the central and western regions; Its technology-intensive manufacturing industries, such as new-energy-vehicle, new-material, and computer industries will prefer the eastern coastal areas for their advantage of efficiency. Second, the progress in transportation infrastructure and the popularization of new technologies will weakened the industries' dependence on cheap labor and lands. During the 14th Five-Year Plan period, the national high-speed railway network will be improved, and the air-rail inter-modal transportation will greatly reduce trade costs. The application of advanced technologies such as big data, artificial intelligence, and IoT will offset the impact brought by the rising labor and land prices, which may slow down the trend of relocating for lower costs of labor and lands. Third, the solicitation of investors based on industrial clusters will become an important development method because it conforms to the trends of market segmentation and division of labor. By introducing supporting projects and related enterprises into the clusters, the positive externalities of clustering will be strengthened through the sharing, matching, and learning mechanisms.

To better adapt to the above-mentioned new changes, China needs to adhere to the principle of respecting local conditions and offering individualized guidance during the 14th Five-Year Plan period. It shall give equal attention to sole proprietorship, joint venture, acquisition, merger, non-equity arrangement, and other methods and scientifically guide the transfer of labor-intensive and capital-intensive industries to the central and western regions, avoiding an excessive industrial outflow to countries and regions in Southeast Asia and South Asia. While enhancing the domestic industrial transfer, it has to constantly optimizing its industrial structure and enhance the manufacturing bases in the eastern region such as the BTH region, the YRD, the PRD, and the Central and South Liaoning, preventing the sudden shift of the productivity centers from the east to the west.

3.3 Advancing New Urbanization and Rural Revitalization

The in-deep integration of urban and rural areas is a crucial part of the coordinated regional development. China's urbanization rate increased from 57.35% to 60.6% from 2016 to 2019, an increase of 3.25 percentage. In the past, China was often described as "rural China," since it took agriculture as its major industry and villages as its basic governance units. At present, however, "rural China" becomes "urban–rural China" with the advancement of urbanization and the close urban–rural interaction. Unfortunately, the urban–rural imbalance has become an obvious problem, which is embodied in the following aspects: the demographic imbalance caused by the concentration of population in urban areas; the spatial imbalance resulted from the expanding urban areas and the shrinking rural areas; the industrial imbalance caused by the transfer of traditional manufacturing industries to rural areas; the institutional imbalance generated by the fragmented urban and rural management system; the imbalance in the urban system caused by the neglect of small and medium-sized cities. These imbalances, together with inter-regional disparities, hinder the coordinated regional development in the new era.

To solve these imbalances, China needs to accelerate its strategic transformation for the in-depth urban–rural integration during the 14th Five-Year Plan period. It shall promote the new-type urbanization and rural revitalization strategies and improve the quality and scale of urbanization. It must attach equal importance to urban and rural areas and enhance the quality of regional economic operation. For these purposes, it is essential to do well in the following three aspects. First, it is necessary to adopt the people-centered philosophy in the new urbanization drive. Fei Xiaotong, one of the foremost Chinese social anthropologists, once observed that China was a typical rural society with village as its basic community. Long influenced by the immobility of lands, its inter-personal relations have become highly dependent on hometown connections and blood relations, which prevented transferred rural population from integrating themselves into urban communities. To facilitate the integration, it is a must to ensure that the transferred rural population can enjoy the rights they deserve in urban areas, such as high-quality child care and education services, reasonable incomes, guaranteed heath and old-age insurances, housing security, and other social insurances. To this end, the supplies of basic public services for child care, education, income distribution, medical care, old-age care, and housing security shall be strengthened, and the the equal access to public services shall be realized. Second, the idea of agricultural industrialization must be implemented. Scientific studies and predictions shall be made on changes in consumer demands and the reconstruction of interpersonal relationship induced by institutional innovations. Leveraging the opportunities brought by the land allocation system, the capital flows to the rural areas, and the urban–rural bi-directional people flows, the agricultural mechanization and modernization can be elevated to a level that is close to the international advanced level. Third, the urban and rural civilizations must be promoted in an integrated way. From the 1980s to the 1990s, when China's urbanization advanced by leaps and bounds, the urban civilization was often regarded as superior than the rural civilization. Later, when China's urbanization rate exceeded 50% and when "urban diseases" such as housing shortage, traffic congestion, and environmental degradation began to plague the economic activities and daily lives of the urban residents, these people took rural life as a mental comfort. In recent years, the thriving rural recreation, rural tourism, and homestay businesses have reflected this trend. To respond to this demand, it is a must to preserve the tangible and intangible heritages in the rural areas and vigorously develop characteristic towns and beautiful countryside. Cultural industries and cultural undertakings with rural characteristics shall also be promoted, and rural environment shall be restored. With improvement in the natural and cultural environments in the rural region, the synergistic development of rural and urban civilizations can be realized.

3.4 Realizing Better Coordination of Land and Marine Development

Occupying the eastern part of the Eurasian continent China faces the Pacific Ocean in the east. It has more than 300 square kilometers of marine territory that is rich in mineral resources, chemical resources, biological resources, and power resources, and its marine GDP has exceeded 8.9 trillion yuan in 2019. In today's world, the development and utilization of ocean resources is widely regarded as an important standard used to judge a country's comprehensive national strength. Back in the early twentieth century, Sun Yat-sen the forerunner of China's modern democratic revolution, noticed the importance of marine development. He said that China should pay equal attention to both the marine and land development to become a great country and in his famous *Jianguo Fanglue* (Founding Strategy), he defined ports as key locations for the development of export-oriented industries and China's international transportation hubs. Since the founding of the PRC in 1949, the five generations of the CPC leadership collective has attached great importance to the marine development and marine economy. In its report to the 19th National Congress of the Party, the CPC Central Committee has incorporated the drive to realize better coordination of land and marine development and accelerate China's marine development in the country's regional development strategy, offering a clear guide for the high-quality marine development during the 14th five-year plan period (2021–2025).

For the better coordination of land and marine development, it is essential to do well in the following four aspects. First, the efforts to safeguard our marine rights and those to promote our international cooperation for our marine development shall be coordinated. Under the guidance of the United Nations Convention on the Law of the Sea, China shall boost international cooperation for marine protection and development while safeguarding its marine rights and interests. This can effectively resolve all kinds of uncertainties in the current international political and economic situations and bring benefits to all mankind. Second, the efforts to pool marine resources and those to protect marine environment shall be combined. The planning of marine functional regions shall be carried out in a timely manner to cooperate with the planning of land functional, achieve mutual-complementing and mutual-reinforcing effects. Third, the marine industrial structure shall be optimized and the marine industrial layout shall be adjusted. Focus shall be placed on the building of a "blue granary," and active efforts shall be exerted to develop new business models featuring the integration of marine shipbuilding industry, marine oil and gas industries, and seaside tourism industry. Fourth, the optimized offshore development and the expansion to ocean space shall be pushed forward together. A coastal economic belt with Shanghai as the center and Beijing, Tianjin, Guangzhou, and Shenzhen as the north and south wings shall be established as a growth role of the marine economy covering offshore and ocean regions. With the implementation of the above-mentioned development ideas, the scale and quality of marine economy will be significantly improved, and the utilization of the marine resources will be optimized. This will provide a strong guarantee for the stable and sustainable development of China's national economy.

3.5 *Pursuing the Green Development*

To ease the contradiction between man and land and build a modernized economic system, it is a must to pursue the green development. In about 100 years from the 18th to nineteenth century, Western capitalist countries quickly realized their industrialization and improved their social productivity, but they generally adopted an extensive-growth model, which was realizing their economic expansion by increasing the input of factors, in the early stage of their industrial civilization. This triggered a series of major environmental disasters, represented by the Sulfurous smog in London and the photochemical smog in Los Angeles. Different from these Western capitalist countries, the PRC spent only about 70 years to develop itself from a backward agricultural economy into an industrial economy, and thus faced even more severe environment constraints caused by resource bottlenecks, environmental pollutions, and ecological destruction. The country urgently need to change this growth model. At the 18th National Congress of the CPC, held in 2012, the CPC Central Committee with Xi Jinping as the core stated that a good ecological environment was equivalent to productivity and that green mountains and lucid waters were economic wealth, proposing the plan to build a resource-saving and environment-friendly society for better lives of the people and more efficient production activities.

To achieve the goal of building a beautiful China, the constraints of ecological capacity and resource carrying capacity shall be taken into consideration in the formulation of all development plans for the 14th Five-Year Plan period. To be more specific, it is a must to do well in the following three aspects. First, the industrial adjustment shall be upgraded for green total factor productivity. The innovation of green technologies shall be accelerated. Technology-intensive industries such new materials, new energy vehicles, and high-end equipment manufacturing shall be developed, and the traditional iron and steel, chemical, and metallurgical industries shall be transformed with green technologies for the realization of the Porter effect. These efforts can inject new impetus into the green development. Besides, green-technology innovations also need the support of high-quality production factors such as talents, information, technologies, etc. To facilitate such innovations, economic regions shall implement preferential policies to realize tax reductions, accelerated depreciation, profit and loss balance, deferred tax payments, and credit support, creating a good policy environment for the development of green industries and realizing the regional coordination in the optimization of green total factor productivity. Second, the industrial transfer shall be carried out in an orderly manner. Guided by policy documents such as the "Guiding Opinions of the PRC State Council on Undertaking Industrial Transfer in the Central and Western Regions" and the "Catalogue for Guiding Industrial Development and Transfer (2018 Edition)," green development standard shall be the primary evaluation criterion for all the industries transferred in or out of a province. Industries with high pollution and high energy consumption must not settle in the development zones. Third, a joint prevention-and-control mechanism of environmental pollution shall be built. For water pollutants, harmless treatment of waste waters must be carried out according to unified standards, and efforts shall be made to improve the water quality of large waters, especially those across administrative regions. For the air pollutants and solid wastes, it is necessary to trace the source, phase out backward production capacities in a timely manner, and enhance the binding effect of environmental regulations, stopping the emissions from their sources. It should be noted that in the environmental protection campaign, the developed areas, the beneficiary areas, and the protected areas often cannot achieve the consistency of interests. To solve this problem, it is a must to resort to systematic thinking and establish a cross-regional ecological compensation mechanism.

3.6 *Establishing a Sound and Open Economic System*

Since adopting the policy of reform and opening-up, China has adhered to the principle of mutual benefit and win–win cooperation and successively established a number of special economic zones, coastal open cities, coastal economic open zones, open cities along the Yangtze River, and open cities along the border, building an all-around, extensive, and muli-level opening-up paradigm. Nevertheless, given all kinds of uncertainties that have emerged since 2018, i.e. frequent Sino-US trade frictions, rising protectionism and unilateralism, and the COVID-19 pandemic, the recovery of the world economy is still facing great uncertainty, and China has to pursue a new development paradigm that has the domestic circulation as the mainstay and the domestic and international circulations promoted each other. It is worth noting that China's focus on the domestic circulation does not mean that it decides to decouple from the global division of labor and the global value chain. China will be more determined to implement its reform and opening-up policy and leverage both domestic and international resources for the vigorous development of its economic regions.

To upgrade its open economy in the 14th Five-Year Plan period, China must do well in the following two aspects. Firstly, the pilot projects of free trade zones shall be further promoted. While increasing the number of the pilot projects, it is essential to optimize the mechanism design for these trade zones: implementing overall opening-up policies for the first-tier zones; exercising safe and efficient management for the second-tier zones; providing one-stop service for online transactions. While developing import and export trade, China shall create a favorable policy environment to attract foreign investors and encourage Chinese enterprises to invest abroad, contract to work on foreign projects or send laborers to work abroad, and introduce international economic and technological cooperation through such activities. Secondly, the field of international cooperation needs to be further expanded. Besides economic and trade cooperation, cooperation in information technology, culture, education, and non-traditional security sectors should also be strengthened to help China better adapt to the increasingly complicated international political and economic environments. For instance, by building a big information channel supported by big data, cloud computing, and IoT technologies, China can increase its efficiency of digital communication with other countries and regions. By actively tapping our characteristic cultural resources and holding cultural exchange activities with other countries and regions, China can enhance mutual cultural-recognition between its people and people from other countries and regions. Typical examples are China-Russia Youth Friendly Exchange Year and International Youth Creative Culture Week. China also needs to enhance its cooperation with other countries and regions in non-traditional security fields such as natural disaster prevention and relief, public health, terrorism, and transnational crimes, nipping the risks brought by its drive to deepen its opening-up in the bud.

References

Chen, Y., Chen, X., & Liu, Z. (2018). China's macroeconomy under structural delevergaing—China's macroeconomic analysis and forecast in 2018 interim report (RUC Research Group of Macroeconomic Analysis and Forecast, Renmin University of China). *Economic Theory and Business Management, 2018*(08), 5–19.

Cheng, Z. (2016). *Jingji xin changtai xia woguo quyu chanye zhuanyi de yindao lujing* (The guiding path for regional industrial transfer in China under the new economic normal). *Economic Review Journal, 2016*(11), 92–95.

Li, G., & Song, C. (2018). The strategic choice of high quality development in Xiong'an New Area. *Reform, 2018*(04), 47–56.

Liu, S., & Long, T. (2020). The political economy of urban and rural transformation. *China Review of Political Economy, 11*(01), 97–115.

Liu, S., & Wang, Y. (2018). Cong xiangtu zhongguo dao chengxiang zhongguo: Zhongguo zhuanxing de xiangcun bianqian shijiao (From rural China to urban-rural China: A perspective on the rural transformation in China). *Management World, 34*(10), 128–146.

Lin, X. (2020). The global ocean economy in 2030: Influencing factors, trends, and suggestions. *Pacific Journal, 28*(01), 50–63.

Liu, Y., Yan, Y., & Liu, X. (2016). The bottom of the Chinese macro economy—China's macroeconomic analysis and forecast in 2015–2016 (RUC Research Group of Macroeconomic Analysis and Forecast, Renmin University of China). *Economic Theory and Business Management, 2016*(01), 5–45.

Liu, Y., Liu, X., & Zou, J. (2018). New normal towards new stage—China's macroeconomic analysis and forecast in 2017–2018 (RUC Research Group of Macroeconomic Analysis and Forecast, Renmin University of China). *Economic Theory and Business Management, 2018*(02), 20–38.

Liu, Y., Liu, X., & Yan, Y. (2019). China's macroeconomic report 2018–2019—China's macroeconomy in the new era of reform and opening up (RUC Research Group of Macroeconomic Analysis and Forecast, Renmin University of China). *Economic Theory and Business Management, 2019*(01), 4–26.

Liu, Y., Yan, Y., & Liu, X. (2020). China's macroeconomic report 2019–2020—China's macroeconomy in the period of structural adjustment (RUC Research Group of Macroeconomic Analysis and Forecast, Renmin University of China). *Economic Theory and Business Management, 2020*(01), 21–41.

Mao, Z., Liu, Y., Yuan, H., & Zhang, Y. (2017). Macro-economy under the dual bottom-line of stable growth and risk prevention—China's macroeconomic analysis and forecast in 2016–2017 (RUC Research Group of Macroeconomic Analysis and Forecast, Renmin University of China). *Economic Theory and Business Management, 2017*(01), 5–25.

Sheng, L., Zheng, X., Zhou, P., & Li, T. (2018). *Woguo jingji fazhan nanbei chaju kuoda de yuanyin* (Analysis of the reasons for the widening north-south gap in China's economic development). *Management World, 34*(09), 16–24.

Sun, J. (2019). Xiong'an xinqu zai jing jin ji xietong fazhan zhong de dingwei (The positioning of Xiong'an New Area in the coordinated development of the Beijing-Tianjin-Hebei region). *Gansu Social Sciences, 2019*(02), 59–64.

Sun, J. (2020). *Xinguan feiyan yiqing dui zhongguo quyu jingji fazhan de yingxiang chuta*n (Initial exploration into the impact of the COVID-19 on China's regional economic development). *Regional Economic Review, 2020*(02), 8–11.

Sun, J., & Zhang, J. (2019). *Changjian jingji dai fazhan de shikong yanbian yu fazhan jianyi* (The spatial-temporal evolution of the Yangtze River Economic Belt and development suggestions). *China Review of Political Economy, 10*(01), 151–171.

Wang, X. (2019). Research on the construction path of free trade ports with Chinese characteristics from the perspective of international experiences. *Economist, 2019*(03), 60–70.

Zhu, D. (2020). Xi Jin-ping's green development concept: Thought foundation, connotative system and contemporary value. *Economist, 2020*(03), 5–15.

Zhang, J. (2013). *Guojia—quyu zhili de chidu chonggou: Jiyu "guojia zhanlue quyu guihua" shijiao de pouxi* (Scale reconstruction of national regional governance: Analysis from the perspective of "national strategic regional planning"). *Urban Studies, 20*(05), 45–50.

Zhang, K. (2018). The new normal of China's regional economy and regional coordinated development in the new era. *Journal of Chinese Academy of Governance, 2018*(03), 102–108.

Zhang, M. (2020). How to understand the new development pattern of "dual circulation" systematically and comprehensively? *Journal of Liaoning University (Philosophy & Social Science), 48*(04), 1–8.

Postscript

The year 2020 was very special. When the COVID-19 epidemic disrupted the development of China's regional economy, the Chinese people, under the strong leadership of the CPC Central Committee, successfully controlled the spread of the virus and achieved economic recovery and progress. This book is a research report focused on the different characteristics of China's economic regions in 2020 and a collation of many research articles.

Here, I would like to express my gratitude to all those, as listed below, who have participated in the collating of this book.

General Report: Sun Jiuwen & Su Xijian
Chapter 1: Gao Yujie
Chapter 2: Li Chengzhang
Chapter 3: Chen Chaojun
Chapter 4: Zhang Hao
Chapter 5: Su Xijian
Chapter 6: Zhang Qian
Chapter 7: Yi Shuchang
Chapter 8: Song Zhun
Chapter 9: Zhang Ao
Chapter 10: Jiang Zhi

It is a great honor for me to edit the entire work, and Su Xijian has provided me with significant assistance regarding the editing, collating, and referencing tasks. I have enjoyed enthusiastic support and financial help from the Research Office of Renmin University and the leadership of the university's School of Applied Economics. China Renmin University Press has also warmly endorsed the value of this book. I would like to express my sincere thanks to all the above parties.

J. Sun, *Regional Economic Development in China in 2020*,
https://doi.org/10.1007/978-981-95-3604-7

About the quotations in this book, we believe that we have annotated the sources of information for most of them, but we cannot arrogantly claim that there is nothing wrong or improper due to the rush of time. Please feel free to comment provided that you find mistakes or improprieties in the book, and for such problems, I hope that you can accept my sincere apologies.

October 13, 2021

Sun Jiuwen

Kolibri 360 GmbH
Ferdinand-Jühlke-Straße 7
99095 Erfurt, Deutschland
produktsicherheit@kolibri360.de